Environmental Accountability Within Corporate Governance and National Systems

Aftab Haider
https://orcid.org/0009-0009-3339-5057
Southwest University of Political Science and Law, China

Zhang Hui
Southwest University of Political Science and Law, China

Muamar Hasan Salameh
https://orcid.org/0009-0003-9489-493X
Prince Mohammad bin Fahd University, Saudi Arabia

IGI Global Scientific Publishing
Publishing Tomorrow's Research Today

Vice President of Editorial Melissa Wagner
Director of Acquisitions Mikaela Felty
Director of Book Development Jocelynn Hessler
Production Manager Mike Brehm
Cover Design Jose Rosado

Published in the United States of America by
IGI Global Scientific Publishing
701 East Chocolate Avenue
Hershey, PA, 17033, USA
Tel: 717-533-8845 | Fax: 717-533-7115
Website: https://www.igi-global.com E-mail: cust@igi-global.com

Library of Congress Cataloging-in-Publication Data

LCCN: 2026038256 (CIP Data Pending)
ISBN13: 9798260036785
Isbn13Softcover: 9798260036792
EISBN13: 9798260036808
British Cataloguing in Publication Data
A Cataloguing in Publication record for this book is available from the British Library.

Table of Contents

Detailed Table of Contents

This chapter explores the evolving legal and policy framework for corporate environmental accountability, a multidimensional concept requiring businesses to take responsibility for their ecological and social impacts throughout their entire life cycles. By employing a doctrinal research methodology, the study analyzes the transition from voluntary ethical obligations to enforceable legal instruments. It distinguishes between responsibility, accountability, and liability, highlighting how international standards like the UN Guiding Principles and the Paris Agreement are shaping corporate governance. The research evaluates the effectiveness of various policy tools, including mandatory due diligence, sustainability reporting frameworks (GRI, TCFD), and market-based mechanisms like the EU Emissions Trading Scheme. Despite challenges such as greenwashing and jurisdictional inconsistencies, the chapter argues for a harmonized global approach that integrates transparency, technological innovation, and robust stakeholder engagement to foster a culture of genuine environmental stewardship.

Environmental reporting is often treated as a problem of measurement accuracy. Credibility, in this view, depends on data quality and the reliability of calculation methods. This chapter adopts a different position. It argues that the main sources of variation arise earlier, during the construction of reporting boundaries. The analysis focuses on the built environment. Environmental performance is studied through decisions that define scope, establish baselines, introduce assumptions, and determine materiality. These elements operate before measurement is finalized. They determine what is included, how impacts are interpreted, and how results are presented. Building on this, the chapter develops a conceptual synthesis drawn from environmental accounting, sustainability reporting, and life cycle assessment. Through this synthesis, disclosure is reframed as an interpretive process rather than a neutral reflection of reality. While reporting outcomes may seem consistent, their underlying conditions may differ. Such differences limit comparability and obscure assumptions.

The chapter evaluates the effectiveness of enhancing ESG-based financial data ecosystems to develop gender-inclusive finance and speed up the achievement of SDG 5 and SDG 8paying particular attention to India and other similar emerging markets. Based on regulatory trends like BRSR framework established by SEBI as well gender-sensitive financing programs facilitated by the World Bank and the International Finance Corporation, the research paper identifies real actions to incorporate social and governance measures in the process of capital allocation. The insights of the case study of fintech-based MSME lending platforms in India shows that alternative data, including digital transaction history and supply-chain records, has helped women entrepreneurs to gain better access to credit and ensure portfolio stability. The chapter suggests a large-scale and open data integration model bridging the ESG reporting, gender analytics, and financial decision-making, consequently promoting inclusive capital flows and enhancing sustainable economic development in emerging economies.

This article analyses the Corporate Sustainability Due Diligence Directive as a shift from disclosure regimes, including Directive 2014/95/EU and Directive (EU) 2022/2464, toward process-based regulation of corporate conduct in global value chains. It asks when due diligence produces substantive environmental outcomes rather than formal compliance. Using doctrinal analysis grounded in Article 114 TFEU, supported by case law and comparative insights from France, Germany, and the textile sector, the article develops a conditional framework. Outcomes depend on enforcement capacity, value chain structure, and corporate response. Where enforcement is credible and supply chains are stable, firms integrate risk into decision-making; otherwise, compliance remains procedural. The Directive also tests the limits of Article 114 by regulating corporate processes and producing extraterritorial effects. Its impact is therefore contingent rather than uniform, highlighting both the potential and limits of process-based harmonization in EU law.

This chapter examines the limits of existing legal frameworks in addressing greenwashing, arguing that its current characterisation as a form of consumer deception or failure to comply with regulations is insufficient. It argues that greenwashing possesses a hybrid nature, combining both commercial and political elements, thereby challenging traditional doctrinal classifications. Relying on Article 19 of the International Covenant on Civil and Political Rights, Article 10 of the European Convention on Human Rights, and UNECE Aarhus convention, this chapter situates greenwashing within the broader freedom of expression framework, while analysing the conditions under which its restriction may be justified. It proposes that greenwashing should instead be viewed as a form of expression capable of producing structural harm within a democratic society and environmental governance as it undermines rights by distorting the informational foundations on which several rights are depended on, not merely by deceiving consumers, investors and the regulator.

Environmental governance is shifting from voluntary principles to enforceable legal obligations. This paper compares frameworks in India, the EU, and international climate litigation, showing a clear move toward hard legal liability. It highlights India's activist judiciary, the EU's advanced corporate accountability rules, and new international legal concepts like ecocide. The study recommends stronger global enforcement, greater corporate liability, and institutionalized climate justice mechanisms.

This chapter examines the evolution of legal accountability for maritime climate harm and marine pollution in the Bay of Bengal. Drawing on doctrinal analysis of the UNCLOS, IMO frameworks, recent advisory opinions of the ITLOS and the ICJ, landmark domestic and regional climate litigation and emerging mandatory corporate sustainability obligations, the chapter argues that a coherent transition from soft law flexibility toward enforceable legal responsibility is now underway. Focusing on Bangladesh as a primary case study, the chapter analyzes progressive hardening of due diligence obligations for flag, coastal and port states through judicial innovation and rights-based litigation and the extension of enforceable environmental obligations to shipping corporations through mandatory due diligence laws. The chapter proposes a three-tier accountability model linking international adjudication, domestic constitutional litigation and corporate due diligence engagement and identifies strategic pathways for transforming persistent soft law commitments into enforceable legal responsibility.

This study examines environmental accountability in coastal fisheries management, focusing on the Black Sea region. Drawing on literature, policy analysis, and stakeholder insights, it reveals how coastal fishing communities—despite their dependence on marine ecosystems—are often excluded from decision-making processes that shape their livelihoods. The findings highlight persistent gaps between sustainability policies and their implementation, including ecological degradation, weak institutional coordination, and unequal power relations. Supported by the TÜBİTAK 2221 Sabbatical Research Program, the study proposes an ecosocial governance approach that integrates ecological sustainability, social justice, participatory decision-making, and fair market conditions. By emphasizing community empowerment and shared responsibility, the research argues that more inclusive and transparent governance systems are essential for achieving sustainable fisheries management.

Pakistan ranks 179th out of 180 countries in the 2024 Yale Environmental Performance Index, yet it possesses some of the most advanced climate jurisprudence in the Global South. This paradox reflects not a lack of environmental laws, but the persistence of a colonial administrative structure designed for extraction rather than conservation. This chapter argues that the Supreme Court's decision in Mehar Badshah v. Government of Khyber Pakhtunkhwa, delivered shortly after the insertion of Article 9-A into the Constitution in October 2024, establishes a doctrine of Administrative Environmental Reasonableness. The doctrine creates a constitutional standard for evaluating climate governance through administrative capacity, transparency, responsiveness, and accountability rather than legislative enactment alone. The chapter further examines the constitutional tension between Article 9-A and the 18th Amendment's devolution of environmental powers, proposing Pakistan as a potential Global South model for climate governance through administrative reform.

This chapter examines the obligations arising for Bangladesh under existing climate law regimes in the post-ICJ Climate Advisory Opinion period. By identifying the applicable treaty and customary international law obligations, the chapter argues that Bangladesh has, in a strict doctrinal sense, acquired a dual legal identity to comply with its climate commitments; simultaneously a climate-vulnerable rights-holder and a climate-violating duty-bearer. The chapter identifies substantive and procedural gaps in Bangladesh's implementation of these obligations, and delineates the legal consequences, including attribution, causation, and the tripartite responsibility framework, that Bangladesh may face for such breaches under the Opinion's post-2025 jurisprudential architecture. The chapter establishes that Bangladesh can no longer evade responsibility under the veil of vulnerability, and must undertake the legislative and institutional action the Opinion requires to discharge its obligations under customary and treaty-based climate law.

Preface

INTRODUCTION

Environmental accountability is now at the centre of debates about how companies are run, how states regulate, and how international law responds to ecological harm. Voluntary sustainability programmes no longer satisfy the demands placed on organisations. Firms are expected to meet legal duties, disclose environmental information, build environmental considerations into strategy, and answer to stakeholders. Legal systems have moved in the same direction: new statutes, court decisions, disclosure rules, due diligence requirements, and international instruments are reshaping what responsibility means in practice.

This book, *Environmental Accountability Within Corporate Governance and National Systems*, gathers work by scholars writing from legal, regulatory, financial, ethical, and governance angles. The chapters treat environmental accountability as something wider than environmental law alone. It discusses corporate governance, international relations, constitutional law, finance, human rights, sustainability reporting, climate litigation, maritime law, fisheries management, religious ethics, and public administration. The authors write from institutions in Asia, Europe, the Middle East, and South America. That range matters, because environmental problems and the responses to them rarely stop at national borders.

The book shows how responsibility is moving from something voluntary to something that can be enforced. Some chapters track how soft-law guidance is giving way to binding duties to identify, manage, disclose, and reduce environmental risks. Others look at constitutional principles, judicial creativity, stakeholder participation, and ideas of climate justice that are changing older governance arrangements. Accountability, on this reading, is not only a matter of compliance. It also depends on transparency, institutional capacity, ethical leadership, and real engagement among governments, businesses, communities, and civil society.

We offer the collection as a contribution to current work on environmental governance and sustainable development. We hope researchers, policymakers, lawyers, corporate practitioners, teachers, graduate students, and others working across related fields will find it useful—especially those interested in how environmental accountability is changing inside national legal systems and in corporate governance worldwide. The chapters mix theoretical argument with policy discussion, and we hope they will be of use to those thinking about research, legislation, institutional reform, and organisational practice.

OVERVIEW OF THE CHAPTERS

The opening chapter, "Legal and Policy Framework for Corporate Environmental Accountability," sets out the legal and conceptual ground for the volume. It traces the shift from voluntary corporate responsibility to enforceable obligations, looking at international frameworks, sustainability reporting standards, mandatory due diligence, and market-based regulation. The chapter shows how corporate governance and environmental law have drawn closer, while noting persistent difficulties: uneven implementation, jurisdictional inconsistency, and greenwashing.

Chapter Two, "The Invisible Lines of Environmental Reporting: Beyond Measurement in Environmental Disclosure," questions the idea that environmental disclosure is mainly a technical exercise in measurement. The authors argue that the important differences often arise earlier—in how reporting boundaries are drawn, what assumptions are made, how materiality is judged, and which interpretive frames are used. Recasting reporting as an interpretive process helps explain why comparability across sustainability reporting systems remains limited.

Chapter Three, "Strengthening ESG-Driven Financial Data Ecosystems for Gender-Inclusive Finance," turns to the link between environmental, social, and governance (ESG) reporting and inclusive economic development. Focusing on India and other emerging economies, the authors consider how ESG data, gender-sensitive finance, and fintech lending can improve women's access to capital, and how those efforts relate to the Sustainable Development Goals on gender equality and economic growth.

Chapter Four, "Conditional Regulation in Global Value Chains," examines the European Union's Corporate Sustainability Due Diligence Directive as a form of process-based regulation aimed at conduct along international supply chains. Combining doctrinal analysis with comparative case studies, the chapter argues that environmental outcomes turn on enforcement capacity, organisational incentives, and the structure of supply chains. On this account, effectiveness depends less on the elegance of the legislation than on how it is put into practice.

Chapter Five, "Climate Advocacy, Corporate Speech, and the Limits of Greenwashing," situates environmental accountability alongside freedom of expression. The authors advance a human rights–based reading of greenwashing as corporate speech that can distort public understanding of environmental performance and, in that way, cause structural harm. Using the international human rights instruments, the chapter considers how corporate speech might be weighed against the needs of democratic environmental governance and informed public participation.

Chapter Six, "A Comparative Study of Environmental Accountability Frameworks in India, the European Union, and International Climate Litigation," compares the broader move from soft-law environmental governance toward enforceable liability. Through judicial activism, European regulatory developments, and emerging international legal concepts, the authors show how climate litigation and related doctrines are tightening accountability, and where further cooperation and institutional reform remain needed.

Chapter Seven, "Recasting Responsibility at Sea: Legal Accountability for Maritime Climate Harm in the Bay of Bengal," takes the discussion into the maritime sphere. Using Bangladesh as a case study, it examines international maritime law, judicial opinions, domestic climate litigation, and corporate due diligence as climate change and marine pollution reshape responsibility at sea. The chapter's three-tier framework—international, national, and corporate—points to ways of strengthening legal responsibility across those levels.

Chapter Eight, "Environmental Accountability in Coastal Fisheries Management and Ecosocial Perspectives from the Black Sea," looks at sustainability, social equity, and participatory governance in coastal fisheries. Using policy analysis, stakeholder views, and existing literature, the authors identify gaps between sustainability policy and practice. They argue for an ecosocial model based on transparency, community participation, and shared responsibility, and for keeping ecological protection in balance with the social and economic needs of coastal communities.

Chapter Nine, "Article 9-A and the Constitutionalization of Administrative Climate Governance in Pakistan," turns to constitutional and administrative reform. Through recent constitutional developments and key judicial decisions, the chapter maintains that climate governance needs accountable administrative institutions as well as legislation. It proposes a doctrine of Administrative Environmental Reasonableness as a possible guide for jurisdictions seeking stronger climate governance through institutional reform and clearer public accountability.

Chapter Ten, "Victim or Violator? Bangladesh's Dual Legal Identity After the ICJ's Climate Change Advisory Opinion," considers the obligations facing climate-vulnerable states under emerging international climate law. Reading Bangladesh's position through treaty duties and customary international law, the chapter argues that a state may be both rights-holder and duty-bearer in the climate regime. That reading complicates simpler stories of climate vulnerability and stresses the need for domestic legal reform and institutional accountability if international commitments are to be met.

The volume closes with Chapter Eleven, "Islamic Environmental Ethics and Climate Justice in Maldives: Integrating Faith, Law, and Community for Sustainable Climate Resilience." The chapter brings ethical and cultural questions into environmental governance. Against the particular risks facing Small Island Developing States, the authors consider how Islamic environmental ideas—including *khalifa*, *mizan*, and *amanah*—can work alongside contemporary legal and policy frameworks. They show how faith-based leadership and environmental education can support climate resilience, and suggest that similar approaches may be relevant in other vulnerable settings.

CONCLUSION

Taken together, the chapters show that environmental accountability has become a busy and unsettled field of governance. The legal, political, economic, and cultural settings differ, but the contributors share a basic point: environmental protection cannot rest on aspiration or voluntary corporate programmes alone. It depends on working legal institutions, open governance, collaboration across disciplines, genuine stakeholder involvement, and mechanisms that can turn sustainability goals into concrete action.

Several themes cut across the book. One is the shift from voluntary responsibility to mandatory legal accountability—through due diligence laws, tighter reporting rules, constitutional reform, judicial innovation, and climate litigation. Another is that rules matter only insofar as they are implemented. Institutional capacity, and the readiness of public and private organisations to treat accountability as part of decision-making rather than a box to tick, remain decisive.

The volume also underlines how interconnected the field is. Legal doctrine, corporate governance, finance, public administration, human rights, environmental ethics, and community participation do not sit in separate boxes. The range of approaches here suggests that serious responses to environmental problems need cooperation across disciplines, institutions, and regions. Experience in one jurisdiction often has something to teach others.

We thank the contributing authors for the work that made this book possible. Their chapters reflect both the difficulty of the subject and a shared interest in practical, evidence-based approaches to sustainable governance. We are also grateful to the reviewers and publishing colleagues whose care improved the manuscript.

We hope *Environmental Accountability Within Corporate Governance and National Systems* will be of use to scholars, policymakers, practitioners, educators, and students concerned with the relationship between environmental stewardship, corporate governance, and public accountability. We hope it encourages more conversation, collaboration, and research to build stronger governance systems in response to environmental risk. The chapters suggest that environmental accountability is not a rhetorical goal. It is a practical condition for resilient institutions, sustainable development, and fairer outcomes.

Aftab Haider
Southwest University of Political Science and Law, China

Zhang Hui
Southwest University of Political Science and Law, China

Muamar Hasan Salameh
Prince Mohammad bin Fahd University, Saudi Arabia

Chapter 1
Legal and Policy Framework for Corporate Environmental Accountability

Pranjal Khare
http://orcid.org/0000-0002-9937-9588
Jindal Global Law School, O.P. Jindal Global University, India

ABSTRACT

This chapter explores the evolving legal and policy framework for corporate environmental accountability, a multidimensional concept requiring businesses to take responsibility for their ecological and social impacts throughout their entire life cycles. By employing a doctrinal research methodology, the study analyzes the transition from voluntary ethical obligations to enforceable legal instruments. It distinguishes between responsibility, accountability, and liability, highlighting how international standards like the UN Guiding Principles and the Paris Agreement are shaping corporate governance. The research evaluates the effectiveness of various policy tools, including mandatory due diligence, sustainability reporting frameworks (GRI, TCFD), and market-based mechanisms like the EU Emissions Trading Scheme. Despite challenges such as greenwashing and jurisdictional inconsistencies, the chapter argues for a harmonized global approach that integrates transparency, technological innovation, and robust stakeholder engagement to foster a culture of genuine environmental stewardship.

DOI: 10.4018/979-8-2600-3678-5.ch001

INTRODUCTION

Corporate environmental accountability (CEA) refers to the accountability of businesses to take responsibility for their operation having an adverse effect on the environment and humanity. It is a developing concept which encompasses both the impact of businesses on the environment, i.e., sustainable environmental practices, and the impact of businesses on people in terms of ethical business practices at the global level. Corporate environmental accountability is also introduced as a framework for understanding business's connection to and involvement in causing environmental, social, political and economic problems as it relates to the global society at large. Further, the concept is intended to be evaluated and measured in a more holistic manner than has been previously used, such as businesses' direct impact on the environment or their use of natural resources, but rather as a part of business's whole life cycle i.e., design, production, distribution, etc. Businesses must understand that they are not just a portion of society, but they have a responsibility to not just recognize the negative impact they are having on the environment, but they also must actively work towards reducing that impact. This chapter analyzes the evolution of corporate environmental accountability from a voluntary ethical system to a defined legal obligation. It makes the argument that although there are now many international guidelines in place that have developed a common language for environmental accountability, these same guidelines leave a large gap between the need and the ability to enforce them properly. Therefore, by examining the transition from soft law reporting to mandatory liability under law, this chapter assesses whether existing regulations can effectively fill the gap to ensure the alignment of corporate goals with results-oriented environmental protection for society.

To evaluate the importance of corporate environmental accountability with the genesis of an ever-increasing spectrum of environmental issues and challenges that cross the boundaries of countries' nation and impact the livelihoods of an ever-growing number of individuals. The environmental issues we are starting to face today are climate change, loss of biodiversity, water scarcity, and pollution, creating an obligation for businesses to take a proactive responsibility in managing their environmental impact. CEA is different from the traditional notions of environmental compliance, which is generally focused on meeting legal requirements or national standards, which is often very narrow in nature. Businesses must evaluate the impact of their entire operations, starting with the extraction of raw materials, through production, distribution, use, and ultimately the disposal of products. Business leaders are expected to include environmental considerations in their business strategies, governance and day-to-day business practices as part of their commitment to conducting business in an environmentally sustainable manner and protecting the environment and natural capital for future generations. (Wahyuni et al., 2024).

Environmental accountability on a global scale has been motivated not only by the ethical obligation to take care of the environment but also because of external pressure from consumers, investors, regulators, and civil society who are becoming increasingly aware of the condition of the environment. If environmental risks are not managed, they may result in reputational damage, operational disruptions, and/or financial loss. In contrast, businesses that are environmentally accountable can gain a competitive edge, access to green financing, and fulfil future market and regulator expectations. Consequently, Corporate environmental accountability is no longer a long-term or optional solution for businesses but has become essential to the continued survival and sustainability of businesses in the long-term. It is the same understanding that multinational corporations use when operating in different jurisdictions with different sets of environmental laws, which underlines the need for a uniform method of managing environmental accountability to create a coherent and effective environmental governance system.

The historical development of CEA towards an international standard is a complex convergence of globalization, environmental crises, and challenges associated with governance. Globally, environmental regulation traditionally relies on national regulatory legislation for the purpose of regulating pollution and managing natural resources. However, because of the evolution of multinational corporations and their extensive and complicated global supply chains, the incapacity of state regulatory authorities to regulate transboundary environmental concerns and business practices across multiple jurisdictions has become evident.

As a result of these realizations, there were multinational efforts to create formal international legal agreements, such as the Stockholm Declaration of 1972, which emphasized the idea of 'common but differentiated responsibility', and the Brundtland Report, a product of the UN Secretary General Waste Commission established in 1983, which created the structure for sustainable development. The Earth Summit of Rio in 1992 furthered these efforts by articulating both environmental protection principles and the principles for businesses operating on a global scale that are committed to sustainability. One of the most important points made in Rio was the recognition of non-state actors as stakeholders for sustainability, most notably corporate entities. However, there remains a significant lack of formal international environmental law to hold corporations accountable for their environmental actions. Thus far, most of the ongoing international efforts to hold corporations accountable for their environmental activities have focused on soft-law approaches, such as voluntary guidelines, codes of conduct, and multilateral initiatives like The Global Compact developed by the United Nations, The OECD Guidelines for Multinational Enterprises, and the United Nations Guiding Principles on Business and Human Rights, that require corporations to be environmentally responsible, conduct due diligence, and interact with stakeholders. These types of frameworks fill the legal vacuum created by the lack of hard law regarding international corporate environmental accountability and create a multi-stakeholder international mechanism to promote corporate responsibility and accountability culture among all sectors globally.

It is crucial to define and differentiate corporate environmental accountability from corporate environmental responsibility and corporate environmental liability prior to discussing corporate environmental accountability. Corporate environmental responsibility generally refers to a corporation's ethical and operational obligation to prevent or reduce environmental harm and promote sustainability, as well as all aspects of voluntary commitment or corporate social responsibility and internal discussions of the development of formal policies designed to reduce environmental harm. Corporate environmental responsibility is an aspirational or normative standard and as such, it reflects the way in which corporations ought to behave reasonably and ethically toward the environment, and therefore reflects what type of conduct corporations ought to exhibit in order to be referred to as acting in compliance with corporate environmental responsibility. (Wang & Li, 2024).

Unlike liability, which is limited to a narrower legal framework, accountability creates answerability and oversight by requiring corporations to report and justify their environmental impact, and rectify any errors made. There will be some basis for environmental accountability that is established through the transparency of environmental disclosures, such as the Global Reporting Initiative's (GRI) and the Task Force on Climate-related Financial Disclosures (TCFD) reports, and through the completion of commitments to science-based targets. Also included in accountability is engagement with the company's stakeholders, government agencies, investors, local communities, and civil society. These parties may bring claims against the company. Additionally, accountability generates expectations for continuous improvement, remediation of any impact, and for fulfilling reporting obligations that go beyond legal compliance with environmental regulations and the compensation of injured stakeholders.

Liability, on the other hand, is defined more narrowly within the context of corporations' potential exposure to liability resulting from harm to the environment due to their actions under international and/or national laws. Possible outcomes of liability include penalties, fines, or orders to remediate any damage caused to the environment. Liability connects parties to enforceable legal mechanisms. Thus, inequitable conduct by a corporation will be discouraged, and injured parties will have a means available for seeking to recover their damages. Whereas responsibility and accountability can be voluntary through the application of internationally accepted best practices, liability employs enforceable legal instruments and will be determined by the applicable jurisdictional authority as such. Therefore, liability offers the strongest coerce tool available in establishing that corporations comply with the legal requirements to remedy any harm caused to the environment. (Morgera, 2020).

Together, these create a continuum of voluntary ethical obligations to legally binding obligations out of the corporate environmental behavior of companies throughout the world to a varying level of degree. In the last several decades and certainly the last several months, there has been a greater focus on expectation, moving from responsibility to accountability to liability, particularly because of increasing environmental crises and growing expectations from stakeholders for corporate transparency and action. The creation of international frameworks increasingly contains components of all three, highlighting the demand for corporate actors to internalize environmental risks and impacts into their core business governance and to be subjected to increasing scrutiny and regulation. Recently, international policy discussions have outlined the multidimensional nature of corporate environmental accountability via a holistic approach. This includes having a broad understanding of environmental footprints including greenhouse gas emissions, resource consumption, waste generation, and ecosystem impacts at multiple points along the value chain. Effective accountability and governance also require extensive stakeholder engagement, where businesses actively involve the communities they impact, their employees, investors and NGOs to monitor corporate decisions and behavior. Transparency and reporting create ways for stakeholders to establish measures and assessments of the corporate actions and performances of all businesses to provide feedback on loop processes that will result in improved environmental outcomes.

Ultimately, the corporate accountability frameworks seek to incorporate proactivity and prevention for all environmental accountability rather than focusing on damage repair. Anticipatory approaches consider risk management, innovation, and pathways to environmentally sustainable technologies and business models as part of the commitment in the Paris Agreement to achieve ambitious functions related to emissions and resource efficiency. Corporations have an obligation to continue transforming their relationship between business and the environment internationally. Therefore, corporate environmental accountability is a developing framework for helping businesses understand their environmental impacts to transactional structure with different stakeholder groups that are interested in the corporation's operations. And apply accountability tools and remediation processes to support business accountability. As the complexities of addressing environmental concerns continue to develop and grow into a more significant international concern, this globally recognized framework will continue to assimilate economic development with ecological sustainability and social equity throughout all the nations (Belal et al., 2015).

The research method applied in this chapter is doctrinal. Therefore, the choice of certain standards for consideration within this research, namely UNGPs, GRI, and TCFD, has been made because they represent the prevailing consensus-driven benchmarks influencing international capital markets today and developing emerging national regulatory frameworks. An analysis of the selected instruments leads to consideration of the limitations of non-binding international law for attaining meaningful compliance by corporations. Furthermore, the chapter utilizes a critical literature review of both primary and secondary legal sources.

LEGAL FOUNDATIONS OF CORPORATE ENVIRONMENTAL ACCOUNTABILITY

Accountability held by corporations regarding their effect on the environment has developed through multiple layers of regulation, such as international environmental legal standards, significant treaties and conventions, soft law, and national and international liability law. Detailed analysis of all applicable environmental laws will follow demonstrating how corporate responsibility and accountability across the globe is regulated through legal structures.

International environmental law related to corporations begins with the concept that environmental protection is global, and therefore shared, concern calling for international order and action beyond a nation's state. The contemporary beginnings of international environmental law can be traced to lodestones in time or law and event, namely, the 1972 Stockholm Declaration, which set out broad principles while clarifying nationally the right of states to explore and use natural resources together with the obligation to not inflict damages outside national claims. This foundation began to develop internationally to establish a legal framework where nations could elevate environmental considerations into domains traditionally considered corporate activities. This event marked the very beginning of an inclusion of non-state actors like corporations into international environmental governance frameworks (Wang, 2023).

This framework continued to advance in 1992 with the Rio Earth Summit, where again the role of corporations was further entrenched in environmental governance, with the introduction of the principle of sustainable development, and comments to implement corporate environmental responsibility. While national governments bear primary international law obligations, the Summit acknowledged that multinational corporations, due to their global reach, play a crucial role in achieving environmental sustainability. International law thus began to develop not just through the creation of binding treaties, but through the creation of non-binding or voluntary soft law mechanisms that articulate recommended corporate behavior.

Examples of binding international treaties having an impact upon corporate environmental accountability include the Convention on Biological Diversity (CBD) that aims to conserve and ensure the sustainable use of biodiversity; the Basel Convention, which regulates the transboundary movement and disposal of hazardous wastes; and the Stockholm Convention, which seeks to eliminate persistent organic pollutants. While the subject of these conventions are state parties, they nevertheless have a limiting effect on corporations, as state parties are required to create implementing legislation in accordance with the conventions, and regulate activities with known environmental impacts. To comply with these conventions, corporations must demonstrate that they are following environmentally sustainable best practices, managing the risk associated with chemical substances, and providing comparisons of their environmental performance in jurisdictions where they operate (Lachenmeier & Montagnon, 2024).

In addition to treaties, there are also internationally endorsed soft-law instruments, such as the United Nations Guiding Principles on Business and Human Rights (UNGPs), which consider environmental due diligence to be part of a corporation's responsibility to respect human rights and the environment. The OECD Guidelines for Multinational Enterprises provide guidance on responsible and effective environmental management with an emphasis on transparency with all stakeholders and continuous improvement. These treaties are meant to be non-binding by their nature, yet they have substantial normative influence as a reference for the expectations of corporate behavior and act to influence the behavior of investors, and act as a reference point for corporate sustainability reporting and benchmarking.

National environmental legislation still plays an essential role in establishing corporate accountability for environmental impact. Most nations have comprehensive environmental laws that regulate pollution, use of natural resources, waste management, and environmental impact. Such laws create obligations for companies to achieve various environmental clearances, meet emissions standards, undertake environmental audits, and develop environmental mitigation or impact management plans, usually in collaboration with an environmental protection or regulatory agency. The scope of these laws relies largely on the ability to administer those laws, the judiciary's willingness to apply the law, and rather importantly, the political will to have those laws followed. Inconsistent levels of legal rigor and legal enforcement across countries complicate the issue of achieving environmental accountability from corporations on a consistent global basis, and the international regimes seek to address that inconsistency by enacting harmonization and dissemination of best practices.

The element of corporate liability, which is the ability to impose a legal consequence for environmental harm, is the fundamental component of accountability. Domestic regimes of liability, entrenched in tort, criminal, and administrative law, allow parties affected by environmental degradation, as well as regulators, to hold corporations accountable for the harm that they have inflicted against the environment. Doctrines based on principles, such as the 'polluter pays' principle and strict liability for inherently hazardous activities, have fundamentally impacted obligations companies owe to each other, the public, and a regulatory scheme in many countries; the obligations have established that the corporation has liability for the consequences even when negligence cannot be established. International legal schemes such as conventions related to oil pollution contain liability provisions that impose duties to clean up and compensation on corporations who cause harm across borders (Zhu, 2023).

As a result, international law has increasingly recognized due diligence obligations that call corporations to positively identify, prevent, and mitigate environmental risks associated with their operations and supply chains. Due diligence is becoming an important part of producing socially responsible sustainability by creating a legislative framework for addressing social and environmental responsibility that includes due diligence processes defined by the OECD Guidelines and the UN Guiding Principles. Due diligence requires organizations to establish organizational policies and processes to identify, assess, track, and report on their environmental impact. It establishes a culture of proactive environmental sustainability rather than reactive environmental sustainability through these due diligence processes. At present, due diligence is embodied primarily in voluntary standards. However, there is evidence from various jurisdictions' legislative and regulatory efforts to move towards an internationally recognized norm for mandatory due diligence at the corporate level. Also, international dispute resolution and arbitration mechanisms increasingly address such complaints from the communities regarding corporate behavior against them, which creates additional opportunities for organizations to be held accountable for their corporate behavior by creating processes for formal accountability in situations where there are no adequate or available domestic remedies. However, to date, these forms of accountability have received insufficient development and are minimally used. (Buhmann & Feld, 2024).

The legal principles of CEA are made up of numerous binding and voluntary agreements, as well as national legislation that applies to liability. As a result, the corporate environmental action regulatory structure is composed of various binding and voluntary elements. However, substantial inconsistencies exist with respect to the enforcement and application of these rules across boundaries. However, legal developments continue to focus on transparency, prevention, and stakeholder engagement.

GLOBAL STANDARDS AND GUIDELINES IN CORPORATE ENVIRONMENTAL ACCOUNTABILITY

At a global level, corporate environmental accountability is fundamentally informed by a variety of standards and guidelines which take the form of non-binding treaties in the sense that they are not legally binding but serve as standards or frameworks that corporations can still voluntarily adopt or are often expected to adopt based on stakeholder pressure and social and regulatory trends. These international standards are the basis of mechanisms to ensure transparency, ethical behavior, sustainability, and associated stakeholder engagement in international corporate environmental practices. There is a limitation on the ability to comprehensively explore the most influential global standards and initiatives impacting corporate accountability, namely the United Nations Guiding Principles on Business and Human Rights (UNGPs), the Accountability Framework Initiative (AFI), and other related international sustainability reporting frameworks. We will examine some consensus-based international mechanisms on ethically based supply chains and transparency in corporate environmental governance.

The UNGPs and other related international normative frameworks are the most impactful and currently updated and contextually relevant frameworks to address the role of business and human rights, including environmental rights. The UNGPs were first adopted in 2011, as the global standard for corporate responsibility to respect human rights, including environmental rights. The UNGPs are structured around three interrelated pillars: the State duty to protect human rights; the corporate responsibility to respect human rights; and access to effective remedies to address human rights abuses. Although they do not center exclusively on the environment, the UNGPs specifically address the connection between environmental harm and rights violations, including the right to health, rights to safe drinking water, and rights to a safe living environment. For corporate entities, this translates into an obligation to conduct a rigorous environmental due diligence process intended to identify and prevent environmental harm and related human rights harm resulting from their operations and their supply chain. The UNGPs have explicit provisions regarding environmental accountability, advocating for the monitoring of risk, transparency and stakeholder engagement. These provisions ensure that corporate human rights compliance includes environmental issues as an agenda item worldwide. The UNGPs have been incorporated into national action plans, investor frameworks and voluntary corporate policies, demonstrating the trend of utilizing the UNGPs to form regulatory and market expectations. (Razzak, 2024).

Practical resources are being utilized through the Accountability Framework Initiative (AFI), to assist businesses aligning their supply chains with the world's zero deforestation, sustainable land use, and biodiversity conservation objectives. The AFI was created via a multi-stakeholder process; thus, companies can use it to develop criteria when they create an environmental due diligence system, risk assessment, and report on sustainability performance. Companies using these criteria will specifically consider the commodities that are most frequently associated with environmental degradation and work toward operationalizing global commitments such as the Paris Agreement within their supply chains. The focus on transparency, traceability, and independent verification of supply chains provides recognition of the complexity of monitoring large multinational companies and their supply chains. The initiative promotes continuous improvement and accountability using measurable targets and public reporting, which will increase the confidence of investors, consumers, governments, and civil society and enhance their willingness to engage. (Repas, 2024).

There are also other various sustainability standards, other than UNGPs and AFI, which have been developed and introduced globally to provide more detailed reporting requirements and sector-specific benchmarks. The Global Reporting Initiative (GRI) is one of the most used standards that permit companies/ corporations to report on their environmental, social, and governance (ESG) impacts using globally accepted indicators. GRI standards connect corporate commitments and disclosures to sustainable development goals (SDGs), leading to greater transparency and comparability across sectors and nations. Likewise, the Sustainability Accounting Standards Board (SASB) promotes ESG disclosure standards that are industry specific, focusing on financial materiality, connecting environmental accountability to the investor's decision-making process. Moreover, the Task Force on Climate-related Financial Disclosures (TCFD), developed under the auspices of the Financial Stability Board, provides a standardized approach to climate-related risk disclosures, centered on governance, strategy, risk management, and metrics. None of these frameworks impose hard legal obligations, yet they have an indirect consequence on capital markets, as they begin to influence how investors assess risk and set lending terms, as well as influence how investors engage, effectively exerting pressure on companies to enhance their environmental reporting and accountability (Bais et al., 2024).

There are certification programs that support these international frameworks, as well as sector-specific initiatives, all of which promote ethical sourcing and responsible resource management. The Forest Stewardship Council (FSC), the Rainforest Alliance, and the Marine Stewardship Council are examples of certifications that promote responsible practices in sustainability for forests, agriculture, and fisheries. The purpose of these independent certification programs is to measure corporate practices against environmental accountability while enabling compliant practices to an internal standard, creating incentives for best practices, and ultimately helping consumers use their purchasing power to hold corporations accountable for their environmental impacts. Establishing consensus-based mechanisms for ethical supply chains and transparency in complex global corporate systems is necessary to ensure accountability. Initiatives such as the Ethical Trading Initiative (ETI), Sustainable Apparel Coalition, and Extractive Industry Transparency Initiative (EITI) bring companies, NGOs, governments, and labor organizations together to create codes of conduct, monitoring systems and grievance mechanisms. (Tysiachniouk & Vorbrugg, 2025).

The collaboration between various stakeholders will create a sense of trust, shared responsibility, and will allow stakeholders to come together as a unit rather than as separate entities. By creating an area in which companies can disclose their environmental performance and the harm done to the environment as a result of their operations, the transparency of the arrangement enables stakeholders, including investors, regulators, customers and the public stakeholders, to hold companies accountable for their supply chain practices. The emergence of global standards and guidelines has resulted in a transformation of corporate environmental accountability as an activity based on voluntary compliance by a small number of companies to now becoming a fundamental aspect of corporate governance. Although these frameworks are not supported by the legally enforceable authority that is available under treaties, the continuing prevalence of these frameworks demonstrates that a new ecosystem of what is referred to as non-legally binding soft law is underway and that they will continue to shape the practice of how the corporate world approaches its standards, practice and expectations from regulators and their communities. The interface between these standards and public policy and stakeholder activism continues to develop and enhance environmental accountability and therefore, support progress towards the achievement of global sustainable development objectives.

Despite this, there are uncertainties related to harmonizing these frameworks and operationalizing them across multiple corporate environments. Greenwashing concerns, inconsistent reporting quality, a relative lack of enforcement, and difficulties related to supply chain traceability remain in areas that merit ongoing focus and change with respect to accountability systems. Research and policy development is increasingly in favor of enhanced harmonization of these standards, deliberative stakeholder engagement, and aligning these standards with binding legal obligations to close gaps and improve global corporate environmental accountability. Developing global standards and guidelines have introduced a new set of tools in the effort to achieve corporate environmental accountability. The emergence of global standards for corporate environmental practices, from the normative framing of the UNGPs, to practical approaches for engagement like AFI, to the scope of ESG reporting standards, and finally, various consensus documents regarding shared supply chains. This work provides a clear example of collective and multiple stakeholder engagement with respect to prompting corporations to be more responsible stewards of the environment around the globe.

COMPLIANCE, MONITORING, AND ENFORCEMENT MECHANISMS

Accountability for environmental issues in business is dependent on compliance, monitoring and enforcement strategies that can effectively transform applicable law or policy into concrete actions. There are multiple levels or layers to the interconnected systems that provide information as to whether businesses have met policy expectations and voluntary standards, national regulatory systems and local internal controls, therefore the systems are very complex and can be unpredictable. It becomes necessary to highlight three categories of factors that affect the level of corporate accountability, including: International processes (IMPLAN) to verify corporate compliance and reporting, Due diligence systems for assessing the environmental impact of corporate operations; and transparency, traceability, and stakeholder consultation across the borders. These factors create the potential for corporate accountability and subsequently the achievement of corporate sustainability.

International processes have developed significantly to promote, enforce, and verify compliance by companies with environmental regulations and standards across borders. Through voluntary reporting standards like the Global Reporting Initiative (GRI), the Financial Disclosure Task Force on Climate Related Financial Disclosure (TCFD), and the Sustainability Accounting Standards Board (SASB), companies can make voluntary reports about their environmental performance. Although there are no legal requirements to use these standards, both the demands of the market and the requirements of regulators will lead to the widespread use of these Types of reporting. The environmental performance indicators used in these frameworks are both qualitative and quantitative, and include reporting on things such as greenhouse gas emissions, use of water, use of energy, use of waste and the impact on biodiversity, which provide a shared language to relay environmental risks to any interested party including investors, government officials and non-profit organizations. Through this harmonization of different types of disclosure frameworks, the ability for stakeholders to compare and benchmark the progress of a corporation toward their environmental goals will be improved, in turn enabling stakeholders to assess the corporation's commitment and verify the corporation's progress towards achieving its environmental objectives. (Joshi, 2012).

Regulatory systems across countries, which have legally enforceable obligations for compliance, supports frameworks for disclosure to satisfy the requirements of applicable regulatory regimes in high-environmental-risk sectors like extractives, chemicals and shipping. Examples of international treaties enforced by the International Maritime Organization (IMO) related to marine pollution and the Basel Convention dealing with transboundary movements of hazardous waste impose obligations on businesses to follow prescribed operational controls and report on those obligations to the IMO or the appropriate national authority. Many multinational corporations have developed an internal environmental management system (EMS) that meets or exceeds the requirements of the ISO 14000 series of standards to establish their commitment to monitoring and rapidly achieving compliance with international laws (Leary, 2023).

Environmental due diligence has become a prominent compliance requirement in almost all global corporate accountability initiatives. Due diligence, according to the United Nations Guiding Principles on Business and Human Rights, involves identifying, assessing, preventing and managing the negative impacts of a corporation's operations and supply chain activities on the environment. Due diligence, also known as environmental due diligence, should consider the entire life cycle of products and services and seeks to proactively assess the risks associated with each corporation, through conducting environmental impact assessments and regular environmental monitoring and management of the impacts of corporations. Through the development of the Corporate Sustainability Due Diligence Directive in various countries, the authority of the law to mandate due diligence now shifts from a voluntary requirement to a requirement that has enforceable legal consequences. The global trend that is emerging, which has already started, is to link the obligation of due diligence with the effectiveness of environmental protection through compliance with statutes or regulations pertaining to both the management of environmental risk and statutory requirements and a proactive management of environmental risk through compliance with laws, is an illustrative example of how corporations will need to approach environmental due diligence not simply as a series of check boxes but as part of the business's overall strategic plan. Consequently, at a minimum, due diligence will require a cross-functional approach that integrates the legal obligations and compliance requirements of corporations with the scientific evaluation of environmental impacts, the consultation with stakeholders, and the reporting of these activities in an open manner within a governance framework. (Corino, 2000).

Monitoring procedures provide an additional means of assuring compliance and continuing oversight of environmental performance. Examples of monitoring methods include government inspections, environmental audits, third-party verifications, and community-based monitoring. Environmental audits, whether conducted internally or externally, are critical to demonstrating compliance with environmental permits, identifying inefficiencies in operations, and reforming operational practices. Government regulators implement risk-based inspection systems around the world, allocating scarce government resources to supervise probable high-risk facilities or industries. Gradually, through more sophisticated technologies (remote sensing, satellite imagery, IoT sensors), real-time environmental monitoring and intervention are becoming economically feasible. New technologies permit regulators and civil society organizations to detect violations while they are occurring and provide a mechanism that allows for a prompter enforcement response (Hu & Xu, 2025).

Enforcement systems also permit standing for breaches of environmental obligations. Enforcement procedures may include administrative sanctions, fines, injunctions, and remediation orders for non-compliance or breaches of harm, and criminal prosecutions for serious breaches of harm or reckless negligence. The capacity to enforce non-compliance will require legally clear enforcement provisions, political will, transparent procedural accountability, and competency of institutions. Moreover, the enforcement of obligations is difficult and can be considerably challenging when multinational corporations engage in environmental practices and policies with relationships across multiple jurisdictions. There may be required bureaucratic and legal cooperation with many borders, and resources may also need to be allocated for domestic and cross-border enforcement through mutual legal assistance treaties and international litigation forums for dispute resolution that hold corporation's accountable outside of national borders. Market-based enforcement provides further legal consequences, but particularly through reputational harm, for non-compliance, as more investors leverage ESG factors to impose consequences on corporate environmental practices. For example, the continued scrutiny of corporate environmental processes from the media, civil society activism, and consumer blacklisting enforce compliance in addition to statutory requirements to meet environmental standards.

Transparency, traceability, and stakeholder engagement are essential tools for connecting compliance, monitoring, and enforcement in global supply chains. Transparency includes timely and factual disclosure of environmental data to help stakeholders examine corporate practices and hold organizations accountable. Reporting platforms and data portals created by entities that have signed onto the Extractive Industries Transparency Initiative (EITI) provide better access to data while increasing confidence in the information. The implementation of traceability systems is crucial for uncovering environmental impacts associated with complex supply chains, as well as for eliminating illegal activities, such as deforestation, and guaranteeing that raw materials have been acquired in an ethical manner. Increasingly, digital ledger technologies, such as blockchain, are being used to provide proof of traceability, to avoid greenwashing, and to support third-party validation. (Yanuardi et al., 2022).

Corporate environmental accountability relies on stakeholder engagement mechanisms to allow communities that are impacted by corporate actions to participate in shaping those actions, as well as enabling investors, employees, and NGOs a voice in the decision-making processes. Grievance mechanisms that are typically defined through international standards are a formalized way for affected stakeholders and parties to report concerns or violations and seek redress from the corporation through an established process. Participatory monitoring, citizen science, and collaborative platforms provide a link between local knowledge systems, thereby enhancing the oversight of legitimacy and effectiveness of these systems. Through authentic stakeholder engagement, it is possible to enhance dialogue, foster partnerships towards achieving sustainable development, reduce the potential for conflict, and improve corporate accountability. (Krebs, 2022).

The integrative processes used to create compliance, monitor compliance, and enforce compliance must use an integrated perspective of environmental accountability within the corporation. These processes can be continuously improved through technology, legal enforceability, and expanded involvement of a variety of stakeholders, thereby meeting new and emerging challenges. There are several core issues that need to be addressed related to enforcement, consistent levels of accountability, and complexity of jurisdiction. The emerging architecture will be a significant step toward placing environmental accountability within the global corporate governance framework. Normative processes that develop to the international standards establish obligations for disclosure and incorporate an anticipatory risk management component. Continuous monitoring can achieve an ongoing monitoring mechanism, while enforcement provisions ensure consequences for moving towards compliance and accountability. Transparency and stakeholder engagement will create adaptive systems of participatory governance. These elements form an integrated framework that progresses environmental accountability from ethical aspirations to the measurable realities of global corporate governance in enforceable terms (Khare, 2026).

POLICY INSTRUMENTS AND MARKET-BASED MECHANISMS

There are several policy tools and market-based mechanisms that encourage corporate conduct to be more sustainable, to uphold environmental standards, and to encourage transparency and stakeholder involvement, all of which affect the progress of corporate environmental accountability. There are three interrelated areas that require additional investigation to provide further insight into the development of corporate accountability related to their environmental impact, that is, environmental standards, certifications & disclosure frameworks; market incentives & economic mechanisms to promote sustainability and; stakeholder engagement & transparency to provide sufficient confidence in corporate accountability.

The backbone of environmental accountability is comprised of environmental standards, certifications & disclosure frameworks as these provide a clear minimum set of requirements and benchmarks for corporate environmental performance. For example, the ISO 14001 environmental management systems standard provides corporate users with a systematic approach for managing their environmental responsibilities, reducing pollution, and continuously improving the company's ecological footprint. These standards focus on organizational processes, including policy, risk assessment, operational control, and training that allow organizations to internalize the notion of environmental accountability in their governance apparatus. Certifications, which are provided by third-party independent issues, demonstrate the entity's compliance with external standards, by showing stakeholders that organizations are meeting externally defined environmental standards. For example, the Forest Stewardship Council (FSC) and Rainforest Alliance certifications demonstrate that an organization is sustainably managing their forestry and/or agricultural product, thus providing additional market value and providing confidence to consumer choices (Gutterman, 2023).

Complementary standards are comprehensive disclosure regimes that require or provide motivation for companies to publicly disclose their environmental impacts. Three organizations, that is, the Global Reporting Initiative, the Sustainability Accounting Standards Board and the Task Force on Climate-related Financial Disclosures, have come together to create a standardized framework, which can be used as a guideline for comparable corporate progress. The standardized framework would allow for comparable corporate sustainability disclosures such as greenhouse gas emissions, resources utilized, water management, and biodiversity impact. Public disclosure helps to establish accountability by providing information to investors, regulators, and the public to make well-informed decisions, which then helps to improve environmental policies within a company. Environmental corporate reporting is evolving to become more consistent with other areas of sustainability. As a result, corporate accountability towards sustainability becomes even more firmly established because of quantifiable corporate sustainability actions across the globe. (Shekhawat & Khare, 2026).

Market-based incentives for sustainable economic behavior apply economic variables to motivate corporations to voluntarily, or in compliance with regulation, enact environmentally responsible behaviors. Market-based environmental policy instruments use price signals, financial incentives, and/or tradable rights to facilitate cost-effective implementation of environmental goals. These instruments differ from command-and-control (or regulatory) approaches by providing flexibility in implementation and rewarding innovation. Price-based instruments for environmental policy include environmental taxes, pollution charges, and subsidies that seek to internalize the environmental externalities, making it more expensive to implement unsustainable practices, while providing rewards for green alternatives (Almaqtari et al., 2023).

Tradable permit systems, or cap-and-trade systems, exemplify quantity-based market mechanisms. A regulatory body establishes a total limit on emissions or resource use and distributes a limited number of permits to firms that enable trade among the firms. Firms that can decrease emissions or pollution below their permit requirements can sell surplus permits allowing for efficient pollution control and integrated cost savings and technological innovation. The European Union Emissions Trading Scheme (EU ETS) is the largest cap-and-trade scheme in the world and has been able to show considerable reductions at lower costs to society by utilizing market-based mechanisms (Dechezlepretre, 2023).

Deposit-refund systems incentivize recycling through fees or tax at the time of purchase and provide a refund upon return, effectively inducing consumers and producers to take ownership of waste management to motivate recycling of the product. Market-based mechanisms also extend to public policy or green public procurement. Governments procure goods and services, environmentally friendly goods or services, for example, which help develop markets for more sustainable products. Environmental labeling and eco-certification mechanisms further provide a market mechanism, since both allow for consumer differentiation and take choice of products based on either the label or certification or the underlying environmental performance, allowing consumers to reward or choose between corporations based on their responsibility to work towards sustainability.

A successful policy will be created by engaging key stakeholders in policy development and maintaining transparency about how policies will be implemented and enforced through legislation and contract law. Multi-stakeholder platforms are able to provide a mechanism for dialogue and cooperation between various stakeholders in the corporate ecosystem, such as corporations, investors, government agencies, NGOs, community groups and consumers, which establishes a common understanding of their roles and functions and fosters cooperative actions toward shared responsibilities in addressing the urgent environmental concerns we face today. Mechanisms for clear, public disclosure of corporate environmental responsibilities and performance offer an easy way for stakeholders to hold corporations accountable for their environmental responsibilities.

Accountability and remedial mechanisms to enable individuals and the broader community to make a formal complaint and seek redress for environmental impacts from corporate conduct, are typically implemented under the UN Guiding Principles on Business and Human Rights, by creating a culture of responsibility and accountability for corporate conduct related to the environment that lies beyond legal compliance, and ensures that corporations have a role in shaping environmental governance through the engagement of affected communities.

Digital technologies support transparency and increase stakeholder engagement through improvements in the access, traceability, and verification of environmental data. Applications of blockchain provide secure and immutable traceability throughout the supply chain, reducing the risk of greenwashing and increasing the level of confidence markets will have in a corporation's commitment to sustainability.

Technologies such as remote sensing, geographic information systems, and online reporting platforms enable real-time monitoring and provide a public good that addresses the information asymmetry inherent in the corporate-regulatory-stakeholder interface. Policy instruments and market-oriented mechanisms are used in concert as an enabling environment for which corporate environmental responsibility is viewed as both economically rational and socially demanded. Environmental standards and certifications establish performance and compliance baselines. Market-based standards and economic instruments shift cost structures to support sustainable innovation and operational performance. And frameworks for stakeholder engagement and transparency create equitable governance processes which foster legitimacy and facilitate continuous improvement. This collection of instruments constitutes an effective toolkit for enabling corporate sustainability, reducing environmental harm, and promoting sustainable development at the global level (Eakpisankit & Pirzada, 2025).

CEA IN PRACTICE AND ASSOCIATED CHALLENGES

The requirement for corporate sustainability has become a global ethical and governance concern, different regionally and with respect to sectors and methods used. Examples of global corporations coupled with regulatory regimes will assist in extracting lessons learned from current international practices. Critical analysis of existing new issues will provide insights on how these influence corporate behavior. Ultimately, the lessons learned serve to distill what may be defined as best practices, along with highlighting existing and emerging issues. This can be seen in the case of greenwashing, the balance of economic growth & environmental protection, and increasing use & importance of technology & innovation in executing corporate governance.

For example: Unilever, Patagonia, Tesla & Nestle are all examples of global corporations that have adopted various approaches toward their commitment to environmental sustainability, like through the integration of sustainability, third party certification, transparency in disclosure and stakeholder engagement. Specifically, Unilever's Sustainable Living Plan, Patagonia's use of recycled materials & active advocacy, Tesla's impact report & carbon report, and Nestle's involvement in water resource stewardship outline specific targets, independent verification through third party audits, and clear communication practices that will provide a genuine difference in the environmental sustainability of their respective policies & practices. (Lawrence et al., 2018).

These corporations' experiences illustrate how corporations can take an environmentally responsible approach to their corporate strategies, innovate, and meet their stakeholders' demands while improving their corporate resiliency and market reputations. In contrast, the case studies from Nigeria's oil and gas sector and Ethiopia's manufacturing industry indicate that there are lower levels of regulatory enforcement, low levels of public participation and engagement, as well as challenges in the implementation of global accountability standards. This illustrates that local conditions must be taken into account when developing voluntary initiatives and that the establishment of normative international standards will require adequate government capacity to comply with and engage with the community. (Wong, 2025).

There are significant differences in how accountability for environmental impacts has been approached in regulatory and market-driven contexts. The European Union (EU) is currently leading in developing binding climate laws that provide clearly defined non-financial reporting and due diligence obligations and create a mechanism for benchmarking jurisdictions. In contrast, the United States employs a fragmented mix of federal statutes, investor ESG requirements, and consumer scrutiny to create a dynamic albeit sometimes inconsistent set of accountability mechanisms. Emerging economies provide many models, with gaps in compliance with and adaptation of international standards requiring better clarity of purpose and signaling harmonization with regulatory, localization of regulation mechanisms, and better capacity development. The comparative analysis demonstrates the merits of transparency, multi-stakeholder governance, and continuous improvement amongst comparative analysis regardless of the governing jurisdiction (Kulovesi et al., 2024).

Case studies and regional differences demonstrate several essential lessons. Initially, mandatory disclosure and third-party validation increase environmental accountability and trust amongst stakeholders. Secondly, collaborative governance, which brings together regulators, communities, civil society, and investors, addresses gaps and promotes ongoing change. Thirdly, organizational responsiveness to changing institutions, changes in policy, and solitude with innovation promotes ongoing accountability. Fourthly, cooperation of global frameworks must be adapted to local contexts with strong support from provincial and State governments. Best practices show that sustainability must be at the core of strategy planning as opposed to a stand-alone responsibility, and that the mixture of accountabilities must demonstrably support measurable environmental impact and resilience.

Greenwashing continues to erode corporate commitment to environmental accountability. Companies may exaggerate or even fabricate sustainability-related claims to satisfy short-term market expectations or similar and superficial compliance to an external standard. The adaptability of verification, independent audit, certification, and public, transparent reporting at the organizational level counteracts this risk across sectors. Regulatory and market oversight functions must be theoretically developed to harness the possibilities of digital technologies and improve civil society engagement and mobilization to protect against companies that exploit ESG assumptions to support false assumptions. Forth, while increased use of blockchain technologies, public disclosure platforms, and tracking devices integrated into the supply chain principles improve and enhance transparency, for many actors, developing new governing and technology knowledge is required (Sneideriene & Legenzova, 2025).

Accountability gaps occur when regulatory systems are weakly established; enforcement of them is weak, or voluntary standards are not rigorously adhered to. In places where governments have limited capacity, companies tend to face higher levels of risk as well as less incentive to engage in sustainable behavior. Thus, international organizations, market actors, and digital infrastructures are especially important in overseeing behavior, setting standards, and promoting best practices. Bridging accountable gaps will require continuous harmonization with respect to local and global approaches and collective commitment from stakeholders.

The balancing act between profit and sustainable stewardship is a source of continued tension among all corporations, particularly those that operate within rapidly developing economies. As corporations navigate through this tension, they are faced with regulatory pressures from various levels of government, and various advocacy groups as well as short- and long-term risk mitigation. Research done by several leading corporations shows that there is great potential to create innovation, enterprise risk management strategies, and competitive advantages by integrating sustainability into their operational model. Therefore, for sustainability to be able to scale, there needs to be a collective goal of adaptive leadership, an ongoing commitment to strategic investments and multiple stakeholder participation to align business interests within the planetary boundaries.

Environmental sustainability is being transformed by digital innovation. Digital revolution supersedes the current environmental corporate responsibility paradigm by diminishing the empirical difference in information available to corporations and the public. However, there are limitations with just having digital transparency alone. Although things like blockchain, cloud computing, and real-time monitoring can improve the accuracy of your information, using these technologies does not guarantee you will hold businesses legally accountable for their actions. Many times, regulatory fragmentation has limited the power of these technologies because there is no single global authority assigned to certify the accuracy of claims made by business. The reuse of products and closed-loop production is both technologically feasible and necessary, as consumers are opting to purchase sustainable products. While there are some companies making progress toward measuring their accountabilities through changes, there is still immense opportunity to create meaningful progress through stimulating alignment between transparent accounting systems, continuous reporting, stakeholder engagement programs, and technology. It is important to note that true accountability is more than meeting a minimum standard.

It is just as important for accountability to be sustainable and responsive at every tier within an organization as well as to all others involved with the organization. As developed countries and developing countries face continuing change and the emergence of future technologies, the direction of corporate accountability will be shaped by strong regulations, informed market-based incentives, verification systems, and cross-sector partnerships.

CONCLUSION AND WAY FORWARD

To effectively strengthen corporate accountability for the environment, there is need to adopt an integrated strategy that incorporates better legal and policy frameworks, better enforcement and compliance mechanisms, changing the corporate culture towards being sustainable, and globally collaborating. One of the essential elements of holding corporations accountable for their environmental impact will be effective laws and policies governing their operations. At international and national levels, existing legal frameworks often have issues with being fragmented, inconsistent, and difficult to enforce. It is necessary to create more integrated and binding legal instruments to establish corporate environmental responsibilities, provide clear due diligence obligations and establish acceptable procedures for remediation in case of an environmental disaster on the part of a corporation. Additionally, countries need to create an enforceable mandatory framework that requires all corporations to perform and disclose comprehensive assessments of environmental risks throughout all their supply chains. Incorporating environmental accountability measures into corporate governance, including incorporating fiduciary obligations of directors and executives, will ensure that there are adequate internal accountability and motivation for continued adherence to these requirements by all levels of the organization. To keep pace with innovation, updating environmental regulations to incorporate new environmental hazards such as biodiversity loss, plastic pollution, and climate change is essential for timely and relevant responses. Regulatory bodies may work to promote and coordinate the harmonization of international treaties and agreements with domestic legislation, thereby eliminating confusion, preventing regulatory arbitrage, and encouraging compliance and accountability of regulated entities.

Legal reform should be accompanied by supporting resources for enforcement and compliance of laws by facilitating the translation of rules into enforceable actions. Many jurisdictions face challenges due to a lack of resources within regulatory agencies, leading to inconsistent enforcement practices, therefore, hindering corporate accountability. Strengthening the capacity of regulatory agencies through additional funding, technical assistance in training staff, and the use of new monitoring technologies such as remote sensing and data analysis can expand the reach and accuracy of regulatory agencies. Requiring that environmental auditing and verification of environmental audits be conducted voluntarily by a third-party entity is an additional means to ensure independent oversight over enforcement and compliance with applicable environmental laws and regulations.

Enforcement strategies should also provide a dual benefit of punitive mechanisms for non-compliance alongside incentives or privileges for exemplary corporate conduct, including preferred access to green finance, and public procurement. Joint government, international, and civil society actors can also collaborate to support transboundary enforcement and monitoring, enough to inform the development of norms for corporations that may act outside jurisdiction or have multinational supply chains. Finally, the integration of investors and financial institutions as enforcers of compliance will support the implementation of environmental risk for lending and investment decisions.

It is critical for a transition to sustainability to engage in a corporate culture change towards permanently embedding environmental accountability into business practice. Legal and regulatory structures provide necessary external pressure but subsequently require businesses to internalize the values of sustainability in their corporate strategy, leadership priorities, and operational practices. Businesses should incorporate environment-focused governance, develop transparency around sustainability goals succinctly linked to the sustainable development goals, and reward innovative sustainable product development, resource efficiency, and circular economy development. Communicating about environmental impacts and actions taken toward them, working with stakeholders, and being responsive to community feedback increases trust and legitimacy. To support employee engagement in ethical behavior, corporate leaders must advocate for and implement sustainable corporate goals. Education, awareness campaigns, and collaboration with institutions of higher education and industry associations provide ongoing opportunities for learning and sharing best practices (Ochoa et al., 2024). Companies that successfully implement sustainability practices in their culture typically achieve a competitive advantage, including improved brand image, increased customer loyalty, and reduced regulatory risk associated with their supply chains.

Lastly, there needs to be a collaborative global approach toward standardizing the environmental impacts that corporations have across national borders and the globalization of supply chains. The increase in voluntary standards and reporting mechanisms signals a desire for greater alignment. However, the number of different standards makes implementation or evaluation complicated. International organizations should lead the convergence of sustainability standards, reporting guidelines, and due diligence, and they should create interoperable frameworks that establish higher standards and greater ambition while not creating unnecessary burdens on businesses. Extending present efforts such as the Accountability Framework Initiative, The United Nations Guiding Principles for Business and Human Rights, and the Task Force for Climate-Related Financial Disclosures, will establish linkage and cumulative benefits.

Capacity-building initiatives in developing nations are an important way to promote equitable distributions of burdens of global accountability on the environment and serve as a means of protecting against the regulatory gaps, which may leave corporations with unregulated opportunities for environmental wrongs. Addressing corporate environmental accountability requires enhanced cooperation from countries, combined with inclusive engagement from all stakeholders, to facilitate the establishment of an international 'level playing field' where environmental accountability is recognized to be a shared responsibility and a business advantage.

In conclusion, for corporate accountability to become environmentally responsible, there needs to be a transition from voluntary disclosure to a requirement of mandatory liability or hard law, as the existing base of soft law has created a legislative framework that serves as a normative benchmark but is weak in its ability to compel international adherence and to achieve real sustainability, future policies will need to emphasize creating uniform standards across jurisdictions in order to eliminate regulatory arbitrage and the creation of an enforceable obligation on companies to engage in adequate environmental due diligence as part of their overall fiduciary duties in corporate governance.

REFERENCES

Almaqtari, F. A., Elsheikh, T., Abdelkhair, F., & Mazrou, Y. S. A. (2023). The impact of corporate environmental disclosure practices and board attributes on sustainability: Empirical evidence from Asia and Europe. *Heliyon*, *9*(8), e18453. DOI: 10.1016/j.heliyon.2023.e18453 PMID: 37560679

Bais, B., Nassimbeni, G., & Orzes, G. (2024). Global Reporting Initiative: Literature review and research directions. *Journal of Cleaner Production*, *471*, 143428. DOI: 10.1016/j.jclepro.2024.143428

Belal, A. R., Cooper, S. M., & Khan, N. A. (2015). Corporate environmental responsibility and accountability: What chance in vulnerable Bangladesh? *Critical Perspectives on Accounting*, *33*, 44–58. DOI: 10.1016/j.cpa.2015.01.005

Buhmann, K., & Feld, L. (2024). Shifting boundaries: A transnational legal perspective on the EU Corporate Sustainability Due Diligence Directive. *Transnational Legal Theory*, *15*(3), 370–396. DOI: 10.1080/20414005.2024.2426961

Corino, C. (2000). Environmental due diligence. *European Environmental Law Review*, *9*(4), 120–124. DOI: 10.54648/268317

Eakpisankit, A., & Pirzada, K. (2025). Comparative environmental disclosure practices: Evidence from the UK and the USA. *Cogent Business & Management*, *12*(1), 2534093. Advance online publication. DOI: 10.1080/23311975.2025.2534093

Gutterman, A. (2023). Certifications and Rating Systems for Assessing Social and Environmental Impact. *ResearchGate*. https://www.researchgate.net/publication/370631183_Certifications_and_Rating_Systems_for_Assessing_Social_and_Environmental_Impact

Hu, B., & Xu, Q. (2025). Environmental regulation penalties and corporate environmental information disclosure. *International Review of Economics & Finance*, *102*, 104344. DOI: 10.1016/j.iref.2025.104344

Joshi, A. B. (2012). Corporate environmental responsibility: A liability or challenge. SSRN *Electronic Journal*. https://doi.org/DOI: 10.2139/ssrn.2157398

Khare, P. (2026). Integrating the SDGs Into Corporate Governance: A Legal Imperative for Global Sustainability. *IGI Global Scientific Publishing*, 391-416. https://doi.org/DOI: 10.4018/979-8-3373-4192-7.ch015

Krebs, D. (2022). Environmental due diligence obligations in home state law with regard to transnational value chains. *Springer International Publishing*, 245-350. https://doi.org/DOI: 10.1007/978-3-031-13264-3_7

Kulovesi, K., Oberthür, S., van Asselt, H., & Savaresi, A. (2024). The european climate law: Strengthening EU procedural climate governance? *Journal of Environmental Law*, *36*(1), 23–42. DOI: 10.1093/jel/eqad034

Lachenmeier, D. W., & Montagnon, C. (2024). Convention on Biological Diversity (CBD) and the Nagoya Protocol: Implications and Compliance Strategies for the Global Coffee Community. *Foods*, *13*(2), 254. DOI: 10.3390/foods13020254 PMID: 38254555

Lawrence, J., Rasche, A., & Kenny, K. (2018). Sustainability as opportunity: Unilever's sustainable living plan. In *Managing Sustainable Business* (pp. 435–455). Springer Netherlands., DOI: 10.1007/978-94-024-1144-7_21

Leary, D. (2023). International maritime organization (IMO). *Yearbook of International Environmental Law*, *34*(1). Advance online publication. DOI: 10.1093/yiel/yvaf050

Morgera, E. (2020). Introduction - In Corporate Environmental Accountability in International Law. *Oxford University Press.* 1-23. https://doi.org/DOI: 10.1093/oso/9780198738046.003.0001

Ochoa, A. G., Khare, P., & Setia, S. (2024). Expanding Environmental Education in Honduras: A Holistic Approach for Sustainable Development. *IGI Global Scientific Publishing*, 153-170. https://doi.org/DOI: 10.4018/979-8-3693-2577-3.ch010

Razzak, M. (2024). The UN guiding principles on business and human rights: A commentary. *Australian Journal of Human Rights*, *30*(1), 125–130. DOI: 10.1080/1323238X.2024.2349322

Repas, L. (2024). Accountability Framework initiative and Canopy: Statement of Alignment. *Canopy*. https://canopyplanet.org/news/accountability-framework-initiative-and-canopy-statement-of-alignment

Shekhawat, V. & Khare, P. (2026). Analyzing the Impact of International Law and Sanctions Mechanisms Towards Sustainable Outcomes. *IGI Global Scientific Publishing*, 327-348. https://doi.org/DOI: 10.4018/979-8-3373-1707-6.ch010

Sneideriene, A., & Legenzova, R. (2025). Greenwashing prevention in environmental, social, and governance (ESG) disclosures: A bibliometric analysis. *Research in International Business and Finance*, *74*, 102720. DOI: 10.1016/j.ribaf.2024.102720

Tysiachniouk, M., & Vorbrugg, A. (2025). The transformation of the Forest Stewardship Council (FSC) Certification Scheme in Russia after the full-scale invasion of Ukraine. *Forest Policy and Economics*, *178*, 103559. DOI: 10.1016/j.forpol.2025.103559

Wahyuni, W., Syamsidar, S., Indasari, F., & Bualemo, L. E. G. (2024). Corporate environmental accountability: A qualitative study of stakeholder expectations and corporate practices. *Advances: Jurnal Ekonomi Bisnis*, *2*(6), 323–336. DOI: 10.60079/ajeb.v2i6.352

Wang, Q., & Li, G. (2024). Research on the effect of corporate environmental responsibility on corporate sustainability and the mediator effect of corporate environmental strategy. *SAGE Open*, *14*(3), 21582440241266115. Advance online publication. DOI: 10.1177/21582440241266115

Wang, X. (2023). The 1972 Stockholm Conference and China's diplomatic response. *Cultura e Scuola*, *6*(2), 146–152. DOI: 10.1177/20966083231184691

Wong, Y. A. (2025). A study on the effectiveness of Patagonia's sustainability-driven brand strategy. *Advances in Economics. Management and Political Sciences*, *180*(1), 302–306. DOI: 10.54254/2754-1169/2025.23726

Yanuardi, Y., Vijge, M. J., & Biermann, F. (2022). Social-ecological reflexivity of extractive industry governance? The case of the Extractive Industries Transparency Initiative in Indonesia. *Environmental Policy and Governance*, *32*(5), 426–437. DOI: 10.1002/eet.1988

Zhu, L. (2023). Some thoughts on application of the polluter pays principle for controlling marine greenhouse gas emissions. *Marine Policy*, *158*, 105877. DOI: 10.1016/j.marpol.2023.105877

Chapter 2
The Invisible Lines of Environmental Reporting:
Beyond Measurement in Environmental Disclosure

Khuloud Ali
Federal University of Rio de Janeiro (UFRJ), Brazil

Ghayth Tintawi
http://orcid.org/0009-0009-9891-9528
IMAGINE Studios, Brazil

Mohamad Khaled Bassma
http://orcid.org/0009-0002-1849-3651
IMAGINE Studios, Brazil

ABSTRACT

Environmental reporting is often treated as a problem of measurement accuracy. Credibility, in this view, depends on data quality and the reliability of calculation methods. This chapter adopts a different position. It argues that the main sources of variation arise earlier, during the construction of reporting boundaries. The analysis focuses on the built environment. Environmental performance is studied through decisions that define scope, establish baselines, introduce assumptions, and determine materiality. These elements operate before measurement is finalized. They determine what is included, how impacts are interpreted, and how results are presented. Building on this, the chapter develops a conceptual synthesis drawn from environmental accounting, sustainability reporting, and life cycle assessment. Through this synthesis, disclosure is reframed as an interpretive process rather than

DOI: 10.4018/979-8-2600-3678-5.ch002

a neutral reflection of reality. While reporting outcomes may seem consistent, their underlying conditions may differ. Such differences limit comparability and obscure assumptions.

ENVIRONMENTAL REPORTING BEYOND NUMERICAL PRECISION

Environmental reporting is often treated as a question of measurement reliability. Under this view, the credibility of disclosure depends on the reliability of the data, the soundness of the calculation methods, and the consistency of the indicators over time. Errors are typically attributed to numerical miscalculation, incomplete datasets, or unsuitable methodological choices. These concerns remain valid, yet they do not fully account for the variation seen within environmental disclosures, even when similar standards and tools are applied. This matters materially in the built environment: Buildings and construction accounted for approximately 34% of global final energy demand and 37% of energy- and process-related carbon dioxide emissions, with embodied carbon contributing roughly 25–28% of total sectoral emissions (Global Alliance for Buildings and Construction and United Nations Environment Programme, 2024).

Looking more closely, divergence frequently appears before any calculation takes place. It begins when reporting boundaries are defined. At that stage, organizations decide which activities to include, which impacts to recognize, and how to interpret relationships among assets, operations, and value chains. These early decisions determine the conditions under which environmental performance becomes visible and measurable. Such an observation is consistent with a wider conception of disclosure as an interpretative process rather than a strictly technical one. Environmental indicators are generally understood to depend on the frameworks that shape them. They are produced through decisions about system boundaries, accounting frames, and estimation rules that translate physical activity into quantified impacts and reported categories (Ascui & Lovell, 2011; Matthews et al., 2008). Two disclosures may therefore present internally consistent data while conveying different representations of performance. The issue is not only whether the numbers are correct. It concerns the logic that defines what the numbers are taken to mean.

A Shift in Focus Toward Boundary Construction

Accepting the role of boundary construction requires a change in analytical orientation. Attention shifts from downstream calculation to upstream design. Environmental reporting can be understood as a sequence of decisions, in which measurement is only one stage within a broader process of representation. Before emissions are calculated or energy use is normalized, reporting teams define the analysis perimeter, establish reference points, and adopt assumptions to fill gaps between available data and required indicators.

These earlier decisions affect disclosure in ways that are less visible than numerical outputs yet more influential in interpretation. Scope determines how impacts are attributed across ownership structures, operational control, tenancy arrangements, and supply chains. Baselines define the reference condition against which change is assessed. Assumptions translate activity into impact through emission factors, modeling parameters, and normalization procedures. Each element introduces interpretation into the reporting structure.

The implications are not minor. When attention remains fixed on numerical outputs, disclosures may appear comparable despite relying on different boundary conditions. Once boundary construction is made explicit, differences in reported performance can be interpreted not only as operational variation but also as outcomes of methodological design. Environmental reporting then appears as a governance practice in which inclusion, exclusion, and interpretation carry analytical weight. This shifts the question from measurement alone to the design choices that make performance visible (De Cristofaro & Gulluscio, 2023; Ng et al., 2022).

To clarify the shift from measurement to construction, Figure 1 illustrates environmental reporting as a process of progressive transformation rather than direct representation. The figure conceptualizes reporting as a funnel in which complex environmental reality is systematically reduced through successive stages of boundary construction, filtering, and representation. At each stage, decisions shape how data are selected, interpreted, and ultimately communicated. As the process advances, abstraction increases while the richness of underlying information decreases. This visualization makes explicit how methodological choices, often treated as technical details, collectively influence the final form of reported performance.

Figure 1. Distortion Funnel illustrating the transformation of complex environmental reality into reported performance through successive stages of boundary construction, filtering, and representation

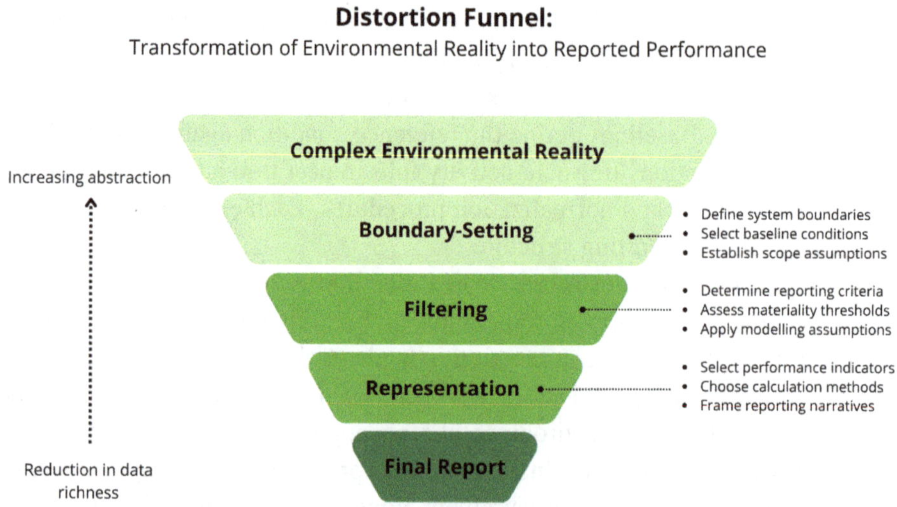

The figure presents the author's conceptual representation of how reporting progressively increases abstraction while reducing data richness. The boundary-setting stage shown in the figure refers to the reporting decisions through which system boundaries, baseline conditions, and scope assumptions are established.

Why the Built Environment Intensifies This Problem

The role of boundary construction becomes more pronounced in the built environment. Here, environmental performance is distributed across multiple actors, extended time horizons, and overlapping spatial scales. Real estate portfolios, construction projects, and infrastructure systems do not operate within clear organizational boundaries. They involve relationships among owners, developers, contractors, operators, tenants, and regulators, each exercising different degrees of control. This disintegration complicates the definition of scope. Decisions must address whether reporting reflects entity-level ownership or project-level activity, whether leased assets are included, and how responsibility is distributed between landlords and occupants. These choices affect both the magnitude of reported impacts and their attribution. In practice, the boundary is never just a technical line.

Temporal structure adds one more level of complexity. Buildings and infrastructure span life cycles that include design, construction, operation, retrofitting, and eventual decommissioning. Environmental impacts are unevenly distributed across these phases. Embodied carbon is concentrated early, while operational emissions continue over long periods. The selection of time frames determines whether disclosures emphasize instant effects or long-term performance.

In addition, the sector relies extensively on modeled data. Energy performance is often simulated, occupancy patterns are generalized, and emission factors are applied using regional averages. These practices are necessary, yet they introduce interpretative layers that are rarely visible in final disclosures. Environmental reporting in the built environment therefore depends heavily on modeling assumptions, system boundaries, and indicator design (Backes & Traverso, 2021; Bragança et al., 2010; Castrillón-Mendoza et al., 2024). Taken together, organizational fragmentation, life-cycle extension, and reliance on modeling create conditions in which boundary-setting decisions strongly influence reported outcomes. Without explicit attention to these decisions, disclosures may appear coherent while masking the assumptions that define their meaning.

Scope and Contribution of the Chapter

This chapter examines how boundary construction shapes environmental reporting in the built environment. It also argues that credibility and interpretability depend not only on measurement accuracy but also on the clarity of decisions that precede measurement. By focusing on the key elements: scope, baselines, and assumptions, this chapter explains how environmental performance is constructed. It also clarifies how environmental performance can be evaluated with greater interpretive rigor.

The discussion adopts a conceptual yet practice-oriented direction. Rather than providing a comprehensive review of reporting standards, it synthesizes recurring patterns identified in the literature on sustainability disclosure, materiality, and greenwashing risk. These patterns are used to examine how reporting frameworks operate in practice and where ambiguity or inconsistency may arise. On this basis, environmental disclosure is presented as an evidence chain that highlights the relationships among boundary construction, data generation, indicator selection, and reported outcomes. This perspective clarifies how disclosures may appear methodologically sound while remaining difficult to compare or verify. Environmental reporting is therefore reframed as a process that constructs visibility. Boundary construction determines what becomes measurable, comparable, and reportable, while also shaping what remains excluded. Making these boundaries explicit provides a basis for more transparent interpretation and supports both academic evaluation and practical decision-making in the built environment.

The chapter contributes to environmental reporting scholarship by introducing the "invisible lines integrity model," which synthesizes and extends existing perspectives by reframing environmental reporting as a constructed process rather than a neutral act of measurement. It also introduces operational tools—including boundary mapping, baseline definition, assumption documentation, and materiality-to-KPI tracing—for evaluating reporting integrity beyond numerical accuracy. The chapter's contribution is therefore both conceptual and diagnostic: it explains how reporting outcomes are produced and provides a structured lens for examining the decisions that shape them.

Methodological Positioning

This chapter adopts an interpretive conceptual synthesis rather than an empirical research design or systematic literature review. Its purpose is not to test hypotheses, measure reporting quality across a sample of organizations, or compile a comprehensive inventory of environmental reporting standards. Instead, it synthesizes selected and conceptually relevant bodies of literature to develop a diagnostic framework for interpreting how the boundaries of environmental reporting are constructed, justified, and communicated.

The synthesis draws primarily from four overlapping domains. The first is environmental accounting and sustainability disclosure, which provide the conceptual basis for understanding reporting as a mechanism of transparency, legitimacy, and stakeholder communication. The second is greenhouse gas (GHG) accounting, which addresses questions of scope, control, consolidation, and value-chain attribution that are central to the representation of environmental responsibility. The third is life-cycle assessment and life-cycle sustainability assessment, which clarify how system boundaries, inventory assumptions, and impact categories shape the interpretation of environmental performance in the built environment. The fourth is materiality and greenwashing literature, which explain how disclosure can become selective through omission, emphasis, framing, or weak linkage between impacts and indicators.

The methodological logic is therefore framework-development oriented. The chapter identifies recurring conceptual tensions across these literatures and reorganizes them into the invisible lines integrity model. This model is not proposed as a replacement for established standards such as GRI, ESRS, or the GHG Protocol. Rather, it is suggested as an interpretive tool that can be used alongside such standards to enhance the visibility of boundary-setting decisions. Its primary contribution is to clarify the upstream decisions that shape reported performance prior to numerical calculation.

The framework is therefore evaluated through conceptual coherence, explanatory usefulness, and practical transferability rather than statistical generalization. Its value depends on whether it helps users identify where reporting judgments enter. It also clarifies how those judgments affect interpretation. Finally, it shows how they can be documented without replacing existing reporting standards or calculation procedures entirely.

CONCEPTUAL FOUNDATIONS OF ENVIRONMENTAL DISCLOSURE

Disclosure as a Response to Information Asymmetry

Environmental disclosure is frequently described as a response to information asymmetry between organizations and their stakeholders. Investors, regulators, and the public do not have direct access to operational data, environmental impacts, or risk exposures associated with corporate activity. In contrast to this lack of access, reporting practices provide structured information to support accountability and decision-making. From this perspective, disclosure is expected to improve market functioning, increase transparency, and enable more informed evaluations of environmental performance.

The reduction of information asymmetry, however, does not occur automatically. Here, a key contrast appears: Disclosure remains controlled by the reporting organization, which determines what information is collected, how it is categorized, and how it is presented. Even when established frameworks such as the Global Reporting Initiative are applied, internal decisions continue to shape the selection and interpretation of information. In this sense, disclosure operates at two levels: as a mechanism for transparency and as a form of communication influenced by corporate priorities (Boiral, 2013).

Further complexity appears in the voluntary-disclosure literature. Firms may disclose more to reduce information asymmetry and signal performance, but disclosure may also serve legitimacy and impression-management purposes, as discussed in prior literature (Healy & Palepu, 2001; Merkl-Davies & Brennan, 2007; Suchman, 1995). These competing interpretations indicate that disclosure is not simply a corrective mechanism. It is a negotiated process in which visibility is actively shaped. Not all disclosures reveal.

Materiality as a Filtering Process Rather Than a Fixed Rule

Materiality is not a purely technical threshold. Under Global Reporting Initiative (GRI) 3: Material Topics 2021, material topics are those representing an organization's most significant impacts on the economy, environment, and people; under the European Sustainability Reporting Standards, disclosure is determined through double materiality, which combines impact and financial materiality (Global Reporting Initiative, 2023). In both systems, judgment enters through stakeholder mapping, significance criteria, and time horizons (De Cristofaro & Gulluscio, 2023; European Financial Reporting Advisory Group, 2024; Global Reporting Initiative, 2023). It is also influenced by context and interpretation. Through this process, environmental impacts are classified as relevant or non-relevant. Risks are prioritized, and selected performance dimensions are translated into indicators.

This filtering process is naturally selective. Determining materiality involves identifying stakeholder groups and assigning weight to different impacts. It also requires defining the time horizon over which significance is assessed. Each of these decisions influences both what is included and what remains outside the reporting scope. Financial materiality focuses on the risks that might affect a company's value, whereas impact materiality considers both environmental and social consequences. When combined, often referred to as double materiality, these perspectives do not remove ambiguity but add additional layers of interpretation. Some absences may be deliberate, as suggested in prior research.

Early empirical work on double materiality reporting shows considerable variation in assessment practice and disclosure design, suggesting that materiality processes remain operationally heterogeneous rather than fixed (De Cristofaro & Gulluscio, 2023). Instead, they tend to adhere to company priorities. Empirical studies suggest that areas supporting strategic positioning are often emphasized, while less favorable impacts receive limited attention or are excluded. Materiality, therefore, acts as a mechanism that defines reporting boundaries before measurement takes place. Through this process, certain impacts become visible. Others do not. Their absence does not indicate irrelevance, but classification within a given framework.

What Becomes Visible, and What Remains Concealed?

The selective character of disclosure creates conditions in which reporting may differ from underlying performance even when data are not falsified. Research indicates that greenwashing in reporting can operate through selective emphasis, omission, framing, and imbalance rather than outright falsification. This is why narrative and boundary choices matter analytically (Downie & Stubbs, 2013). Positive initiatives, improvements, or innovations are emphasized, while less favorable aspects are minimized or omitted. This pattern appears consistently in both conceptual discussions and empirical studies. Analyses of corporate reports show that favorable disclosures are often highlighted, whereas negative events receive limited attention. Counter-accounting research comparing official reports with external sources has demonstrated that significant adverse events may be absent from disclosure, despite their relevance to environmental performance (Boiral, 2013).

Economic perspectives provide an additional explanation. Firms may disclose information selectively when the expected reputational benefit exceeds the potential cost of scrutiny. External monitoring can influence this behavior, yet it does not eliminate it. In some cases, increased scrutiny raises the perceived risk of transparency, resulting in more controlled disclosure practices rather than fuller reporting (Lyon & Montgomery, 2015).

These dynamics indicate that greenwashing develops through the interaction between selective visibility and stakeholder views. Reporting frameworks commonly emphasize completeness and balance while relying on self-reported data. This leaves organizations with considerable discretion in how principles are applied. The outcome can appear methodologically sound, while still reflecting selective structuring of information. What is disclosed, and what remains unseen?

Reporting as a Constructed Representation

Taken together, these observations support an interpretation of environmental reporting as a constructed representation. Reporting systems do not simply record environmental performance; they contribute to how it is constructed. They influence the definition, measurement, and communication of that performance. This shaping occurs through several layers, including boundary construction, materiality assessment, indicator selection, and narrative framing.

Critical perspectives have described sustainability reports as representations that could differ from underlying conditions. In some cases, they resemble projections of an idealized image of performance instead of a direct account of impacts. The accumulation of indicators, narratives, and visual elements can create coherence without ensuring full correspondence with reality (Boiral, 2013).

This does not reduce the value of environmental reporting. It reframes how it should be interpreted. Evaluating disclosure requires attention to the processes through which it is constructed. Boundary construction, the operationalization of materiality, and the selection and presentation of information all influence the credibility of reported outcomes. By laying down these conceptual foundations, this section prepares the ground for examining how scope, baselines, and assumptions function as core mechanisms in the construction of environmental disclosure.

SCOPE: WHERE RESPONSIBILITY BEGINS AND ENDS

From Entity Reporting to Asset-Level Representation

Scope defines the area in which environmental impacts are recognized and reported. In conventional practice, this area is consistent with the reporting entity. It reflects legal ownership and financial consolidation structures. The GHG Protocol Corporate Standard requires organizations to set organizational boundaries and then operational boundaries before compiling the inventory, using either the equity-share or control approach for consolidation (World Resources Institute and World Business Council for Sustainable Development, 2004).

Within the built environment, such entity-based boundaries often fail to reflect how impacts are generated in practice. Buildings and infrastructure are operated through fragmented arrangements that involve developers, owners, operators, and occupants. Under these conditions, aggregated reporting at the entity level can conceal variation in environmental performance among individual assets. This limitation corresponds to a broader observation in carbon accounting research, where different accounting frames lead to different representations of the same underlying conditions (Ascui & Lovell, 2011).

In response, attention has shifted toward asset-level representation, where performance is associated with individual buildings or projects. This strategy improves spatial resolution. It does not remove ambiguity. Decisions remain necessary regarding inclusion, aggregation, and the treatment of shared systems. Scope continues to function as a constructed boundary, even when defined with greater precision.

The Question of Control and Operational Responsibility

Defining scope requires assigning responsibility. Reporting frameworks rely on the notion of control to determine whether emissions should be included. The GHG Protocol distinguishes between equity-share and control-based consolidation approaches, requiring organizations to define organizational boundaries and then operational boundaries before emissions are inventoried (World Resources Institute and World Business Council for Sustainable Development, 2004).

In application, these differentiations lead to different outcomes. In the built environment, property owners are often distinct from those who control their day-to-day operations. A landlord may own a building while having limited influence over how tenants consume energy. Occupants, in contrast, shape consumption patterns without direct control over building systems. Emissions attribution, therefore, depends on the boundary logic that is selected rather than on the physical origin of impact.

This condition becomes more complex as value-chain accounting expands. The Science Based Targets Initiative (2021) called for companies to assess Scope 3 emissions and, when Scope 3 emissions exceed 40% of total Scope 1, 2, and 3 emissions, to set ambitious Scope 3 targets covering at least two-thirds of Scope 3 emissions. Such an extension shows a broader view of accountability. It also introduces uncertainty, as responsibility becomes associated with influence rather than ownership (Brander et al., 2018).

Extending or Limiting the Life-Cycle Perspective

Scope also determines how environmental accounting engages with the life cycle of assets. In the built environment, impacts occur across material production, construction, operation, and end-of-life phases. The inclusion or exclusion of these stages alters how performance is represented. Life-cycle approaches show that narrow boundaries capture only a portion of total impact. For the average sector in Matthews et al. (2008), Tier 1 plus Tier 2 emissions capture only 26% of total Tier 3 emissions, indicating that narrow footprints omit most supply-chain-related impacts (Matthews et al., 2008). A large share of environmental impact, therefore, lies beyond usual reporting limits. Expanding boundaries to include full life-cycle impacts increases completeness. It also introduces methodological demands. Analytical structures based on input–output relations and extended environmental accounting reveal the distribution of emissions across interconnected systems. These approaches highlight the significance of considering broader boundaries. This understanding emphasizes the need for comprehensive environmental assessment (Kitzes, 2013; Lenzen et al., 2013). At the same time, reporting standards allow flexibility in defining these limits. Variation in practice follows.

To illustrate how scope expands across the life cycle, Figure 2 presents a multi-layer representation of environmental boundaries. The model shows how reporting progresses from direct emissions to increasingly comprehensive system coverage, extending to value chain and full life cycle impacts. Each outward layer captures additional sources of environmental impact that are typically excluded under narrower reporting definitions. At the same time, this expansion is accompanied by increased analytical complexity and uncertainty, as indirect impacts require assumptions, modeling techniques, and aggregated data structures. The figure, therefore, makes explicit the central trade-off in life-cycle assessment: broader boundaries improve completeness but also introduce interpretive challenges. This tension explains why reporting practices vary and why comparisons across organizations remain difficult when scope definitions differ.

Figure 2. Multi-layer environmental reporting boundary model illustrating the conceptual expansion from direct emissions to indirect value-chain emissions and broader whole-life or cradle-to-grave assessment boundaries

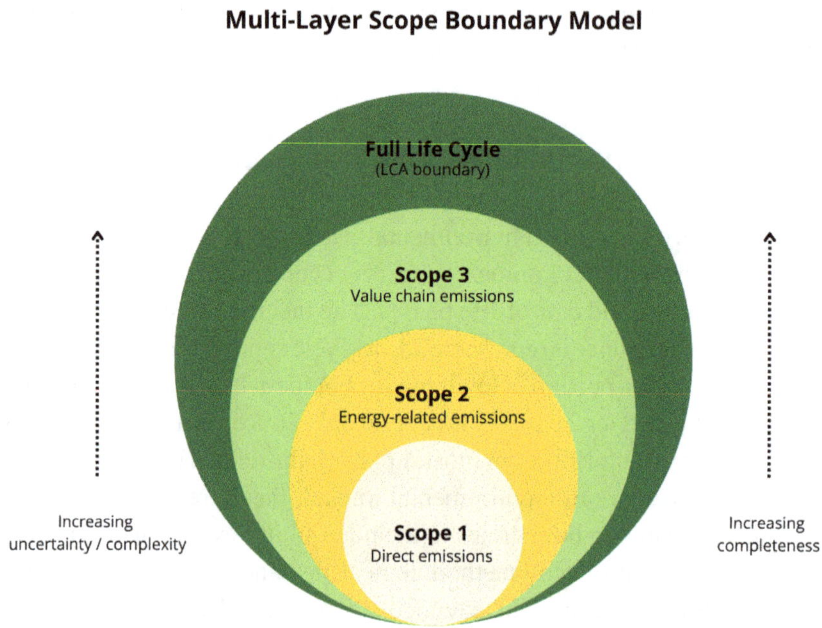

The figure is the author's conceptual representation based on greenhouse gas (GHG) accounting and life-cycle assessment literature. It is intended to show increasing reporting completeness and methodological complexity rather than to present a strict equivalence between corporate GHG accounting scopes and life-cycle assessment boundaries.

What Remains Outside the Boundary?

Any definition of scope produces exclusion. Impacts that fall beyond the selected boundary do not disappear. They are removed from representation. These exclusions often cover emissions from the supply chain, indirect processes, and downstream effects. A significant portion of the total environmental impact arises from these factors. Evidence from Australian companies shows uneven Scope 3 assessment practices, with implications for comparability; similar comparability concerns also motivate the formalization of Scope 3 accounting in the GHG Protocol (Brander et al. discuss this explicitly in relation to Scope 2 market-based accounting, Brander et al., 2018; Downie & Stubbs, 2013). Critical work also shows that methodological choices in carbon accounting can materially alter reported outcomes.

From a systems viewpoint, these exclusions lead to important consequences. Environmental impacts are distributed across connected activities. Limiting boundaries simplify reporting but reduce completeness (Lenzen et al., 2013). Scope, therefore, functions not only as a technical parameter. It defines how environmental performance is made visible. Figure 2 already visualizes this structure through concentric layers of inclusion, showing how narrower boundaries capture less of the system-wide impact.

BASELINES: HOW PERFORMANCE ACQUIRES MEANING

The Role of Starting Points in Defining Progress

Environmental performance is often presented as evidence of improvement over time. However, this improvement is not inherent in the data itself; it is established only relative to a defined reference condition. Baselines establish the reference point used to evaluate change. Without such a reference, environmental indicators become isolated values. This isolation makes their meaning impossible to interpret. Research in environmental accounting shows that variations in boundary construction and reference conditions occur. These variations may lead to different interpretations of the same system's performance (Matthews et al., 2008). Performance, therefore, should not be treated as an intrinsic property of systems. It emerges through methodological structure. The baseline in this narrative is key to the process. This point is often overlooked in practice.

This reliance is more pronounced within the built environment. Performance is assessed over prolonged time horizons and under variable operating conditions. In building energy analysis, the energy baseline serves as the reference against which efficiency gains are calculated. Its definition influences the scale of reported savings and the interpretation of system performance (Castrillón-Mendoza et al., 2024). A building may appear efficient under one baseline. However, under another, the same building may show limited improvement. The physical system remains unchanged, but the reference shifts. This condition demonstrates a broader feature of life cycle assessment methodologies. Performance is interpreted across system stages extending from resource extraction to end-of-life, each defined within a structured boundary and reference framework (Backes & Traverso, 2021). How the result is read depends on how the reference condition was built. The baseline not only provides context.

It shapes how performance is understood. What changes is not the system, but the lens through which it is assessed.

Interpreting Change Through Alternative Scenarios

Baselines are often constructed through modeled scenarios rather than derived from historical observation. These scenarios may represent business-as-usual conditions, projected demand, or alternative system states. Under this approach, environmental reporting relies on modeling. Interpretation follows from assumptions embedded in scenario design. This reliance signals a shift in how performance is produced. Results increasingly emerge from simulation rather than direct measurement. In carbon accounting, different analytical frameworks can provide distinct representations of the same system even when the underlying data remain unchanged (Ascui & Lovell, 2011). Measurement alone does not determine performance. Assumptions guide interpretation.

In building performance analysis, the selection of variables determines the baseline trajectory. Climatic conditions, occupancy patterns, and operational schedules each influence how change is perceived (Castrillón-Mendoza et al., 2024). A scenario under extreme conditions may yield substantial efficiency gains. A scenario based on average conditions may not. The difference lies in construction. It does not lie in the system alone. Life cycle assessment reinforces this pattern. System perimeters, inventory data, and impact assessment methods bring variability into results (Backes & Traverso, 2021). Comparisons across studies become difficult when baseline construction differs. Performance can be understood as not only observed but also configured.

The Effect of Normalization on Reported Outcomes

Normalization is applied to adjust environmental indicators for external conditions. These conditions may include floor area, occupancy, or production levels. This adjustment improves comparability but also introduces interpretive choices in determining which variables are relevant and how they are treated. Rather than simply clarifying performance, normalization reshapes it. Extending this line of reasoning, evidence from building energy analysis indicates that the choice of independent variables influences estimates of efficiency and performance (Castrillón-Mendoza et al., 2024). When variables such as weather or occupancy are included or excluded, the resulting metrics change. These shifts influence how performance improvements are interpreted.

Research in sustainability assessment points to a related issue. Indicator frameworks can produce ambiguous or confusing assessment outcomes when selection logic and conceptual design are weakly specified (Backes & Traverso, 2021; Michalina et al., 2021). Frameworks used to define indicators do not always provide clear justification. As a result, variation emerges in how performance is measured and compared. Indicators should not be treated as neutral outputs. They are products of methodological selection. Similarly, in life cycle sustainability assessment, normalization interacts with the integration of environmental, economic, and social dimensions. Combining indicators across these categories adds extra complexity. Aggregation can reduce transparency. Trade-offs may become less visible (Backes & Traverso, 2021). Improvements in one dimension may coincide with a decline in another. Interpretation depends on how normalization and aggregation are structured. Normalization, therefore, operates as more than a technical adjustment. It defines how data are transformed into indicators. It also defines how those indicators are read.

Revisions, Recalculations, and Shifting Targets

Baselines evolve over time. They change as methods develop, data improves, and system boundaries are revised. Recalculations may follow updates in emission factors, changes in modeling techniques, or shifts in organizational structure. These adjustments can alter the interpretation of performance trends. Physical performance may remain constant. Reported outcomes may not. This pattern reflects broader characteristics of sustainability reporting. Disclosures are shaped by institutional practices and evolving frameworks rather than by measurement alone (Boiral, 2013). Baseline revision forms part of this process. It is not only technical. It is interpretive. Within life cycle approaches, interpretation is closely linked to system definition and impact assessment. When these elements change, results change. The underlying system may remain the same (Backes & Traverso, 2021). Performance is therefore contingent on methodological structure.

In the construction sector, this phenomenon becomes more clearly observable and structurally influential. Assets extend across long life cycles. Multiple sustainability dimensions must be considered within a single framework. Studies indicate that variation in indicator selection, incomplete coverage, and shifting system boundaries lead to fragmented interpretations of performance (Backes & Traverso, 2021). Stability becomes difficult to maintain. Taken together, these observations show that baselines function as evolving instruments of interpretation. Their definition and revision shape environmental disclosure. They determine how performance is measured. They also determine how it is understood. In this sense, baselines connect data to interpretation.

ASSUMPTIONS AND MATERIALITY: FROM DATA TO INTERPRETATION

Assumptions Embedded in Environmental Quantification

Environmental metrics are often presented as objective representations of system performance. In practice, they emerge through layers of assumptions that convert activity data into quantified impacts. These assumptions operate across several levels. They appear in emission factors, choices about boundary construction, allocation rules, and in the use of proxy data when direct measurement is unavailable. Quantification, therefore, cannot be understood as a purely technical procedure. It involves interpretation shaped by methodological choice (Ascui & Lovell, 2011; Matthews et al., 2008). What appears as a precise numerical result frequently reflects a sequence of decisions that remains largely hidden.

In carbon accounting and sustainability reporting, these assumptions are frequently embedded within standardized methods. This can create the impression that results are neutral and comparable. Yet such methods depend on generalized coefficients and simplified system representations. These may not reflect local or project-specific conditions. The use of emission factors illustrates this point. Outcomes vary depending on geographic scope, temporal resolution, and data source. Even a small adjustment in the assumed electricity mix or operational profile can alter results. Precision is presented. Uncertainty remains.

Such assumptions are necessary. They allow complex systems to be analyzed at scale. Their influence, however, is rarely made explicit. In many reporting contexts, assumptions are treated as background conditions rather than as active components of analysis. This limits interpretability. It also reduces transparency for decision-makers. Making assumptions visible is a core condition for credible and interpretable reporting, particularly where secondary data, modeled inventories, or proxy factors are used (Ascui & Lovell, 2011; Science Based Targets Initiative, 2021).

Modeling Choices Under Conditions of Uncertainty

Environmental assessment frequently depends on models when direct measurement is incomplete or impractical. These models incorporate assumptions about system activity, operational conditions, and future states. Modeling enables the analysis of complex systems. It also introduces uncertainty. That uncertainty is not peripheral. It is intrinsic to how environmental performance is constructed. Several sources contribute to this uncertainty. Data gaps remain common. Input parameters vary. Modeling approaches differ. In input–output assessments, impacts are estimated across interconnected economic systems. This requires assumptions about sectoral relationships and aggregated data structures (Lenzen et al., 2013). In energy modeling, results depend on assumptions about occupancy, climate, and system performance. These assumptions are necessary. They also cause variability that is difficult to resolve fully.

The presence of uncertainty does not invalidate modeling. It changes how results should be read. Modeled outputs are conditional estimates. They reflect underlying assumptions rather than fixed outcomes. This distinction is often overlooked. Results are frequently presented as definitive indicators without sufficient acknowledgment of their limits. The consequence can be unwarranted confidence in precise-looking outputs when their underlying assumptions, choices about boundary construction, and scenario logic remain underexplained (Ascui & Lovell, 2011; Backes & Traverso, 2021; Kitzes, 2013). A different approach is possible. Results can be expressed as ranges or alternative scenarios rather than as singular values.

Materiality as a Process of Inclusion and Exclusion

Materiality determines which environmental impacts are included in reporting and which are excluded. It has traditionally been defined in relation to financial relevance. Sustainability frameworks have expanded this definition to include environmental and social dimensions. Despite this expansion, materiality remains a selective process. It filters complex system interactions into a limited set of indicators. What is not selected does not appear in the reporting narrative.

This process is interpretive. It is also strategic. Decisions about materiality depend on stakeholder expectations, regulatory pressures, and organizational priorities. Different entities may therefore report different aspects of performance, even under similar conditions. Empirical studies show that disclosures typically emphasize favorable indicators while downplaying less favorable ones, resulting in a selective representation (Janik & Ryszko, 2025; Lyon & Montgomery, 2015).

The introduction of double materiality adds complexity. It brings together two perspectives: the organization's impact on the environment and the environment's impact on the organization. This expansion broadens the scope of analysis. It also increases variability in how material issues are defined and prioritized (De Cristofaro & Gulluscio, 2023). In practice, consistency is difficult to achieve. Criteria for inclusion remain flexible. Materiality does not simply identify relevance. It constructs visibility.

The Transformation of Impacts Into Indicators

The final stage of reporting translates quantified impacts into indicators that can be communicated and compared. This transformation requires abstraction. Complex system interactions are reduced to a limited number of performance measures. Indicators make communication possible. They also remove context.

The construction of indicators involves multiple decisions. Metrics must be selected. Data must be aggregated. Results must be normalized. Each step influences how performance is represented. Indicator construction is shaped jointly by technical measurement choices and by reporting frameworks that determine what counts as material, comparable, and decision-useful (European Financial Reporting Advisory Group, 2024; Global Reporting Initiative, 2023). Comparability is often prioritized. Context may be reduced.

This reduction introduces risk. Important aspects of performance may be lost or simplified. Organizations may focus on indicators that demonstrate improvement while excluding those that reveal decline. Such practices increase the risk of greenwashing, in which reporting presents a favorable image that is not fully supported by underlying data (Janik & Ryszko, 2025; Lyon & Montgomery, 2015).

Indicators play a significant role in shaping how performance is defined and interpreted. They determine what is measured, what is compared, and what is treated as meaningful. Understanding this transformation is necessary to interpret environmental disclosures thoroughly and assess their credibility.

TOWARD REPORTING INTEGRITY: MAKING THE INVISIBLE VISIBLE

The Invisible Lines Integrity Model

The previous sections show that environmental reporting does not operate as a neutral act of measurement. It develops through a sequence of decisions that define boundaries, set reference conditions, introduce assumptions, and filter impacts through materiality. These decisions do not act at the margins. They determine how performance is formed and understood. Environmental performance should therefore be read as an outcome of methodological construction rather than as a direct representation of reality.

A gap follows from this condition. Reporting frameworks emphasize transparency and comparability, yet they tend to focus on outputs such as indicators and disclosures. The decisions that precede these outputs remain less visible. These upstream elements shape inclusion, measurement, and interpretation. They can be described as the invisible lines of reporting. They structure what becomes visible.

The invisible lines integrity model is introduced to address this gap and to offer a diagnostic approach to environmental reporting. Unlike conventional frameworks that can be read as checklists of technical requirements, this model presents reporting as a layered process that makes decision points explicit. Each stage in the model transforms the appearance of performance, enabling scrutiny of how outcomes are produced and providing distinctive means to evaluate integrity. This focus on visibility complements existing approaches.

The model does not replace existing standards. It functions alongside them as a conceptual extension. Its role is diagnostic and interpretive. It reveals the decision points that shape reporting and provides a basis for assessing their transparency and internal consistency. Attention shifts from reported outputs to the processes that generate them.

To clarify this structure, Figure 3 shows the invisible lines integrity model as a layered sequence of decision stages. These stages connect system inputs to final disclosure. The model demonstrates how environmental performance is gradually built through several layers such as scope definition, baseline creation, assumptions, and materiality assessment before being translated into indicators and reported results. Instead of illustrating a simple linear process, the figure emphasizes increasing levels of interpretation and transformation as reporting develops. It also highlights the role of feedback, where reported performance can influence future boundary-setting decisions. In this way, the model offers a visual framework for understanding how reporting outcomes are shaped and where scrutiny should be focused.

Figure 3. Decision Layer (Invisible Lines), illustrating the layered and iterative process through which environmental reporting inputs are transformed into reported performance and disclosure narratives

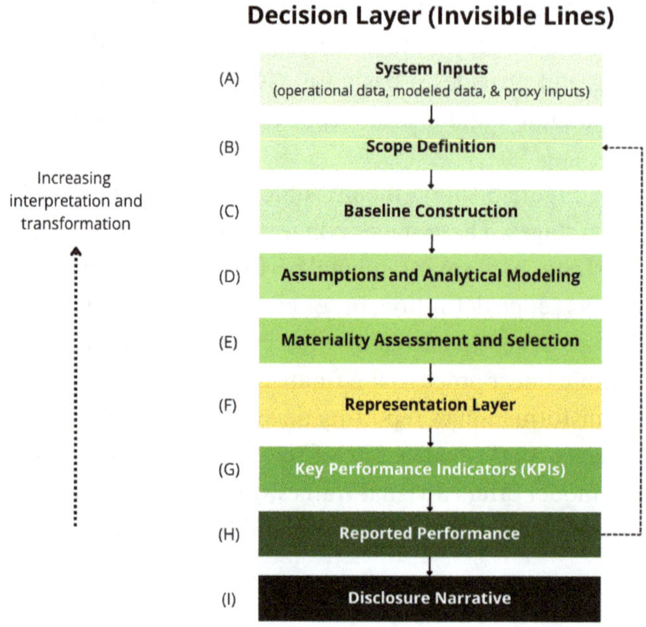

The figure presents the author's conceptual representation of the invisible lines integrity model, showing how scope definition, baseline construction, assumptions, materiality assessment, representation, and key performance indicators (KPIs) progressively shape the interpretation of environmental performance.

Making Scope Explicit Through Boundary Mapping

The first component of the model addresses scope. Scope defines what is included in reporting and what remains outside it. These decisions determine responsibility and influence completeness (Matthews et al., 2008). In practice, scope is often presented as fixed, although it is not.

Boundary mapping is proposed as a structured way to make these decisions visible. Instead of presenting scope as a single definition, this approach depicts it as a set of layers. These layers distinguish among direct and indirect emissions, operational impacts, and those embedded in materials, and stages of the asset life cycle. When these layers are mapped, the structure of inclusion becomes clear. So do the points of exclusion.

A tension emerges between completeness and complexity. Expanding boundaries increases coverage. It also increases uncertainty and reliance on assumptions. This tension should not be hidden. Making it explicit allows stakeholders to evaluate whether scope definitions are consistent with the purpose of reporting. In practice, boundary mapping can be implemented through visual or tabular formats attached to disclosures. No additional data is required. Existing information is reorganized. The result is greater clarity regarding what is included and what is not.

Stabilizing Interpretation Through Baseline Definition

The second component addresses baselines. Baselines define how performance is interpreted. Their selection involves decisions about reference years, scenarios, normalization practices, and recalculation rules (Castrillón-Mendoza et al., 2024). These decisions influence meaning. To manage this, the model implements a baseline charter. This is a structured declaration that defines how baselines are constructed, maintained, and revised. It sets conditions for change. It also clarifies how continuity is preserved across reporting periods.

The primary function of the baseline charter is to stabilize interpretation. When baselines change without explanation, reported performance shifts in ways that are difficult to interpret. With clear rules in place, variation can be traced. Changes in results can then be linked either to physical system performance or to methodological adjustment. Baseline choice is not merely technical; it determines the counterfactual against which change is judged and therefore shapes the reported magnitude of improvement (Backes & Traverso, 2021; Matthews et al., 2008). The charter also supports comparison. Different organizations often apply different baseline definitions. Full standardization may not be achievable. Documentation, however, allows differences to be understood. Interpretation becomes more precise. Flexibility is not removed. It is structured. Baseline selection remains an interpretive act. The charter makes that act explicit.

Documenting Assumptions and Strengthening Accountability

The third component focuses on assumptions. Assumptions are embedded across the reporting process. They appear in data collection, modeling, and calculation. They allow complex systems to be analyzed. They also influence outcomes (Kitzes, 2013). The model proposes an assumption ledger. This ledger records the key assumptions used in reporting. It includes emission factors, modeling parameters, allocation rules, and proxy data. Each entry is accompanied by a justification. Where possible, sensitivity to change is indicated.

The aim is not to remove uncertainty. It is to expose it. When assumptions are documented, users can understand how results are derived. They can also evaluate robustness. This creates a basis for accountability. Assumptions can be reviewed. They can be questioned. They can be revised. Transparency in sustainability reporting depends on more than data accuracy. It depends on the openness of the processes that produce that data (Boiral, 2013). Making assumptions visible reduces the risk of misinterpretation. It also supports more well-informed decision-making.

In practical terms, the assumption ledger improves comparability across reporting contexts. It does so by clarifying the conditions under which results are produced. Outcomes often differ because assumptions differ. Direct comparison is not possible without understanding these differences. When assumptions are documented, interpretation becomes more precise. Users can identify how results were derived and where variation originates. This reduces ambiguity in cross-case evaluation. It also allows differences to be attributed to methodological choices rather than to performance alone.

The effect is cumulative. Transparency extends beyond reported values to the structure that generates them. Analytical reliability increases because results can be examined in context. Comparisons become more grounded. Environmental reporting, in this sense, becomes easier to interpret and more consistent in its application.

Linking Materiality to Indicators: From Selection to Representation

The final component addresses the link between materiality and indicators. Materiality determines what is included. Indicators translate those elements into measurable form. This link is often treated as technical. It is interpretive. It involves decisions about relevance, measurement, and communication (Janik & Ryszko, 2025; Lyon & Montgomery, 2015). The model introduces a materiality-to-key performance indicator (KPI) trace. This trace connects identified impacts to the indicators used to represent them. It documents how each indicator relates to a specific impact and explains why it was selected. Indicators are no longer presented in isolation. They are tied to a defined process.

This connection reduces the risk of selective reporting. When indicators are detached from materiality, favorable metrics can be prioritized while others are excluded. The trace makes such choices visible. It allows evaluation. It also improves interpretation. Indicators are understood within context. They reflect underlying impacts rather than functioning as isolated values. Reporting becomes more aligned with its purpose. The components of the invisible lines integrity model deliver a structured way to make hidden aspects of reporting visible. Scope, baselines, assumptions, and materiality are not peripheral. They define how reporting operates. When these elements are made explicit, attention shifts from outputs to processes.

This shift affects both practice and research. For practitioners, it offers a way to improve transparency and strengthen credibility. For researchers, it provides a structure for analyzing reporting systems and identifying sources of variation. Environmental reporting will continue to involve complex data and evolving methods. Complexity is not the problem. Interpretation is. Making the invisible lines visible is an important step toward more reliable and meaningful disclosure.

Applying the Invisible Lines Integrity Model in Built Environment Reporting

The practical value of the invisible lines integrity model can be illustrated through a hypothetical built environment reporting case. Consider a real estate organization that owns and manages a portfolio of ten commercial office buildings. During a reporting cycle, the organization announces that portfolio-wide operational carbon emissions have been reduced by 28% over the past five years. Although this reported reduction appears clear and positive, its interpretation depends on several upstream decisions that are not immediately visible in the headline figure. These include whether tenant energy consumption was included, whether the baseline year was representative, whether retrofitting materials were considered, which emission factors were applied, and how material topics were linked to selected key performance indicators. Such decisions are central to environmental accounting because system boundaries, accounting frames, and estimation assumptions shape how environmental impacts are translated into reported figures (Ascui & Lovell, 2011; Matthews et al., 2008).

A conventional disclosure may report the final carbon reduction figure and provide supporting energy data. The invisible lines integrity model asks a deeper question: how was the reported figure constructed? The first step is boundary mapping. If the organization reports only landlord-controlled electricity and excludes tenant electricity, the result may reflect the performance of base building systems but not the environmental impact of the fully occupied asset. If embodied carbon from façade replacement, HVAC upgrades, or interior refurbishment is excluded, the disclosure may emphasize operational improvements while leaving retrofit-related impacts outside the reporting boundary. This issue is especially relevant in the built environment, where operational and embodied impacts are distributed across ownership structures, tenant behavior, supply chains, and life-cycle stages (Backes & Traverso, 2021; World Resources Institute and World Business Council for Sustainable Development, 2004). The issue is not that one boundary is inherently wrong, but whether the selected boundary is explicit, justified, and consistent with the stated purpose of the report.

Table 1 presents a simplified application of the model to this hypothetical portfolio. The table is not intended as a reporting template, but rather as a means to illustrate how the model translates implicit decisions into visible documentation points.

Table 1. Illustrative Application of the Invisible Lines Integrity Model to a Hypothetical Building Portfolio

Model component	Example reporting decision	Disclosure risk if unclear	Documentation response
Boundary mapping	Include landlord-controlled electricity, common-area energy, tenant energy, purchased cooling, and retrofit-related embodied carbon either fully or selectively.	Reported reductions may appear stronger if tenant or retrofit impacts are excluded.	Provide a boundary map showing included and excluded asset-level, operational, and life-cycle categories.
Baseline charter	Use a single pre-retrofit year, a multi-year normalized baseline, or a modeled business-as-usual scenario.	Improvements may be overstated if the baseline year had unusually high consumption or low occupancy.	State the baseline year, normalization rules, recalculation triggers, and justification for the reference condition.
Assumption ledger	Apply grid emission factors, occupancy assumptions, operating-hour schedules, and floor-area normalization.	Precise results may conceal uncertainty or dependence on proxy data.	Record assumptions, data sources, rationale, and sensitivity where relevant.
Materiality-to-KPI trace	Select KPIs such as energy use intensity, operational carbon, embodied carbon, water use, and thermal comfort.	Favorable indicators may be emphasized while less favorable impacts remain unreported.	Link each KPI to a material impact, stakeholder concern, or reporting requirement.

The same portfolio could therefore yield different interpretations depending on how decision layers are configured. Under a narrow operational boundary, the organization may report a large reduction in landlord-controlled emissions after upgrading lighting and HVAC systems. Under a broader boundary that includes tenant electricity and embodied carbon from retrofit materials, however, the reduction may be smaller, delayed, or accompanied by a short-term increase in total life-cycle emissions. Although both results may be internally consistent, they do not communicate the same environmental performance, due to differences in the construction of the reporting boundary. This reflects a wider challenge in carbon accounting and life-cycle assessment: while broader boundaries can improve completeness, they also increase methodological complexity and uncertainty (Backes & Traverso, 2021; Matthews et al., 2008).

The baseline charter is particularly important in this example. If the organization uses the year before retrofit—which had unusually low occupancy efficiency or abnormal cooling demand—as the baseline, the reported improvement may appear larger. A multi-year normalized baseline, by contrast, may yield a more stable comparison. A modeled business-as-usual baseline may be appropriate if the portfolio changed significantly owing to acquisition, renovation, or occupancy turnover. Thus, in building energy analysis, baseline selection directly influences the interpretation of efficiency gains and reported performance improvements (Castrillón-Mendoza et al., 2024). The charter does not eliminate judgment, but records how the reference condition was selected and when it should be recalculated, enabling users to distinguish between actual performance improvement and methodological change.

The assumption ledger will document the technical choices underlying the carbon calculation. For example, it will specify whether location-based or market-based electricity factors were applied, whether national or local grid factors were used, how missing tenant data were estimated, and whether cooling loads were based on measured consumption or modeled energy use. It will also clarify whether floor area was measured as gross floor area, net lettable area, or conditioned floor area. Although these details may appear minor, each one can affect comparability and interpretation. Prior work on carbon accounting shows that methodological choices—including emission factors, accounting frames, and allocation procedures—can materially influence reported outcomes (Ascui & Lovell, 2011; Brander et al., 2018). By making these assumptions visible, the ledger reduces the risk of numerical precision being mistaken for methodological certainty.

The materiality-to-KPI trace completes the process by linking selected indicators to the impacts they are intended to represent. In the office portfolio example, energy use intensity and operational carbon may be material because of regulatory requirements and investor expectations. Embodied carbon may be material owing to retrofit activity. Thermal comfort may be material because energy reductions achieved through higher cooling setpoints could affect occupant well-being. Water use may be material in water-stressed regions but less central in other contexts. This logic aligns with reporting approaches that require material topics and sustainability indicators to be connected to significant impacts, risks, opportunities, and stakeholder concerns (European Financial Reporting Advisory Group, 2024; Global Reporting Initiative, 2023). The trace prevents indicators from appearing as isolated metrics and instead anchors them in the underlying materiality logic.

Implementation constraints should also be acknowledged. Many organizations lack complete tenant data, consistent embodied carbon inventories, or internal expertise in life-cycle assessment. Reporting teams may face time constraints, fragmented asset information, incompatible software systems, and uncertainty about how much methodological detail users will be interested in. These barriers do not invalidate the model, but shape how it should be introduced. A phased approach is therefore advisable. Organizations can begin by documenting existing boundaries and assumptions without altering the calculation method. In subsequent reporting cycles, they can expand the assumption ledger, formalize baseline rules, and add a materiality-to-KPI trace.

This phased use is important because the model is not designed to increase reporting burden but to enhance interpretability. Even when data remain incomplete, explicit documentation enables users to understand what the disclosure includes, what it excludes, and how uncertainty has been managed. In this way, the invisible lines integrity model functions as a practical bridge between technical reporting requirements and interpretive transparency, helping readers evaluate not only the reported outcome but also the decisions that made the outcome visible.

CONCLUSION

Environmental reporting cannot be reduced to a technical exercise in measurement. Accuracy, consistency, and methodological rigor remain necessary. They do not fully account for variation across disclosures. The analysis shows that the most significant differences arise earlier, before calculation takes place. They emerge in the decisions that define how performance is constructed. Scope, baselines, assumptions, and materiality are not secondary parameters. They determine what becomes visible and what can be reported.

Viewing reporting in this way shifts attention. Outputs no longer stand alone. Indicators and disclosures depend on the frameworks that produce them. Those frameworks define system boundaries, establish reference conditions, and translate environmental activity into simplified representations. Two organizations may present internally consistent results. Their interpretations of performance may differ. The divergence does not necessarily indicate error. It reflects differences in design.

This pattern becomes especially significant in the built environment. Impacts are distributed across fragmented ownership structures and extended life cycles. Value chains remain interconnected. Data are often modeled or estimated through proxy relationships. Under such conditions, boundary-setting decisions carry analytical weight. Choices about inclusion, control, and baseline construction directly influence reported outcomes. Without attention to these choices, disclosures may appear coherent. Their underlying assumptions remain hidden.

Interpretation operates throughout the reporting process. Baselines define how change is understood. Assumptions translate activity into impact. Materiality determines what is included. Indicators convert these elements into communicable form. At each stage, complexity is reduced. Abstraction increases. This reduction is necessary. It also introduces the possibility of selective representation. Reporting does not only describe performance. It constructs it. Recognizing this condition clarifies how reporting should be evaluated. The objective is not to remove complexity or uncertainty. It is to make the processes that manage them visible. This is the context in which the invisible lines integrity model becomes useful. The model presents reporting as a sequence of decision stages. It traces the progression from system inputs through scope definition, baseline construction, assumptions, materiality, and indicator selection. Each stage shapes interpretation. The model makes these stages explicit. It also reveals their interdependence. Outputs can influence subsequent design decisions. Reporting becomes iterative.

Four components support this structure. Boundary mapping represents layers of inclusion and exclusion across organizational and life-cycle dimensions. The baseline charter defines how reference conditions are established and maintained over time. The assumption ledger records the parameters that shape quantification and clarifies uncertainty. The materiality-to-key performance indicator (KPI) trace links selected impacts to reported indicators. Representation remains connected to the process that defines it. These components do not introduce new standards. They operate alongside existing frameworks. Their role is interpretive. They strengthen transparency and improve comparability. They also support more precise evaluation by exposing decision points that are often implicit.

The implications go beyond reporting practice. For practitioners, the framework delivers a way to improve clarity without requiring additional data. For regulators, it underscores the need to address decision logic, not just reported outputs. For researchers, it provides a basis for examining reporting systems as constructed representations. It opens a path for analyzing how boundary choices influence outcomes.

Limitations remain. The chapter develops a conceptual approach. It does not include empirical testing across reporting contexts. The application will depend on organizational capacity and data conditions. Further work is needed to evaluate the model's performance in practice and its interaction with evolving standards.

Environmental reporting is not only about measuring impacts; it also defines the conditions under which those impacts are recognized. Making the boundaries of scope, baselines, assumptions, and materiality explicit allows differences in outcomes to be interpreted as differences in reporting design rather than as numerical variation alone.

This shift is necessary. Environmental decisions now carry economic consequences and regulatory implications that extend beyond individual organizations. Reporting outcomes influence investment flows, compliance strategies, and public trust. Under these conditions, reliability cannot depend on numerical accuracy alone. Data may be correct. Interpretation may still vary. What matters is how that data is framed, bounded, and translated into indicators.

Reliable reporting requires explicit processes. These processes define how system boundaries are drawn, how baselines are established, and how assumptions are introduced into analysis. When such elements remain implicit, differences in outcomes are difficult to interpret. Transparency becomes limited to outputs rather than extending to their construction. This weakens comparability and reduces confidence in reported performance. Positioning reporting as a system of constructed visibility addresses this gap. It clarifies how performance is produced rather than only how it appears. The framework developed in this chapter supports this objective. It enables closer examination of decision points and allows reporting practices to be applied with greater consistency and interpretive control.

In this sense, the contribution of this chapter extends beyond conceptual clarification. It establishes a structured lens through which environmental reporting can be interrogated, compared, and refined across different institutional and operational contexts. As sustainability reporting frameworks continue to evolve, particularly with the increasing formalization of double materiality and life-cycle thinking, the need for interpretive transparency becomes more pronounced. The invisible lines integrity model responds to this need by repositioning reporting as a system of accountable construction rather than passive disclosure. Its value lies in its adaptability: it can be applied across scales, from single-asset evaluations to portfolio-level reporting, without requiring fundamental changes to existing data infrastructures. More importantly, it introduces a shift in evaluative logic. Instead of asking whether reported numbers are correct, it enables an examination of how those numbers were produced. This distinction is critical for advancing both methodological rigor and institutional trust in environmental reporting systems.

Future development of this approach should focus on operational validation across diverse reporting contexts, particularly within large-scale real estate portfolios and infrastructure systems where boundary complexity is most pronounced. Empirical application would allow testing of how the model performs under varying regulatory, data, and organizational conditions, and whether it improves interpretability in practice. In parallel, integration with emerging digital tools, including reporting assisted by artificial intelligence and automated data pipelines, presents an opportunity to embed boundary transparency within reporting systems themselves, particularly where algorithmic systems increasingly influence environmental data interpretation, decision processes, and sustainability governance (Ali & Tintawi, 2026).

AI DISCLOSURE

Any AI-assisted tools used during manuscript preparation were limited to language refinement, structural editing support, proofreading, and editorial drafting assistance. The conceptual framework, argument, analysis, interpretation, and final scholarly decisions remain the responsibility of the author.

ORIGINALITY AND FIGURE ATTRIBUTION

All figures, models, and conceptual frameworks presented in this chapter are the original work of the author unless otherwise indicated. Where existing concepts from the literature have informed the development of these representations, they have been interpreted and synthesized into original forms for the purpose of this analysis.

METHODOLOGICAL NOTE

As clarified in the methodological positioning section, this chapter presents an interpretive conceptual synthesis rather than an empirical study or systematic review. The framework is intended to support analytical clarity, not to prescribe universally applicable measurement standards.

FUNDING AND CONFLICT OF INTEREST

The author declares no financial support or conflicts of interest related to this work.

INTERPRETIVE SCOPE

The arguments presented in this chapter reflect analytical interpretations of environmental reporting practices based on existing academic literature. They are not intended to make claims about the conduct of specific organizations or to imply misconduct in any particular case.

REFERENCES

Ali, K., & Tintawi, G. (2026). Governing environmental decisions with artificial intelligence: Algorithmic sustainability for accountability and lifecycle impact, a policy review. *Preprints*. DOI: 10.20944/preprints202602.1746.v3

Ascui, F., & Lovell, H. (2011). As frames collide: Making sense of carbon accounting. *Accounting, Auditing & Accountability Journal*, *24*(8), 978–999. DOI: 10.1108/09513571111184724

Backes, J. G., & Traverso, M. (2021). Application of life cycle sustainability assessment in the construction sector: A systematic literature review. *Processes (Basel, Switzerland)*, *9*(7), 1248. DOI: 10.3390/pr9071248

Boiral, O. (2013). Sustainability reports as simulacra? A counter-account of A and A+ GRI reports. *Accounting, Auditing & Accountability Journal*, *26*(7), 1036–1071. DOI: 10.1108/AAAJ-04-2012-00998

Bragança, L., Mateus, R., & Koukkari, H. (2010). Building sustainability assessment. *Sustainability (Basel)*, *2*(7), 2010–2023. DOI: 10.3390/su2072010

Brander, M., Gillenwater, M., & Ascui, F. (2018). Creative accounting: A critical perspective on the market-based method for reporting purchased electricity (scope 2) emissions. *Energy Policy*, *112*, 29–33. DOI: 10.1016/j.enpol.2017.09.051

Castrillón-Mendoza, R., Rey-Hernández, J. M., Castrillón-Mendoza, L., & Rey-Martínez, F. J. (2024). Sustainable building tool by energy baseline: Case study. *Applied Sciences (Basel, Switzerland)*, *14*(20), 9403. DOI: 10.3390/app14209403

De Cristofaro, T., & Gulluscio, C. (2023). In search of double materiality in non-financial reports: First empirical evidence. *Sustainability (Basel)*, *15*(2), 924. DOI: 10.3390/su15020924

Downie, J., & Stubbs, W. (2013). Evaluation of Australian companies' scope 3 greenhouse gas emissions assessments. *Journal of Cleaner Production*, *56*, 156–163. DOI: 10.1016/j.jclepro.2011.09.010

European Financial Reporting Advisory Group. (2024). *European sustainability reporting standards (ESRS) 2024*. https://www.efrag.org

Global Alliance for Buildings and Construction and United Nations Environment Programme. (2024). *Global Status Report for Buildings and Construction: Beyond foundations—Mainstreaming sustainable solutions to cut emissions from the buildings sector* (I. Hamilton, H. Kennard, O. Rapf, J. Amorocho, S. Steuwer, J. Kockat, & Z. Toth, Trans.; Nos. 978-92-807-4131-5). DOI: 10.59117/20.500.11822/45095

Global Reporting Initiative. (2023). *GRI 3: Material topics 2021.* https://www.globalreporting.org/standards/

Healy, P. M., & Palepu, K. G. (2001). Information asymmetry, corporate disclosure, and the capital markets: A review of the empirical disclosure literature. *Journal of Accounting and Economics*, *31*(1–3), 405–440. DOI: 10.1016/S0165-4101(01)00018-0

Janik, A., & Ryszko, A. (2025). Greenwashing in sustainability reporting: A systematic literature review of strategic typologies and content-analysis-based measurement approaches. *Sustainability (Basel)*, *18*(1), 17. DOI: 10.3390/su18010017

Kitzes, J. (2013). An introduction to environmentally-extended input–output analysis. *Resources*, *2*(4), 489–503. DOI: 10.3390/resources2040489

Lenzen, M., Moran, D., Kanemoto, K., & Geschke, A. (2013). Building Eora: A global multi-region input–output database at high resolution. *Economic Systems Research*, *25*(1), 20–49. DOI: 10.1080/09535314.2013.769938

Lyon, T. P., & Montgomery, A. W. (2015). The means and end of greenwash. *Organization & Environment*, *28*(2), 223–249. DOI: 10.1177/1086026615575332

Matthews, H. S., Hendrickson, C. T., & Weber, C. L. (2008). The importance of carbon footprint estimation boundaries. *Environmental Science & Technology*, *42*(16), 5839–5842. DOI: 10.1021/es703112w

Merkl-Davies, D., & Brennan, N. (2007). Discretionary disclosure strategies in corporate narratives: Incremental information or impression management? *Journal of Accounting Literature*, *26*, 116–196.

Michalina, D., Mederly, P., Diefenbacher, H., & Held, B. (2021). Sustainable urban development: A review of urban sustainability indicator frameworks. *Sustainability (Basel)*, *13*(16), 9348. DOI: 10.3390/su13169348

Ng, A. W., Yorke, S. M., & Nathwani, J. (2022). Enforcing double materiality in global sustainability reporting for developing economies: Reflection on Ghana's oil exploration and mining sectors. *Sustainability (Basel)*, *14*(16), 9988. DOI: 10.3390/su14169988

Science Based Targets Initiative. (2021). *Science-based target setting manual (Version 1.1).* https://sciencebasedtargets.org/resources

Suchman, M. C. (1995). Managing legitimacy: Strategic and institutional approaches. *Academy of Management Review, 20*(3), 571. DOI: 10.2307/258788

World Resources Institute and World Business Council for Sustainable Development. (2004). *The greenhouse gas protocol: A corporate accounting and reporting standard* (Revised Edition). World Resources Institute.

KEY TERMS AND DEFINITIONS

Assumption Ledger: A structured record of key assumptions used in environmental reporting, including their justification and influence on results.

Baseline: A defined reference condition against which environmental performance and changes over time are evaluated.

Boundary Construction: The process of defining what is included or excluded in environmental reporting, shaping how performance is represented.

Constructed Representation: The view that environmental reporting reflects methodological and interpretive choices rather than a direct depiction of reality.

Environmental Reporting: The process through which organizations measure, interpret, and disclose information about environmental impacts and performance.

Greenwashing: The selective presentation of environmental information to create a misleadingly favorable impression of performance.

Invisible Lines Integrity Model: A framework that makes explicit the decision layers (scope, baselines, assumptions, and materiality) shaping reporting outcomes.

Materiality: The process of determining which environmental impacts are significant enough to be included in reporting.

Normalization: The adjustment of environmental indicators to account for external variables to improve comparability.

Scope (Environmental Reporting): The defined boundary that determines which activities and impacts are included in reporting.

Chapter 3
Strengthening ESG-Driven Financial Data Ecosystems for Gender-Inclusive Finance:
Advancing SDG 5 and SDG 8 in India and Emerging Economies

Kanika Arora
http://orcid.org/0009-0007-5420-5190
Mittal School of Business, India

Nancy Sahni
http://orcid.org/0000-0002-7748-3632
Mittal School of Business, India

ABSTRACT

The chapter evaluates the effectiveness of enhancing ESG-based financial data ecosystems to develop gender-inclusive finance and speed up the achievement of SDG 5 and SDG 8paying particular attention to India and other similar emerging markets. Based on regulatory trends like BRSR framework established by SEBI as well gender-sensitive financing programs facilitated by the World Bank and the International Finance Corporation, the research paper identifies real actions to incorporate social and governance measures in the process of capital allocation. The insights of the case study of fintech-based MSME lending platforms in India shows that alternative data, including digital transaction history and supply-chain records, has helped women entrepreneurs to gain better access to credit and ensure portfolio stability. The chapter suggests a large-scale and open data integration

DOI: 10.4018/979-8-2600-3678-5.ch003

model bridging the ESG reporting, gender analytics, and financial decision-making, consequently promoting inclusive capital flows and enhancing sustainable economic development in emerging economies.

1. INTRODUCTION

Over the recent years, the global financial system has experienced a dramatic change between the emergence of sustainable finance and Environmental, Social and Governance (ESG) investing. Globally, sustainable investment assets have reached USD 30 trillion, and it is an indicator of the increasing interest shown by investors, financial institutions, and policymakers to introduce environmental sustainability, social responsibility, and accountability of governance to the decision-making process of finance (Global Sustainable Investment Alliance, 2023). ESG frameworks are becoming more influential in corporate strategies, capital allocation choices, and financial risk analyses in the developed and emerging economies. Nevertheless, even though there is the accelerated growth in ESG-based investments, there has been continued inequality in the allocation of financial resources, especially regarding the inclusion of gender into the access to finance. A significant gap in financing of women entrepreneurs is one of the most acute issues in the global financial systems. Due to structural barriers, women-owned small and medium enterprises (SMEs) have a global credit gap of more than USD 1.7 trillion, especially in emerging and developing economies where women are facing a variety of structural barriers that restrict the access to formal financial services (World Bank, 2024). Other limitations that women-led businesses face include low level of collateral ownership, poor financial literacy levels as well as biases in the classical credit assessment frameworks. These predicaments are further aggravated by the fact that there is no disaggregated financial information by gender in banking systems and investment models which therefore complicates financial institutions to effectively determine the creditworthiness and economic potential of businesses that are managed by women.

The increasing focus on ESG investing is a special chance to overcome these systemic financial inequalities. The social aspect, within the ESG approach, focuses on such problems as diversity, inclusion, labour rights, and fair economic inclusion. Incorporating indicators of gender in ESG reporting and financial reporting frameworks can thus be important in ensuring gender-inclusive finance. Financial institutions can define the lack of representation of certain segments of entrepreneurs through the inclusion of gendered financial data in investment decision-making and build more welcoming lending approaches. Studies have also indicated the existence of better governance systems, greater innovation rates, and better long-term financial results in firms and economies that support gender equity (Friede et al.,2015). The topicality of gender-inclusive finance is also strictly correlated with the United Nations Sustainable Development Goals (SDGs). Specifically, SDG 5 (Gender Equality) aims at removing obstacles to economic engagement of women, whereas SDG 8 (Decent Work and Economic Growth) focuses on inclusive economic development, effective jobs, and sustainable entrepreneurship (United Nations, 2015). To ensure these are achieved, it is necessary to have financial systems that can facilitate equitable access to capital especially women run micro, small and medium enterprises (MSMEs). Women-led MSMEs occupy a considerable portion of entrepreneurship activity in most emerging economies and make a considerable contribution in relation to the creation of employment and economic growth. However, such businesses still experience an unequal disadvantage of lower access to formal financing than male-owned companies.

India is a valuable case study concerning the social-driven interaction of the ESG frameworks, financial data ecosystems, and gender-inclusive finance. Within the last ten years, India has taken some regulatory and policy measures to enforce sustainable finance as well as enhance financial inclusion. The implementation of the Business responsiveness and sustainability reporting (BRSR) framework by the Securities and exchange board of India (SEBI) is one such notable development that requires ESG disclosures in large listed companies. The BRSR framework has indicators connected to the workforce diversity, gender ratio, and social sustainability benchmarks, thus prompting corporations to incorporate the gender-related considerations into the sustainability approaches (Securities and Exchange Board of India [SEBI], 2023). Meanwhile, digital financial infrastructure in India, such as Unified Payments Interface (UPI), Aadhaar identification systems, and account aggregator structures have made new opportunities to alternative credit assessment models and inclusive financial services.

One of the primary sources of this change is the fintech innovation. Online lending websites are becoming more open to alternative forms of data including transaction history, supply chain records and digital patterns of payment history to evaluate the creditworthiness of a borrower. These technologies allow financial institutions to appraise entrepreneurs with no conventional collateral or credit background and therefore create more access to finance to underserved population, such as women entrepreneurs (Frost et al., 2019). Such digital financial ecosystems have the potential to be very important in closing the gender financing gap, especially in those emerging economies where the substantial part of the population works in informal sectors. Regardless of these favorable changes, however, there are a number of challenges. The ESG reporting systems in different countries are still disjointed and there is still lack of incorporating gender-disaggregated financial information in investment regimes. In most instances, the ESG disclosures focus on the environmental indicators (Ahmad et al., 2023; Haider et al., 2024; Hui et al.,

2025), and the social indicators, especially the gender related financial indicators, are poorly developed. To build ESG-oriented financial data ecosystems, the concerted actions of regulators, financial institutions, fintech sites, and international development agencies are needed.

It is in this context that the chapter reviews how gender-inclusive finance can be facilitated through ESG-oriented financial data ecosystems and enhanced in developing economies, such as India and other developing countries in relation to SDG 5 and SDG 8. Based on the international trends in ESG investment, regulatory policy, including the BRSR framework in India, and a case study of female-led enterprise lending platforms facilitated by fintech, the chapter demonstrates how capital could be more effectively allocated to women-led businesses through integrated data systems. Connecting ESG reporting, gender analytics, and digital financial infrastructure, the research points at the ways of enhancing inclusive economic growth and sustainable financial development in emerging markets.

1.1 ESG and Gender-Inclusive Finance Literature Review

In the last twenty years, Environmental, Social, and Governance (ESG) investing became a key element of sustainable finance. ESG frameworks promote environmental responsibility, social inclusion, and governance accountability to be set in financial decision-making processes by investors and corporations. The rise of ESG investing is a wider trend toward responsible capital allocation, in which financial performance is being assessed more and more concerning social and environmental concern. The global sustainable investment assets have surpassed USD 30 trillion, as it is stated by Global Sustainable Investment Alliance, and it is easy to understand that ESG-oriented financial markets develop fast (Global Sustainable Investment Alliance, 2023).It has also been noted in academic literature that LEVA has a positive correlation with financial performance and ESG performance. Friede et al.(2015) have provided a detailed meta-analysis of more than 2,000 empirical studies and discovered that most studies had indicated a positive correlation between ESG practices and corporate financial performance. In line with that, Broadstock et al. (2020) established that companies that performed better in ESG-related factors were more resilient in economic shocks, such as the COVID-19 pandemic. These conclusions imply that when ESG considerations are part of financial decision-making, the risk management and the creation of long-term value can be improved.

The growing significance of ESG investing has resulted in the emergence of several sustainability reporting frameworks that are focused on enhancing transparency and accountability of corporate reporting. These are the Global Reporting Initiative (GRI), the Sustainability Accounting Standards Board (SASB), the Task Force on Climate-related Financial Disclosures (TCFD), and the United Nations Principles for Responsible Investment (PRI). These frameworks do motivate businesses to publish the non-financial performance measures on environmental sustainability, social responsibility, and corporate governance practices. Nevertheless, even after the extensive spread of ESG reporting standards, there is still substantial fragmentation of the structure and comparability of ESG disclosures across jurisdictions (Boffo & Patalano, 2020).One of the most significant weaknesses of current ESG reporting models is a somewhat underdeveloped approach toward incorporating gender-specific financial indicators in the sustainability disclosures. Although most ESG models incorporate diversity and inclusion indicators, detailed gender-disaggregated financial information, e.g. the proportion of credits given to female entrepreneurs or gender-based investment flows are not adequately measured. Consequently, financial institutions do not have the data infrastructure that is required to integrate gender considerations in methods of capital allocation (Afzal & Haider, 2025; Ali et al., 2026).

1.2 Gender Aspects ESG Finance and the Global Credit Gap

The gender-inclusive finance literature has indicated the continued gap in access to financial resources between male-owned and female-owned enterprises. Women entrepreneurs are also very vital in the economic growth especially in the emerging economies where the contribution of micro, small and medium enterprises (MSMEs) is high in employment production and generation of income. Women entrepreneurs are still hindered by structural constraints in accessing formal financial services although they are of economic significance.The estimates of women owned SMEs in the world show that there is a gap of about USD 1.7 trillion in financing, which portrays that access to credit, venture capital and institutional financing is rampantly constrained (World Bank, 2024). Conventional credit assessment models tend to apply strong emphasis on collateral property and official financial records both of which are unfavorable to women entrepreneurs as a result of institutional disparities in property holdings and money integrate. Consequently, women owned enterprises often rely on informal sources of funds or self-financial systems, restricting their growth and innovation abilities.

It has also been found that economic gains can be made by enhancing gender diversity and financial inclusion. Research indicates that companies that have gender balanced leadership structures have better governance performances and better financial results. According to McKinsey Global Institute (Woetzel et al., 2018), the participation of women in the economy through improving gender equality could have a significant boost in the global GDP, and productivity in the industry. Gender equality in the labor markets and entrepreneurial ecosystem should be encouraged to ensure an inclusive economic growth, especially in the framework of the Sustainable Development Goals (UN Women, 2022; OECD, 2021).

The social pillar of ESG can offer a significant chance in combating gender disparities in economic systems. Gender-related metrics are the way to ensure that financial institutions can recognize underrepresented segments of entrepreneurs and distribute resources better through the use of ESG reporting and investment strategies. The introduction of gender-responsive financial instruments, like gender bonds, women-centered investment funds, and others, has become an innovative way of targeting capital toward women-led businesses. Indicatively, the International Finance Corporation has also launched gender bonds which are specifically aimed at financing women-owned businesses, hence fostering financial inclusion and sustainable economic growth (World Bank, 2023).

1.3 Fintech, Digital Finance, and Access to Inclusive Credit

The last trends in digital financial technologies have brought additional opportunities to the process of financial inclusion and minimization of gender inequalities regarding access to credit. Fintechs are starting to use alternative data points more and more, including digital transaction history, supply chain history and mobile payment history, to determine the creditworthiness of borrowers. Such technologies allow financial institutions to assess entrepreneurs with no historic collateral or formal credit history, so they are able to reach more underserved populations with finance (Frost et al., 2019).The emergence of digital financial ecosystems is especially prominent in the emerging economies in which significant parts of the population work in informal sectors. In India, the growth of digital financial infrastructure containing Unified Payments Interface (UPI), Aadhaar digital identity systems, and account aggregator systems has changed the availability and effectiveness of financial services. The platforms have the ability to generate high amounts of transactional data that can be used to create alternative credit scoring models and promote financial inclusion.

Empirical evidence has shown that fintech lending services can help small businesses and entrepreneurs to obtain finance in a much better way. Gomber et al. (2017) emphasize that digital finance innovation lowers information asymmetry levels between borrowers and lenders through mechanisms of assessment of credit using data. These systems are especially useful with women in businesses, who might have no traditional credit history, but have a digital credit history in terms of digital transaction history on e-commerce websites or mobile payments. In addition, ESG-based financial frameworks can be complemented by fintech innovations in the form of data infrastructure of gender-inclusive investment strategies. The combination of fintech data and ESG reporting systems will help financial institutions to follow the gender-disaggregated financial flows and assess the social impact of the investment portfolio. This integration may reinforce ESG-based financial ecosystems and improve the performance of initiatives of gender-based finance.

1.4 ESG Finance and the Sustainable Development Goals

The overlap of ESG investing and the United Nations Sustainable Development Goals (SDGs) has become the focus of more and more scholarly and policy discussion. ESG frameworks offer financial processes whereby the private capital would play the part in meeting the sustainable development goals. Specifically, the SDG 5 (Gender Equality) and SDG 8 (Decent Work and Economic Growth) have a strong correlation with gender-inclusive financial systems. SDG 5 is expected to end gender-based discrimination and ensure more women participate fully in economic and social life. To realize this objective, there must be financial systems that are equally accessible to credit, investment capital, and entrepreneurial resources. In the meantime, SDG 8 focuses on inclusive economic development, productive work, and sustainable business. Women-led businesses are a fundamental aspect in these goals especially in developing economies where MSMEs have been a significant element in job generation.

Although there is increasingly close correlation between ESG models and SDG goals, there are major huddles in converting sustainability promises into quantifiable financial results. A significant weakness is the unavailability of integrated data systems that could be able to connect ESG reporting to gender-disaggregated financial metrics. The financial institutions can find it difficult to assess the effect of their investments on gender equality and inclusive economic development without this data infrastructure. As a result, the reinforcement of ESG-based financial information systems has become a significant concern among policy-makers, financial institutions, and developmental organizations. Such ecosystems can help enable more inclusive capital allocation by incorporating ESG disclosures, gender analytics, and digital financial data and contribute to the attainment of global development objectives(United Nations Development Programme [UNDP], 2024)

2. ENVIRONMENTAL, SOCIAL AND GOVERNANCE ISSUES

2.1 ESG-Based Financial Data Ecosystems: Theoretical Foundations

The increased adoption of Environmental, Social and Governance (ESG) issues within financial markets has led to the realization of the need to have a strong financial data ecosystem which can facilitate sustainable investment decision making. A financial data ecosystem describes the network of relationship of financial institutions, regulatory frameworks, digital platforms and analytical systems that create, process, and use both financial and non-financial data to determine the allocation of capital. ESG-based financial data ecosystems, in the view of sustainable finance, are important in facilitating investors and financial institutions to evaluate the social and environmental consequences of financial practices. Nevertheless, irrespective of the fact that ESG reporting systems have been evolving rapidly, the inclusion of gender-disaggregated financial information into the system is low.

An effective ESG financial data ecosystem is based on various components that are connected to each other. To begin with, the ESG disclosure frameworks offer consistent reporting systems through which corporations can report on their performance in terms of environment, social, and governance. The use of global reporting systems like Global Reporting Initiative (GRI), Sustainability Accounting Standards Board (SASB) and the United Nations Principles of responsible investment (PRI) have helped to enhance transparency in corporate sustainability performance. The Securities and Exchange Board of India (SEBI) in India has proposed a framework of Business Responsibility and Sustainability Reporting (BRSR), which requires disclosure of ESG to large listed companies. The BRSR model also incorporates social indicators concerning workforce diversity, gender representation, and community engagement, thus, providing the basis on which the metrics concerning gender were to be addressed in the report on corporate sustainability (NITI Aayog, 2021).

Second, gender-disaggregated financial information is one of the significant components of sustainable financial ecosystems. Gender-disaggregated data are the financial data classified by gender, which allows policymakers and financial institutions to discover the differences between gender access to credit and access to investment capital and financial services. In the absence of this data, it will be hard to assess the practical use of the financial inclusion programs to assist female entrepreneurs. Both the World Bank and the International Finance Corporation have highlighted that to improve the financing gap that women-owned businesses face across the world, it is important to enhance gender-disaggregated data systems. In the emerging economies, where micro, small and medium enterprises (MSMEs) owned by women are reported to play a key role in generation of employment opportunities, the availability of gender specific financial information can be improved in order to improve policy design and financial inclusion strategies.Third, the digital financial infrastructure is increasingly becoming relevant to the empowerment of ESG-driven data ecosystems. Financial technology (fintech) apps, digital identity systems and digital payment platforms have been producing significant amounts of transactional data which can be analyzed to assess creditworthiness and investment opportunities. Programs like Unified Payments Interface (UPI), Aadhaar digital identification and the Account Aggregator framework in India have redefined the financial data structure of the country with their ability to facilitate secure sharing of data between financial institutions. Such digital systems are useful sources of alternative data that may supplement conventional credit assessment models and increase access to finance in underserved populations.

The artificial intelligence (AI) and sophisticated analytics application to the credit assessment and investment decisions is another essential element of ESG-led financial data ecosystems. Conventional credit assessment frameworks are very dependent on the possession of collateral and past credit documents, which in most cases do not favor women entrepreneurs who might not have proper records in the form of a formal financial history. AI-based credit scoring models can use other data sources like electronic payment records, supply chain data and e-commerce activities to create more comprehensive credit ratings. Such models have become more popular in fintech lending sites in emerging economies that use them to assess MSME borrowers to minimize information asymmetries between lenders and entrepreneurs (Frost et al., 2019). The combination of these elements into a unified financial data ecosystem can greatly help to improve the efficacy of ESG-powered financial systems. Once ESG reporting, gender-disaggregated financial statistics, digital infrastructure, and improved analytics are interlinked, financial organizations can create more well-rounded investment plans, which would focus on inclusive economic growth. This can be supported by such integrated systems so that investors can track the gender effects of their financial portfolio, underserved entrepreneurial groups, and invest in businesses that enhance gender equality and sustainable economic development.

The theoretical framework suggested in this chapter explains how financial data ecosystems that take into account ESG can be used to support gender-inclusive finance and work toward the realization of the Sustainable Development Goals. The framework initiates with ESG reporting mechanisms that are strategies of capturing corporate sustainability performance. These revelations are subsequently added to gender-disaggregated financial data and online financial records that have been created on fintech systems and digital payment systems. The integrated datasets help financial institutions to perform AI-based credit checks and investment decisions. The investment decisions arrived at are inclusive in allocating capital to women owned businesses and gender sensitive business ventures. Finally, the processes lead to more widespread development in accordance to SDG 5 (Gender Equality) and SDG 8 (Decent Work and Economic Growth).Enhancing gender-inclusive finance in emerging economies is a major avenue, thus, strengthening ESG-driven financial data ecosystems is a crucial direction.

Through enhancing data transparency, advancement of digital financial infrastructure, and inclusion of gender analytics in investment procedures, policymakers and financial institutions can create financial systems that facilitate fair access to capital and also contribute to sustainable economic growth.

2.2 Conceptual Framework Figure

Figure 1. ESG-Driven Gender-Inclusive Financial Data Ecosystem

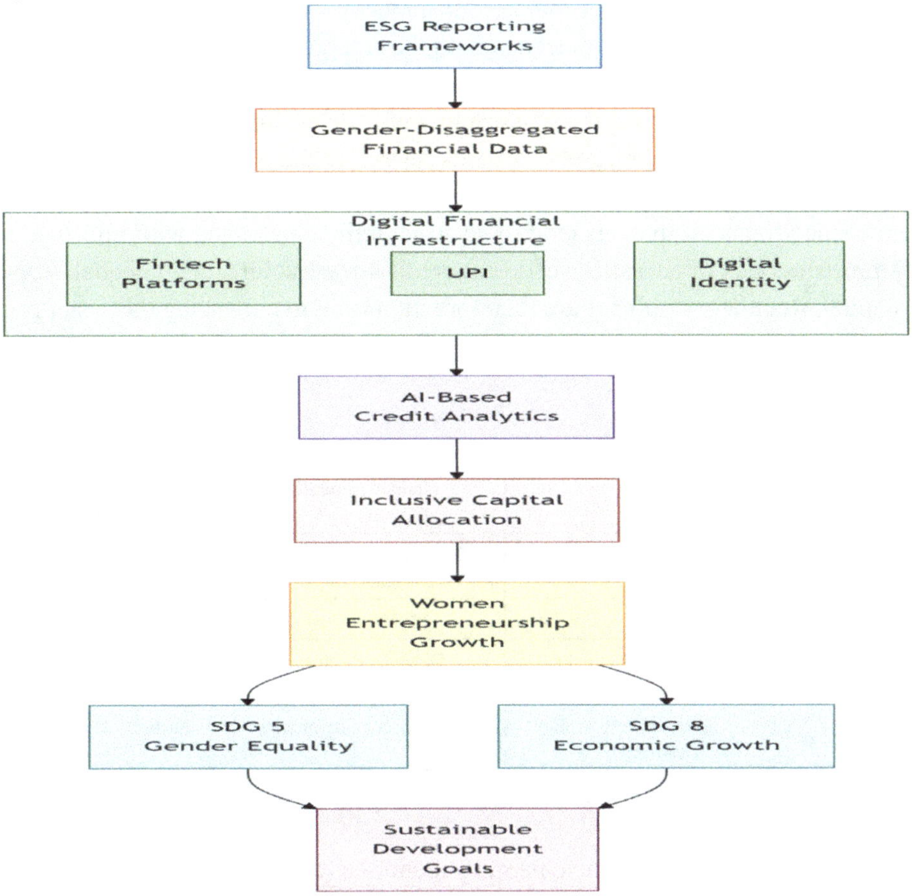

Source: *Author's conceptual framework.*

The **figure 1** demonstrates the interactions between ESG reporting systems, female-disaggregated financial data, and digital financial infrastructure to facilitate the use of AI in credit analytics and equitable capital distributions. These processes enhance the development of the women-owned businesses and help to reach SDG 5 (Gender Equality) and SDG 8 (Decent Work and Economic Growth).

3. REGULATORY LANDSCAPE: INDIA AND EMERGING ECONOMIES

Various laws that help to promote transparency, accountability, and inclusive financial access are directly associated with the emergence of ESG-based financial ecosystems and gender-inclusive finance. The world governments and financial regulators have been coming up with sustainability disclosure policies and financial inclusion policies to make capital markets reflect sustainable development goals. Regulatory efforts are especially in the emerging economies as shown in **Table 1** below since financial markets tend to be structurally challenged with information asymmetries, lack of access to sufficient credit infrastructure, and unequal access to capital. Regulatory policies are therefore important in enhancing the systems of ESG reporting and advancing gender-sensitive financial practices.

Table 1. Comparison of ESG Regulatory Frameworks in Emerging Economies

Country	ESG Framework	Gender Indicators	Implementation Focus
India	BRSR (SEBI)	Workforce diversity, gender leadership	Corporate ESG reporting
Indonesia	OJK Sustainable Finance Roadmap	MSME inclusion	Responsible lending
Brazil	CVM Sustainability Guidelines	Diversity disclosures	Corporate transparency
Global	UN Principles for Responsible Investment	Gender equality metrics	Sustainable investing

Source: Author's compilation based on official ESG regulatory frameworks

The **Table 1** depicts that the emerging economies are now conceptualising sustainability and social inclusion measures into financial regulations.

3.1 India's ESG and Gender-Inclusive Finance Framework

The ESG and Gender-Inclusive Framework of Finance in India. India has been one of the most successful rising economies in implementing the ESG-type of regulatory provisions to increase the corporate transparency and sustainable finance practices. One of the major achievements in this direction is the introduction of Business Responsibility and Sustainability Reporting (BRSR) framework by the Securities and Exchange Board of India (SEBI). The BRSR framework, which is introduced as a compulsory reporting system to the top 1,000 publicly listed companies by market capitalizations, is intended to streamline sustainability reporting and enhance better access to information on ESG-related topics in capital markets

(SEBI, 2023). The BRSR framework is a developmental process of the previous corporate social responsibility reporting systems, as it incorporates the environmental, social, and governance indicators in their entirety as a reporting framework. Notably, the model assigns gender-based metrics in the social aspect of the ESG reporting. The companies must also report on matters of workforce diversity, gender representation at the managerial levels, equal remuneration policies, and employee welfare policies. Such disclosure helps investors and regulators to determine the corporate performance based on gender equality and social sustainability. Gender indicators are also significant because of how they are incorporated into ESG reporting, with implications on gender-inclusive finance. The BRSR framework will enhance the transparency in the corporate governance by releasing gender-related corporate data to the public, which will allow financial institutions to assess how the intended investment decisions impact society. According to the latest studies of BRSR disclosures, Indian companies are starting to report more on gender diversity metrics, which is evidence of the rising interest in the significance of inclusive corporate governance.

Besides the ESG reporting policy, India has also introduced various financial inclusion policies to assist women entrepreneurs and small firms. Collateral-free loans to micro enterprises and small enterprises, known as the Pradhan Mantri Mudra Yojana (MUDRA) programme, was initiated by the Government of India, and most beneficiaries in this scheme are women entrepreneurs. Likewise, the Stand-Up India program is an initiative that promotes the loaning of money by banks to the women and marginalized business people to start a new business. Such initiatives have helped to increase access to formal credit by women-owned firms especially in areas where women businesspeople had previously been disadvantaged by lack of financial opportunities. The other key institutional player in ensuring gender-inclusive finance in India is the Small industries development bank of India (SIDBI). SIDBI has put in place a number of programs to cater to women entrepreneurs such as special credit lines, financial literacy programs, as well as capacity-building programs of women-led businesses (Small Industries Development Bank of India [SIDBI], 2024) These programs are effective supplements to the overall financial inclusion efforts, as they help deal with financial and knowledge gaps among women entrepreneurs. The digital financial infrastructure in India has also enhanced inclusive finance regulatory environment. The high growth of digital payment systems like Unified Payments Interface (UPI) has also raised the level of financial transparency and access to digital financial services. Financial institutions are also using digital identity systems like the Aadhaar system to check the identity of borrowers and also allow them to provide credit delivery systems that are more efficient. The combination of these efforts will lead to the creation of a digital financial environment that will be able to sustain gender-inclusive lending and investment practice.

3.2 ESG and Sustainable Finance Regulations in the Emerging Economies

Outside India, a number of emerging economies have launched regulatory efforts to incorporate sustainability issues in financial markets. These frameworks are indicative of the increased awareness that inclusive economic growth can be promoted through sustainable finance and ways of addressing environmental and social problems. In Indonesia, sustainable finance has been enforced by the Financial Services Authority (Otoritas Jasa Keuangan, or OJK), which has the mandate of making financial institutions incorporate environmental and social factors in their loan and investment decisions. The OJK Sustainable Finance roadmap focuses on sound lending and encourages financial inclusion of micro and small business, such as women owned businesses. Another instance is the Indonesia endorsement of fintech-based microfinance solutions that use digital financial information to reach more underserved communities with credit. On the same note, Brazil has also instituted sustainability disclosure by its securities authority, the Comissao de Valores Mobiliarios (CVM). Sustainability policies in Brazil promote the reporting of information concerning the environmental and social risks, such as the workforce diversity and corporate governance practices. Such disclosures allow investors to assess the sustainability performance of any companies that operate in the Brazilian market. Governance projects in emerging economies are also more focused on the incorporation of ESG reporting into the wider financial inclusion approach. These efforts have been facilitated by multilateral development organizations like the World Bank and the International Finance Corporation in the form of programs that encourage gender-inclusive financial instruments. Indicatively, gender bonds have come up as novel financing systems aimed at redirecting the investment capital to women-owned ventures and gendered development projects.

In spite of these developments, a lot challenges are yet to be resolved in the process of making ESG reporting standards consistent and comparable across nations. Sustainability reporting systems are usually fragmented by differences in regulatory requirements, data availability and institutional capacity. Consequently, investors and policymakers might struggle with the issue of ESG performance comparison across jurisdictions or the analysis of the gender effect of financial investments. Enhancing regulatory alignment and enhancing the inclusion of gender-specific financial information in ESG frameworks, in turn, can play a critical role in the process of advancing gender-inclusive finance in emerging economies.

4. EMPIRICAL EVIDENCE AND CASE STUDIES

Empirical data is essential in the perception of the way ESG-based financial systems and digital financial innovations affect gender-inclusive access to money. In emerging markets, there is a large but underrepresented portion of the business ecosystem that is constituted by women entrepreneurs. Although the number of women-owned businesses has played a key role in the creation of jobs and locally boosting economic growth, women still experience considerable challenges in accessing credit facilities and investment capital in form of formal credit. Most recent studies have shown that ESG-founded financial data ecosystems and digital lending systems can positively contribute to the achievement of financial inclusion among women entrepreneurs significantly.

4.1 Global Trends in Gender-Inclusive Finance

International financial statistics indicate that there are always imbalances in the ability of women owned businesses to access credit and invest. The World Bank and the International Finance Corporation state that a global credit gap among women entrepreneurs owning small and medium enterprises (SMEs) is estimated at about USD 1.7 trillion, especially in the emerging economies in which financial business organizations are usually based on borrowing on collateral (World Bank, 2024). Although there has been a growing awareness of the economic value of women entrepreneurs, institutional financing frameworks have always been constituted by risk assessment schemes that disadvantage the business that does not follow a detailed financial history or the provision of collateral assets. The ratio of women-owned businesses to all businesses globally is about 20 percent, but they receive much less of venture capital and formal credit financing. According to research based on world investment reports, less than 10 percent of venture capital worldwide goes to women entrepreneurs, and this represents system-wide biases in entrepreneurial funding systems. Such restrictions on financing not only reduce the potential growth of women owned ventures, but also economic productivity and innovation, in general. Meanwhile, ESG-oriented investment assets have been developing at an exceptional pace over the last several years, and in some financial markets, the assets under global sustainable investments have grown by over 25 percent annually. The increasing size of ESG investment opens the possibility of incorporating gender-responsive investment practices into the new frameworks of sustainable financial investments. Gender-based investment funds and gender bonds are new financial instruments to overcome these financing imbalances. These tools are aimed at facilitating more general development goals by channeling capital to women-run and gender-based undertakings based on SDG 5 and SDG 8.

Table 2. Global Gender Financing Gap for Women-Owned Enterprises

Indicator	Global Estimate	Source
Women-owned SMEs worldwide	~20% of total SMEs	World Bank
Global financing gap for women entrepreneurs	USD 1.7 trillion	IFC
Share of venture capital received by women-led firms	<10%	OECD
Women participation in formal credit systems (emerging economies)	~30%	Global Findex
Women-led MSMEs receiving institutional finance in India	~12%	RBI

Source: Author's compilation based on reports from the World Bank, International Finance Corporation, Organisation for Economic Co-operation and Development, World Bank Global Findex Database, Demirgüç-Kunt et al., 2022, and Reserve Bank of India.

Table 2 points out the gender financing disparity that remains constant in the financial systems of the world. Although they contribute to a large share of the world in terms of entrepreneurship, the list of enterprises owned by women is still under-funded by institutional capital.

4.2 Gender Finance and MSME Credit in India

India also offers a good example to analyze gender-inclusive financial systems in emerging eco-systems. There is a significant contribution of micro, small and medium enterprises (MSME) in the Indian economy that have been the basis of creation of employment and industrial production. The data given by the Reserve Bank of India show that in India, around 30 percent of the GDP and more than 40 percent of exports is done by the MSME sector. In this ecosystem, women entrepreneurs are a booming business ownership segment. Nonetheless, there is an unequal access to institutional finance. According to the recent RBI data, women-led MSMEs are being given sizeable portions of overall institutional credit that the MSME sector is taking about 12 percent of the total credit, which is significantly lower than the international average of about 27 percent. This disparity underscores the institutional dilemmas that women entrepreneurs are still struggling with to gain access to formal financial systems. Such disparities are generally caused by limited ownership of collateral, absence of formal credit records, and sex-related prejudices of the conventional lending paradigms. Government financial inclusion programs like the MUDRA loans, Stand-Up India as well as the women entrepreneurship programs of SIDBI have tried to overcome these problems by increasing access to credit amongst women entrepreneurs. Through such programs, the number of women-led enterprises that are involved in the formal financial system has increased. Nonetheless, even with such policy initiatives, access to finance remains to be a restrictive factor to many entrepreneurs working in informal sectors using models of credit evaluation which are still traditional.

Table 3. Credit Access for Women-led MSMEs

Country	Share of Women Entrepreneurs	Access to Institutional Credit	Key Programs
India	~20% of MSMEs	12% credit share	MUDRA, Stand-Up India
Indonesia	~23% of SMEs	~15% credit share	Microfinance initiatives
Brazil	~25% of SMEs	~18% credit share	Financial inclusion programs
Global Average	~20%	~27%	Gender finance initiatives

Source: Author's compilation based on World Bank, 2024; IFC,2021; OECD, 2021; RBI,2025)

Table 3 shows that women entrepreneurs have different access to credit depending on their emerging economies, and therefore, there should be gender sensitive financial policies.

4.3 Fintech Lending Platforms and Alternative Credit Assessment

The swift growth of financial technology (fintech) sites has opened up new potential avenues of enhancing financial inclusion, especially of underrepresented entrepreneurs. Alternative data sources, such as digital payment history, supply chain transactions and e-commerce activity, are used by Fintech lending platforms to assess the creditworthiness of borrowers. These platforms minimize the usage of the traditional collateral-based lending models and rather, make use of the data-based credit assessment algorithms to measure the business performance and repayment capacity. In India, a number of fintech startups have become key participants in the growth of credit access to MSMEs and women entrepreneurs. Lendingkart, Indifi and Lentra among others have built digital lending systems that use machine learning algorithms and transaction-based analytics to assess loan applications. Through the process of assessing digital financial records as opposed to basing their operations on collateral assets alone, these platforms can determine the creditworthiness of entrepreneurs who do not have a formal credit history.

Case Study: Lentra Fintech Platform

Lentra is an Indian digital lending infrastructure firm and has created AI-based credit assessment solutions, which combine various data sources to assess MSME borrowers. The platform uses transactional information, online payment records, and business performance metrics to create risk images of loan applicants. As reported in the industry, Lentra digital lending systems have helped financial institutions to process loan application in about 30 percent less time compared to normal credit assessment systems. Moreover, Lentra has been able to finance over 50,000 women-owned businesses between 2023 and 2025 with the help of its digital lending eco-system, which illustrates the possibilities of credit models led by fintech to enhance gender-focused finance. The data provided by the portfolio performance show that digital lending models are also capable of enhancing the credit risk assessment which leads to the growth of the portfolio stability and lower rates of defaults.

Microfinance Innovation: The Case of Amartha in Indonesia

Another valuable lesson that can be learned regarding digital financial ecosystems and gender-inclusive finance is the example of Indonesia. Amartha is a microfinance platform that is currently working on a peer-to-peer lending platform in digital format that aims to assist women in rural settings. The site leverages supply chain data and digital transaction history to assess the performance of borrowers and deploy micro-loans to businesses owned by women. The lending model introduced by Amartha involves the impact evaluation mechanisms based on ESG that monitor the social outcomes of women contributing to the economic development, their income earned, and development of their community. The platform is reported to have increased the rate of borrower retention by about 22 percent which proves the effectiveness of the digital credit assessment systems in aiding the sustainable operations of the microfinance systems.The case of Amartha shows that fintech-enabled models of microfinance have the potential to further enhance access to credit by women in informal economic markets to grow their businesses. Such platforms can empower the financial ecosystems based on ESG by merging digital data analytics and social impact monitoring.

4.4 Gender Bonds and World Sustainable financing efforts

Another current innovation in gender-inclusive finance is the formulation of gender bonds, which are financial instruments, whose purpose is to raise funds to finance projects that empower women economically. Multilateral bodies like the International Finance Corporation have been in the frontline to encourage the idea of gender bonds as a tool of drawing the private investment to gender-inclusive development.The IFC has initiated gender bond programs amounting to more than USD 2 billion which focus on financial institutions and development projects, which promote women entrepreneurs and the participation of women in the workforce. The programs allow investors to invest in social impact-generating projects without compromising their financial gains. Research indicates that gender-based investment portfolio has generated returns as high as 15 percent above the average emerging market investment and this fact indicates the economic potential of gender responsive investment strategies. Gender ties show how the financial instruments that are motivated by ESG can steer the investors towards the same direction as the sustainable development agenda. These tools empower the development of ESG frameworks to be connected with gender-inclusive financial endeavors by interconnecting financial performance and quantifiable social results.

5. OPPORTUNITIES AND PROBLEMS IN ESG-BASED GENDER-INCLUSIVE FINANCE

Although the role of ESG frameworks and digital financial innovations as tools of facilitating an inclusive finance is increasingly becoming acknowledged, various structural issues still curtails the efficacy of gender-inclusive financial ecosystems. The major challenges associated with these are the absence of coherent ESG data systems, the unavailability of gender-disaggregated financial data, and the existence of informal economic systems that limit access to formal credit markets. Simultaneously, the new digital financial systems and data-oriented financial technologies present an important chance to exceed these constraints and reinforce equitable capital distribution.Among the critical issues of ESG-based financial ecosystems, the disintegration of ESG reporting systems should be mentioned. Even though corporate transparency has been enhanced by global sustainability reporting standards like the Global Reporting Initiative (GRI), Sustainability Accounting Standards Board (SASB), and national frameworks including Business Responsibility and Sustainability Reporting (BRSR) in India, inconsistency in ESG reporting across jurisdictions is likely because of the absence of harmonized reporting frameworks. Such fragmentation renders it hard to have the investors and financial institutions to compare sustainability performance and determine the social impact of investments. Specifically, indicators of gender are still underrepresented in the majority of ESG disclosure systems, which restricts the possibilities of investors in assessing the gender equality performance within their corporate and financial portfolios.

The other urgent issue is associated with low access to gender-disaggregated financial information in banking and investment systems. Numerous financial institutions lack a systematic monitoring of lending/investment undertakings in gender terms regarding the owners or entrepreneurs of business. Consequently, policy makers and financial regulators do not have quality data on how effective gender-oriented financial inclusion efforts are. It is hard to formulate specific financial policies that respond to the credit limits of women-led business without such data infrastructure.The unofficial character of economic operation in most developing economies continues to are an impediment in the quest to increase gender inclusive finance. The large percentage of women-owned businesses are found in informal sectors that do not involve any financial transactions recorded in formal banking institutions. It has been estimated that in the majority of cases, almost 60 percent of women-led businesses in the emerging economies do not have formal financial records, which restricts their access to institutional credit or investment funds. Conventional lending procedures are based on the ownership of collateral, financial reports, and credit reports- all of which many small business owners cannot fulfill.

Along with these structural issues, smaller enterprises can also face obstacles due to compliance expenses, which are related to ESG reporting and regulatory frameworks. Although big organizations usually have the resources and organizational capabilities to address the requirements of the ESG disclosures, micro and small businesses might be overwhelming by these reporting requirements. The growth of ESG frameworks without simplified ESG reporting systems to fit smaller companies will have the counter-effectual impact of excluding portions of the entrepreneurial ecosystem that most require financial assistance.Although these issues exist, a number of new opportunities provide good avenues towards the empowerment of ESG-based financial ecosystems and the enhancement of gender-diverse finance performance. Among the most remarkable ones is the fact that the digital public financial infrastructure in emerging economies is growing at an unprecedented pace. On Unified Payments Interface (UPI), Aadhaar digital identity and Account Aggregator architecture has provided a complete digital financial ecosystem in India whereby the secure sharing of data between financial institutions takes place. Such systems result in massive amounts of online financial information that may be harnessed to construct alternative credit evaluation systems.

The digital financial technologies also facilitate application of artificial intelligence and data analytics in credit assessment, which can contribute greatly to the financial inclusion. Credit scoring systems based on AI use digital payment history, supply chain operations, and e-commerce to measure creditworthiness of the borrower. It is possible that such models can assist financial institutions in measuring entrepreneurs who do not have a standard credit history and show a steady economic activity by conducting transactions online.Additionally, the combination of fintech systems with ESG reporting systems offers the possibility to create gender-sensitive financial analytics. Incorporating ESG disclosures with digital financial information will enable financial institutions to track gender-related financial flows, characterize underserved segments of the entrepreneurship, and create investment strategies that help to facilitate the inclusion of women in the economy.All in all, the issues that relate to the ESG data fragmentation and informal financial systems can be resolved through the joint efforts of regulators, financial institutions, fintech innovators, and development organizations. The emerging economies can create more inclusive financial ecosystems by enhancing gender-disaggregated financial data availability and building digital financial infrastructure through which sustainable economic development can be promoted.

6. POLICY RECOMMENDATIONS FOR STRENGTHENING ESG-DRIVEN GENDER-INCLUSIVE FINANCE

To enhance the financial data ecosystems that are based on ESGs, policy efforts need to be spent on mobilizing all stakeholders (regulators, financial institutions, investors, development organizations). Although new steps towards sustainable finance and financial inclusion have provided valuable opportunities to access more capital, the underlying gender financing gap needs specific regulatory changes and institutional solutions. Policymakers need to work on enhancing the data transparency, enhancing digital financial infrastructure, and gender responsive investment strategies.

Figure 2. Integrated ESG Financial Data Architecture

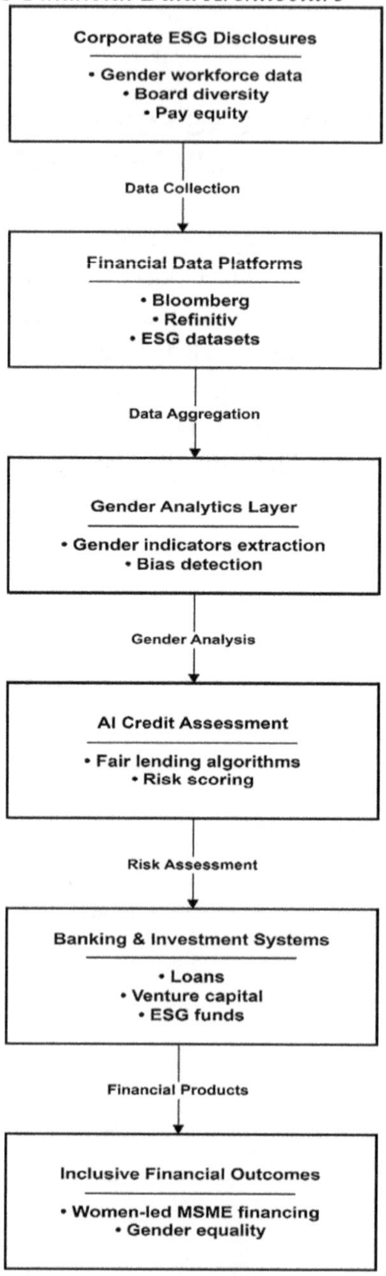

Source: *Author's conceptual framework developed for this study*

Figure 2 illustrates the integration of ESG reporting systems with digital financial infrastructure and gender analytics to improve capital allocation decisions.

6.1 Regulatory Measures

Regulatory bodies are at the center stage of creating the structures that will see the consideration of gender in ESG-related financial systems. Compulsory reporting of gender-disaggregated financial information as a part of ESG reporting is one of the priority policies. Although sustainability reporting mechanisms like the Business Responsibility and Sustainability Reporting (BRSR) framework in India have incorporated indicators on workforce diversity and gender representation, financial regulators need to further extend such requirements on gender specific metrics on giving credit, supplier diversity and support entrepreneurship. Regulators can increase the disclosure of gender-disaggregated financial indicators by legally obligating companies and financial institutions to disclose their capital allocation and help investors assess the social impact of financial activities. Standardized reporting of ESG across jurisdictions would also contribute to the comparability of reports and encourage the creation of globalised gender-sensitive finance standards. The seventh strategy is financing, which deals with financing the organization's operations through a financial institution.

6.2 Financing Strategies

This involves funding the activities of the organization via a financial institution. Financial institutions also have a major role in ensuring gender inclusive finance through development of lending practices that respond to structural obstacles that women entrepreneurs have. Financial institutions (banks and non-banking) must include gender-sensitive credit assessment schemes which consider other forms of business performance other than the conventional lending models based on collateral. A case in point is in providing valuable information about business viability and repayment capacity, digital transaction histories, e-commerce records and supply chain relationships can serve as valuable sources of information. Also, financial institutions could be increased to develop specialized financial products that specifically target women entrepreneurs. These can be gender bias credit lines, mentorship and financial literacy programs that are directed at enhancing entrepreneurial capacities.

6.3 Investor and Market Initiatives

The role of investors in creating gender-inclusive financial ecosystems is becoming even more significant as a result of embracing gender-lens investing approaches. Gender-lens investing is the types of investments specifically designed to give capital to companies and projects that enhance gender equality and economic empowerment of women. Including gender-related metrics in ESG investment strategies, institutional investors will be able to harmonize financial gains with the larger social impact goals. Gender bonds and sustainability related loans are examples of financial tools that offer a novel way of directing capital to women-owned businesses and gender-sensitive development initiatives. Such instruments have already shown promise of mobilizing the private investment towards gender-oriented projects as illustrated by Multilateral development institutions such as the International Finance Corporation and the world bank.

6.4 Building Digital Financial Infrastructure

The other policy priority that is critical is the use of digital financial infrastructure to promote financial inclusion. The emerging economies governments must still invest in the digital identity technology, payment platforms, and open-banking infrastructure allowing the safe exchange of data among financial institutions. Computer systems like the Unified Payments Interface (UPI) in India and Account Aggregator can serve as useful examples of how to merge financial information in the banking, fintech, and regulatory sectors. Such digital infrastructures have the ability to create massive datasets to serve AI credit scoring models and gender-disaggregated financial analytics. The combination of ESG reporting and digital financial platforms would help regulators and financial institutions monitor more efficiently the gender-inclusive capital flows and draft specific financial inclusion policies.

6.5 Implementation Roadmap

Policymakers can use a gradual strategy to make sure that gender-focused finance models based on ESG are successfully implemented. Phase One (2026-2028) may be based on pilot initiatives in nations with developed digital financial systems, i.e., India and Indonesia. These pilot projects would involve trying out integrated ESG reporting systems, which would bring together sustainability reports and disaggregated financial reporting on gender. Phase Two (2028-2030) may be the expansion of the models to involve more regional financial systems in emerging economies especially in ASEAN and other developing regions. Setting regional standards of reporting ESG and sharing data would also ease the flows of investment across the borders and enhance the global programs of gender-embracing finance. With such regulatory reforms, institutional approaches and technological advancement policymakers will be able to establish financial ecosystems that are more accommodating to women entrepreneurs and inclusive economic development. These attempts will not just reinforce ESG-driven financial markets, but will also enhance faster speeds in achieving Sustainable Development Goals 5 and 8.

CONCLUSION

The increasing popularity of Environmental, Social, and Governance (ESG) investing has radically changed the financial markets in the world and provided the new opportunities to the promotion of the inclusive and sustainable economic development. With the growing trend of sustainable investment assets globally, there has been an increasing regard of the integration of social impacts in decision making concerning finance. Improving access to women entrepreneurs is also one of these social priorities that pose a major challenge to the policymakers, financial institutions and development organizations. Although there has been a rapid development of ESG oriented financial markets, a significant source of financing gap on women led enterprises remains especially in the developing economies where structural barriers remain in place to restrict access to formal credit systems. This chapter reviewed how reinforcing ESG-based financial data ecosystems can help finance based on gender and speed up achievement of Sustainable Development Goal 5 (Gender Equality) and Sustainable Development Goal 8 (Decent Work and Economic Growth). The review has indicated that even though ESG models have enhanced transparency in corporate sustainability practices, there is still a lack of gender-disaggregated financial information in the financial systems. In absence of strong data infrastructures to record the gender-related financial indicators, financial institutions have a challenge in assessing the economic potential of businesses led by women and develop inclusive investment policies. The chapter also revealed that digital financial innovations and fintech platforms offer good opportunities to meet these constraints. Other credit scoring models that use digital transaction histories, supply chain information and e-commerce have the ability to increase access to credit to entrepreneurs who do not have traditional collateral or credit histories. The examples of fintech lending platforms and microfinance projects demonstrated how digital financial ecosystems can decrease information asymmetries involving both borrowers and lenders as well as enhance credit risk assessment. Such developments especially apply in the emergent economies where a significant part of business activity is in the informal sectors.

Regulations are also important in enhancing gender-friendly financial ecosystems. Such initiatives as Business Responsibility and Sustainability Reporting (BRSR) in India show that the mandatory disclosure of ESG can prompt corporations to incorporate measures of social sustainability, such as gender diversity and inclusion in the workforce, in their reporting. Equally, financially inclusion efforts by the government and digital financial infrastructure efforts would help increase access to formal financial services to underserved populations. Nevertheless, there are still a number of issues, which hamper the performance of gender-inclusive finance based on ESG. The weaknesses of the ESG reporting standards, a lack of access to gender-disaggregated financial data, and the presence of informal economic systems are the major challenges. To solve these issues, the work of the regulators, financial institutions, fintech innovators and international development organizations must be coordinated. Enhancing data integration between ESG reporting regimes and digital financial framework will be a condition of enhancing transparency in gender-based financial flows and conditional to the creation of evidence-based policy. In terms of policy, the chapter focused on the necessity of obligatory financial disclosures on gender issues on the base of ESG reporting systems, the introduction of gender-sensitive models of credit evaluation by financial institutions, and the development of gender-sensitive investment strategies by institutional investors. The ability of financial systems to promote inclusive capital allocation can also be increased by tapping into digital public financial infrastructure and credit analytics, which are based on artificial intelligence. In general, enhancing ESG-based financial data ecosystems is an essential direction of enhancing gender-sensitive finance and developing sustainable economic growth in emerging economies. Through the combination of ESG reporting systems, digital financial systems, and gender-disaggregated financial systems, policymakers and financial institutions can establish more inclusive financial systems, which can enable women entrepreneurs to prosper and enhance overall economic growth. The next step of the research ought to be to assess the effects of ESG motivational integration of financial data in the long run on female entrepreneurship, MSME development, and job creation. Comparison analysis between the emerging economies would also be a useful source of information on how regulatory frameworks and digital financial infrastructures affect the success of gender-inclusive finance programs. As the world financial systems keep on changing, the intersection between ESG systems and gender-responsive financial approaches will only need to be strengthened in the future as a necessary factor of attaining sustainable and inclusive development outcomes.

REFERENCES

Aayog, N. I. T. I. (2021). *SDG India Index and Dashboard 2020–21: Partnerships in the decade of action*. Government of India. https://www.niti.gov.in/reports-sdg

Afzal, J., & Haider, A. (2025). Environmental and Resource Protection Law by Ke Zhou, Luozhi Yi, Xinjian Su and Youhai Sun. *International Journal of Law and Legal Advancement, 1*(2). Advance online publication. DOI: 10.64060/IJLLA.v1i2.3

Ahmad, I., Haider, A., & Zeb, B. (2023). In the name of nature: The legal frontiers of environmental preservation. *Journal of Asian Development Studies, 12*(4), 401–411. DOI: 10.62345/jads.2023.12.4.30

Ali, K., Tintawi, G., Bassma, M. K., & Haider, A. (2026). Algorithmic sustainability: Governance conditions for AI-driven environmental decision-making. *Preprints.org*. https://doi.org/DOI: 10.20944/preprints202602.0733.v2

Boffo, R., & Patalano, R. (2020). *ESG investing: Practices, progress and challenges*. OECD Publishing. https://www.oecd.org/content/dam/oecd/en/publications/reports/2020/09/esg-investing-practices-progress-and-challenges_8a73d683/b4f71091-en.pdf

Broadstock, D. C., Chan, K., Cheng, L. T., & Wang, X. (2020). The role of ESG performance during times of financial crisis: Evidence from COVID-19 in China. *Finance Research Letters, 38*, 101716. Advance online publication. DOI: 10.1016/j.frl.2020.101716

Demirgüç-Kunt, A., Klapper, L., Singer, D., & Ansar, S. (2022). *The Global Findex Database 2021: Financial inclusion, digital payments, and resilience in the age of COVID-19*. World Bank. https://doi.org/DOI: 10.1596/978-1-4648-1897-4

Friede, G., Busch, T., & Bassen, A. (2015). ESG and financial performance: Aggregated evidence from more than 2,000 empirical studies. *Journal of Sustainable Finance & Investment, 5*(4), 210–233. DOI: 10.1080/20430795.2015.1118917

Frost, J., Gambacorta, L., Huang, Y., Shin, H. S., & Zbinden, P. (2019). BigTech and the changing structure of financial intermediation. *Economic Policy, 34*(100), 761–799. DOI: 10.1093/epolic/eiaa003

Global Sustainable Investment Alliance. (2023). *Global sustainable investment review 2022*. https://www.gsi-alliance.org/wp-content/uploads/2023/12/GSIA-Report-2022.pdf

Gomber, P., Koch, J., & Siering, M. (2017). Digital finance and FinTech: Current research and future research directions. *Journal of Business Economics*, *87*(5), 537–580. DOI: 10.1007/s11573-017-0852-x

Haider, A., Mathlouthi, N., & Ahmad, I. (2024). Beyond the books: Real world challenges in implementing environmental laws in Pakistan. *Remittances Review*, *9*(1), 1383–1405. DOI: 10.33282/rr.vx9il.93

Hui, Z., Haider, A., & Khan, A. (2025). International trade and plastic waste in oceans: Legal and policy challenges. *Frontiers in Marine Science*, *12*, 1627829. Advance online publication. DOI: 10.3389/fmars.2025.1627829

International Finance Corporation. (2021). *Sustainable bonds: Bridging the gap for gender equality*. World Bank Group. https://www.ifc.org/en/insights-reports/2021/sustainable-bonds-to-bridge-the-gender-gap

Krüger, P. (2015). Corporate goodness and shareholder wealth. *Journal of Financial Economics*, *115*(2), 304–329. DOI: 10.1016/j.jfineco.2014.09.008

OECD. (2021). *Gender and the environment: Building evidence and policies to achieve the SDGs*. OECD Publishing., DOI: 10.1787/3d32ca39-

Reserve Bank of India. (2025). *Report on trend and progress of banking in India 2024–25*. https://www.rbi.org.in/Scripts/AnnualPublications.aspx?head=Trend%20and%20Progress%20of%20Banking%20in%20India

Securities and Exchange Board of India. (2023, July 12). *BRSR Core – Framework for assurance and ESG disclosures for value chain* (Circular No. SEBI/HO/DDHS/DDHS-RACPOD2/P/CIR/2023/120). https://www.sebi.gov.in/legal/circulars/jul-2023/brsr-core-framework-for-assurance-and-esg-disclosures-for-value-chain_73854.html

Small Industries Development Bank of India. (2024). *Annual report 2023–24*. https://www.sidbi.in/annualreport/AnnualReport202324/index.php

United Nations. (2015). *Transforming our world: The 2030 agenda for sustainable development*. https://sdgs.un.org/2030agenda

United Nations Development Programme. (2024). *UNDP annual report 2023*. https://www.undp.org/publications/undp-annual-report-2023

Woetzel, J., Madgavkar, A., Sneader, K., Tonby, O., Lin, D.-Y., Lydon, J., Sha, S., Krishnan, M., Ellingrud, K., & Gubieski, M. (2018). *The power of parity: Advancing women's equality in Asia Pacific*. McKinsey Global Institute. https://www.mckinsey.com/mgi

Women, U. N. (2022). *Progress on the Sustainable Development Goals: The gender snapshot 2022*. https://www.unwomen.org/en/digital-library/publications/2022/09/progress-on-the-sustainable-development-goals-the-gender-snapshot-2022

World Bank. (2023). *Women Entrepreneurs Finance Initiative (We-Fi)*. https://www.worldbank.org/en/programs/women-entrepreneurs

World Bank. (2024). *Women, business and the law 2024*. https://www.worldbank.org/en/publication/women-business-and-the-law

Chapter 4
Conditional Regulation in Global Value Chains:
The Corporate Sustainability Due Diligence Directive and the Limits of Article 114 TFEU

Naim Mathlouthi
http://orcid.org/0000-0003-3506-5341
Prince Mohammad bin Fahd University, Saudi Arabia

Muamar Salameh
http://orcid.org/0009-0003-9489-493X
Prince Mohammad bin Fahad University, Saudi Arabia

ABSTRACT

This article analyses the Corporate Sustainability Due Diligence Directive as a shift from disclosure regimes, including Directive 2014/95/EU and Directive (EU) 2022/2464, toward process-based regulation of corporate conduct in global value chains. It asks when due diligence produces substantive environmental outcomes rather than formal compliance. Using doctrinal analysis grounded in Article 114 TFEU, supported by case law and comparative insights from France, Germany, and the textile sector, the article develops a conditional framework. Outcomes depend on enforcement capacity, value chain structure, and corporate response. Where enforcement is credible and supply chains are stable, firms integrate risk into decision-making; otherwise, compliance remains procedural. The Directive also tests the limits of Article 114 by regulating corporate processes and producing extraterritorial effects. Its impact is therefore contingent rather than uniform, highlighting both the potential and limits of process-based harmonization in EU law.

DOI: 10.4018/979-8-2600-3678-5.ch004

1. INTRODUCTION

The Global Value Chain has altered the nature of economic and environmental impacts, making it difficult to manage through traditional regulatory regimes. The production process is no longer territorial; it is networked and distributed, not hierarchical, and contractual (Salminen et al., 2022, p. 1068). This continues the asymmetry in which control over economic and strategic matters is left to key firms, while environmental impacts are exported to areas with less stringent controls. It might be tempting to say this is a coordination problem, but this is not the case. This is a problem of separation of ownership and control, and the absence of the traditional legal mechanisms at the global level is an added challenge (Schilling-Vacaflor & Lenschow, 2021, p. 680).

In the EU, the initial regulatory efforts focused on disclosure. The attempts to enhance disclosure comprise the directives 2014/95/EU and 2022/2464 (on the disclosure of environmental and social risks). The idea was that more disclosure would create indirect pressure on companies from their investors, consumers, and other stakeholders. This has been partially successful: there is more reporting, and the information is available. There is little connection between reporting and improvement. Knowing the risks can have only a limited effect on processes. This seems to be a characteristic of global value chains (Wilhelm, 2024, p. 462).

The Corporate Sustainability Due Diligence Directive breaks with this. Rather than emphasizing information sharing, it emphasizes corporate governance. Companies need to identify, prevent, and mitigate negative effects across their operations, subsidiaries, and supply chains, including those outside the Union. (Friesenbichler et al., 2024, p. 3) Structurally, this reorientation is not remarkable, but it is important: it shifts attention from internal risk management to external risk communication. Its ability to deliver meaningful environmental outcomes is questionable. Process-oriented obligations could, in theory, incorporate risk considerations into organizational structures. Nevertheless, such reliance on open-textured standards invites interpretive flexibility, which may allow the surface to accommodate minimal changes in fundamental practices (Evans & Silbey, 2021, p. 2048).

The current literature covers aspects of this problem, which is often abstract. Some of these analyses support the idea of instilling responsibility within corporate governance. (Cuomo et al., 2015, p. 255) Others emphasize the procedural nature of due diligence and the dangers of symbolic compliance. (Villiers, 2022, p. 561) Other studies examine the extraterritorial impact of EU regulation, assuming that access to the internal market is one of the incentives for harmonization at the international level (Koos, 2024, p. 162). These views both intersect and cannot fully account for the divergent practical outcomes occasioned by similar obligations. Variation in implementation warrants deeper scrutiny.

From a constitutional perspective, this regulatory shift raises a more specific question concerning the limits of Union competence. The Directive is based on Article 114 TFEU, which allows harmonization measures to facilitate the establishment and functioning of the internal market. The Court of Justice has found that these measures must remove actual or potential barriers to trade or distortions of competition (Case C-376/98, Germany v Parliament and Council (Tobacco Advertising I). On the other hand, the Court has acknowledged that harmonization may have incidental aims, such as public health and consumer protection, as long as a sufficient link with the functioning of the market is maintained (Case C-380/03, Tobacco Advertising II, EU: C:2006:772; Case C-58/08, Vodafone, EU: C:2010:321). The Directive is located at the fringes of this jurisprudence by regulating processes rather than product features and by effectively going beyond EU territory. This raises the question of whether such process-based obligations can be fully reconciled with the logic of the internal market (Article 114) or constitute a more ambitious form of regulatory integration.(Blažo, 2025)

The study brings together a doctrinal analysis of the Directive and its basis in Article 114 TFEU, a comparative study of French and German due diligence regimes, and a sector analysis of the textile industry. It does not rely on exhaustive coverage but instead provides an analytical framework to explain variations across sectors and jurisdictions. Therefore, the importance of the Directive lies more in the empowering conditions than in the homogeneous effects.

2. THE CONCEPTUAL POSITIONING AND LITERATURE

2.1 Due Diligence in Pre-Existing Scholarship

The academic literature on corporate due diligence in the European Union has taken different paths, each shedding light on the regulatory changes introduced by the Directive. There is a spectrum of inquiry that considers due diligence as a tool for bringing responsibility to corporate governance (Sinnig & Zetzsche, 2025, p. 1). Based on the UN Guiding Principles on Business and Human Rights, due diligence is defined as a process that facilitates the incorporation of environmental and social risks into companies (Taylor, 2020). The focus is on incorporation: recognizing and minimizing risks, and integrating them into standard decision-making processes rather than ad hoc or marginal responses. This method has analytical value but assumes a change worth examining, because implementing procedures does not always alter old practices, particularly when organizational incentives remain unchanged.

A second view is more critical, emphasizing the procedural nature of due diligence and its potential to degenerate into formalism. In this perspective, the indeterminateness of due diligence conditions, as expressed through terms such as "appropriate measures," allows for superficial compliance. Businesses may develop policies and codes of conduct and implement auditing mechanisms that indicate compliance on paper, but essential production arrangements are largely unaffected (Assche & Narula, 2022, p. 166). This has resonated in academic and political debates, yet it can be misinterpreted as implying that proceduralism is inherent and neglects situations in which further commitments are required.

A third stream of research places due diligence in the broader context of the EU's externalities. It implies that market access to the internal market may oblige companies to harmonize their practices at the extraterritorial level and export EU standards through market power, rather than territorial jurisdiction. This reasoning aligns with Anu Bradford's argument about the export of regulations, market size, and regulatory capacity. It means that, in the context of due diligence, multinational corporations may align their operations with EU standards for compliance (Elbel et al., 2023, p. 103250). However, such consistency is unlikely, given variations in companies' dependence on the EU market and their capacity to restructure operations, which can lead to sporadic, inconsistent, or "convenient" adjustments.

2.2 The Conditional Understanding of Due Diligence

These perspectives are not contradictory but tend to focus separately on mechanisms of internalization, proceduralism, and extraterritorial proliferation, ignoring contextual factors that lead to dominance. This means that the literature offers plausible explanations of the due diligence process but little on the contextual differences in impact. The problem is not to identify mechanisms but to develop a model of difference.

In this article, the literature gap is addressed by refocusing on the interactions among mechanisms, structures, and institutions rather than on mechanisms in isolation. It selectively draws on literature on regulatory governance, such as Julia Black's emphasis on situational factors in the interpretation and application of norms. The argument advanced in the present article is more modest: due diligence is a contingent regulatory mode; its impacts depend on three interrelated factors: enforcement capacity, the value chain structure, and the corporate response (Black, 2001). This form of conceptualization is a small departure from the narratives of inherent or structural regulatory effects. It provides subtlety and therefore refrains from inferring legislative intent. The same mandate can receive different degrees of attention through interpretive, enforcement, and organizational processes. It can change the environment in some instances and be ritualistic in others. This is less about regulatory problems than about context (Cohn, 2001). This perspective also qualifies theories of external regulatory reach. It means that rather than EU norms "spreading", convergence depends on firm-level factors and constraints. The Directive may, but may not, result in convergence; it is contingent on factors similar to those in national implementation. This contingency defines, not detracts from, the model.

3. LEGAL BASIS: ARTICLE 114 TFEU AND THE INTERNAL MARKET FRAMEWORK

3.1 Prerequisites to the Application of Article 114 TFEU

The Directive, under Article 114 TFEU, falls within the constitutional genre, as such measures harmonize national laws that constitute barriers to the establishment or operation of the internal market. This is technically accurate: Before the Directive, different Member States had adopted distinct due diligence measures, such as those in France and Germany, with varying reach, enforcement, and liability. This gives rise to confusion and distortions for non-domestic actors (Wilhelm, 2024, p. 459). This is shaded out by judicial precedent, though: divergence is not enough. Indeed, in Germany v Parliament and Council, the Court found that Article 114 requires a link to the internal market. Harmonization must address actual or predictable trade barriers or distortions of competition that require substantive justification rather than mere empirical evidence (Tobacco Advertising I).

Later jurisprudence has relaxed this rigor. Indeed, in British American Tobacco, the Court supported actions that promoted ancillary objectives, such as public health, to the extent that the market connection was substantive. Similarly, in Vodafone Ltd v Secretary of State for Business, the Court affirmed the need for preventative steps to prevent fragmentation. The shift in existing barriers toward potential disintegration gives the Union the ability to preempt rulemaking (Vodafone, 2010). In the case of the Directive, such jurisprudence provides plausible, though disputed, grounds for harmonization. Multinational firm organization is intruded by national due diligence divergences. Transiting through multifaceted regimes increases costs and enforcement burdens, influencing strategies and thereby distorting cross-jurisdictional competition (Sinnig & Zetzsche, 2025, p. 24). However, the connection to market functioning is more attenuated than in traditional internal market harmonization measures. Due diligence controls operational procedures rather than product characteristics; in this context, extending these to traditional obstacles strains the analogy (Bartl, 2022, p. 969). Harmonization so appeals to a broad conception of an internal market, including regulatory preconditions for cross-border activity, a breadth the Court tolerates, without exhausting latent doctrinal tensions (Schmidt, 2018, p. 70).

3.2 Tensions and Limitations Within the Internal Market Framework

The extra purposes and impacts of the Directive pose significant doctrinal challenges. It goes beyond harmonizing market access by including governance obligations that extend beyond the EU and prioritize the environment and human rights (Peters et al., 2023, p. 450). This raises an important question about its foundation in the internal market (Article 114 TFEU): can measures this wide be justified under the Treaty? On the upside, the different ways countries manage due diligence clearly impact international trade and investment by imposing different costs. That provides a strong impetus for harmonization. Crucially, the Court has confirmed the breadth of Article 114 to encompass a variety of objectives, including organizational and procedural measures, as long as they achieve positive market integration gains (Cherednychenko, 2023, p. 21). This accords with the Court's jurisprudence since Tobacco Advertising.

But proportionality? That is the rub and potentially the fatal flaw. The Directives are broad, covering all stages of the global value chain, and risks upstream and downstream of the EU. This is no small internal-market irritant, such as minor variations in administrative and labeling practices across member states. In Omega Spielhallen, the Court invalidated a measure that went too far, imposing public morals on free-market principles, suggesting that the environment cannot wholly devour trade. Viking Line further suggested that trade cannot be bogged down by excessive social goals. Sure, legislators can choose how far to harmonize, but not too far: measures cannot go beyond the goal of removing market distortions. This one is right on the edge of Article 114's traditional limits, or over the line. The courts will have to tweak it to keep it in line, perhaps to limit it to EU risks (Omega Spielhallen, 2004). The clash becomes more acute with its obvious extraterritorial ambitions. Non-EU firms get caught in the net if they do big business in the EU and then have to follow it across the globe, changing their supply chains. This is not the traditional way to liberalize the market. Rather, it is very interventionist, which clashes with traditional notions of territory and precedent, which admitted only incidental effects outside the EU. These were really side effects, not inherent(Kalimo et al., 2023, p. 32). Nonetheless, the Directive's preventive stance is consistent with the more lenient policy after Tobacco Advertising. The Court permitted measures to prevent future trade barriers. However, the focus on procedures and the de facto extraterritoriality stretch this link too far. The link with the internal market becomes, at best, tenuous and, at worst, requires court intervention. It invites close judicial scrutiny and potential invalidation(Kübek & Mancini, 2023, p. 535). These core ideas have bigger effects. Article 114 imposes the same general obligations on all member states but leaves it to them to implement them. This guarantees a variety of results. Such issues, including the application of due diligence, affect its success. Thus, it highlights the balance and tension in Article 114 between unity and national autonomy, despite the size of the Directives potentially affecting that balance. (Shulman, 2024, p. 515).

4. LEGAL DESIGN OF DUE DILIGENCE OBLIGATIONS

4.1 Scope and Structure of Obligations

The Directive establishes a due diligence process that mirrors the nature of modern production, which extends beyond the firm itself to include its operations, subsidiaries, and the "chain of activities" (Bueno et al., 2024, p. 296). This is no coincidence; it seeks to better fit the law to the economy, in which power is also relational rather than merely hierarchical. For example, in long garment supply chains, a European brand may not make the cloth, but it should have the final say on pricing and specifications. However, it introduces some uncertainty. As responsibility shifts from a firm in a narrow sense, the scope of responsibility becomes unclear, forcing firms to navigate uncertainty until regulators issue guidance (Boström et al., 2014, p. 5).

This is an aspect of the chain of activities, as it covers upstream processing and, to some extent, downstream activities, such as customer distribution. Crucially, it is not all business links but those related to the firm's economic activities. The question is: what is the connection? It may be contractual, economic (e.g., suppliers with a single buyer), or the firm's power (e.g., audit and tech standards). This is not clear-cut under the Directive, as it uses generic language that can apply across different industries, such as cobalt extraction for mobile phones or cocoa harvesting for chocolate. Thus, flexibility is good, but may lack certainty, with the risk of litigation to establish precedent. (Sellare et al., 2022, p. 862) The structure's requirements are designed in a process-based manner. Businesses are required to integrate due diligence into their policies, map negative impacts, take preventive and corrective measures, monitor performance, and publicly disclose. These are not only cumulative but a circle rather than a list, meaning they are ongoing rather than a "one and done" approach (Ford & Nolan, 2020, p. 31). On paper, this approach allows due diligence to adapt to global value chains and circumstances, such as sudden realignments in geopolitical conditions that affect sourcing. However, it also makes assessment more difficult, since there is no clear end to compliance and it involves an element of interpretation, particularly when impacts are gradual and take years to manifest (Schilling-Vacaflor & Lenschow, 2021, p. 690).

One key element of this design is risk-based management, in which firms prioritize their actions according to the risk and likelihood of negative consequences. This allows firms to decide which risks are worth managing and how many resources to devote to them, such as ramping up on child labor concerns while scaling back on small environmental adjustments. While this is real-world, given constrained resources and extensive supplier chains, it involves internal, non-observable processes. It thus raises concerns about peer consistency and strategic myopia, thereby posing systemic risks (Assche & Narula, 2022, p. 166).

4.2 Legal Character and Open Standards

Due diligence obligations have a legal nature defined by the difference between the conduct and the result (Shulman, 2024, p. 512). Companies must take reasonable steps to identify and address risks, but they do not face absolute liability for all damage resulting from their actions. This reflects an acknowledgment of practical limits. In complex, geographically decentralized production systems, outcomes are not always controllable, even when firms are diligent. The emphasis, however, is on the sufficiency of actions rather than the attainment of certain outcomes (Wilhelm, 2024, p. 462).

The open-ended standards support this approach. Words such as " appropriate measures or reasonable steps do not have a fixed meaning. Their contents are context-dependent, such as the nature of the risk, the firm's position in the value chain, and the level of influence the firm can exert (Freitas, 2024, p. 3). This indeterminacy serves a functional purpose by enabling context-sensitive application across heterogeneous sectors. It enables the Directive to be applied across sectors with differing characteristics and avoids the inflexibility associated with more prescriptive rules. Simultaneously, it brings about interpretative uncertainty. Due diligence is not precisely defined in advance, but rather emerges from supervisory practice and, ultimately, judicial interpretation(Sinnig & Zetzsche, 2025, p. 24). *The more difficult question is how the phrase should be fleshed out by the courts and supervisory authorities without creating legal uncertainty. In practice, it is likely that appropriate measures will be evaluated against what a reasonably diligent firm in the same sector, size, and position in the value chain would have done, based on the severity and likelihood of the risk rather than the outcome. This makes the standard contextually responsive but also implies that its meaning is accretionary on a case-by-case basis: until a stock of supervisory decisions and judgments accumulates, firms have no ex ante information on where the boundary between adequate and inadequate effort lies. The open texture that enables the standard to be adaptable is the same attribute that delays its definition until after the compliance decisions have been made.* There is, however, an underlying tension. Open standards provide an opportunity for substantive engagement with environmental risk, enabling firms to adjust their responses to specific conditions. However, they also make space for more formal means to be used. It remains possible to comply with the structure of due diligence policies, procedures, and reporting without significantly altering how production is organized(Rezarta & Richard, 2025). The difference between these results is not simply predetermined by the law's text, but by the interpretation and application of the standards in practice. The Directive also creates aspects that go beyond immediate risk management. Some businesses must implement transition plans aligned with climate goals, making due diligence a longer-term strategic priority. The degree of this association is also unclear. Transition planning implies a shift towards incorporating sustainability into business models, yet it is presented as a procedural requirement. The ability to affect substantive outcomes rests on the operationalization of firms and the evaluation of supervisory authorities (Crovini et al., 2025, p. 334).

Together, these attributes make the Directive distinctive. It expands the range of responsibility and arranges obligations to facilitate adaptation. Flexibility and determinacy are difficult to combine. Obligations may be difficult to apply when overly flexible. Also, they are more accurate, where they run the risk of not being able to capture the complexity of global value chains(Scheper, 2025). The Directive avoids fully resolving this tension, and its practical implications will depend on the circumstances under which it is applied.

5. ENFORCEMENT AND LIABILITY

5.1 Administrative Supervision and Enforcement Capacity

The implementation of the Directive is left, to a large extent, to national enforcement agencies. National authorities must appoint supervisory authorities with the power to investigate, request information, and sanction. Formally, this is not uncommon in EU law, where the Union sets the standards and enforcement is decentralized, a doctrine known as subsidiarity (Art. 5 TEU). However, due diligence makes this system much more complex. Oversight of compliance does not consist of validating actions, but evaluating organizational processes that are often opaque, iterative, and hard to monitor. This stems from the relational and anticipatory nature of due diligence: it involves global supply chains and requires probabilistic risk calculations rather than retrospective fact-finding (Guye et al., 2025). One might be tempted, then, not to rely solely on the distribution of formal powers for effective supervision. This is confirmed by comparative regulatory theory: in complex areas, like environmental due diligence, there is a risk of "implementation slippage" from decentralized enforcement, whereby national differences reduce the supranational ambitions, as evident in case studies of EU environmental law, including the Industrial Emissions Directive(Sinnig & Zetzsche, 2025, p. 25).

The competence of oversight bodies becomes critical not only in financial resources but also in technical and multi-agency expertise and learning. When institutions have these traits, as is the case in Denmark with its strong environmental authorities, they can grapple with the content of due diligence(Foerster et al., 2025). This may include scrutinizing how companies identify risks through advanced materiality analyses, how they prioritize risks through scenario modeling, and the credibility of risk mitigation strategies over time (including stakeholder consultations and third-party verifications). These circumstances allow supervision to go beyond paperwork and influence behavior. Companies must not only provide evidence of their policies but also demonstrate that these policies effectively create a "compliance culture" that instills the right corporate incentives to deliver sustainable outcomes (Rodríguez-Gómez et al., 2025). The scenario would seem more constrained in situations of resource scarcity, a situation faced by many mid-tier Member States amid fiscal stringency and competing demands. Even if formal powers exist, resource limitations may drive regulators to focus on more visible indicators of compliance: policies, reporting mechanisms, and audit trails all offer tangible evidence of compliance, even if they may not reflect actual practices. The distinction between compliance and engagement is important here, and evokes Hart's positivist critique of the "internal point of view" of law. A company may develop a sophisticated framework while operating with a business-as-usual approach to production, taking advantage of informational asymmetries that overwhelm regulators and lacking sophisticated data-mining and whistleblower mechanisms.

Thus, differences among Member States are inevitable and reflect path dependency. With technocratic legacies, the Nordics may develop technical expertise and take more regulatory initiatives, deploying integrated information systems for monitoring. The South and East, in turn, might have more fragmented bureaucracies and less scrutiny and oversight by civil society (Barbieri et al., 2022, p. 768). The Directive provides co-ordination arrangements, such as the European Network of Supervisory Authorities. However, these cannot fully address differences in government capacities and cultures, nor can they address the search for the lowest common denominator. Regulation is not uniform across all areas, and companies are not scrutinized equally across jurisdictions and sectors. This is made worse by the risk-based approach to due diligence: authorities are likely to prioritize sectors with obvious environmental risks (e.g., textiles, agriculture) over those with less obvious risks (e.g., electronics, services). In short, this approach could be a cover for a "variable geometry" of sustainability governance regimes, through which global corporations "forum shop" regulatory jurisdictions, all the while subverting the normative impetus of the Directive towards harmonized accountability (Schilling-Vacaflor & Lenschow,

2021, p. 690). The Directive is reliant on national regulators. Each member state must designate supervisory bodies with investigatory, information-gathering, and sanctioning powers. This is part of the EU's decentralized enforcement model of subsidiarity. However, due diligence is dispersed and iterative across value chains, with risk probabilities that make it hard to supervise, and it focuses on assessing invisible organizational practices rather than actions. Comparative regulatory theory warns of "implementation slippage" in complex areas such as environmental due diligence, as evidenced by precedents such as the Industrial Emissions Directive, in which national differences undermine supranational objectives (Yuan & Mähönen, 2024, p. 17).

The Capacity of financial, technical, interagency coordination and flexibility is crucial, as effective agencies, like Denmark's environmental protection agencies, examine companies' risk assessment (materiality exercises), prioritize risks (scenario analysis) and verify the credibility of mitigation measures, which include consultations and verification by stakeholders, producing a "culture of compliance" to reward sustainability(Marco & Cagliano, 2026). Capacity constraints in many middle-tier Member States, amid fiscal austerity, on the other hand, lead policymakers to rely on visible symbols such as policies and audits, conflating formal and substantive compliance. The distinction between the two types of compliance becomes particularly relevant in this situation, as businesses can have "paper tiger" policies while leaving production practices largely unchanged. Variations between Member States reflect institutional legacies: technocratic Nordic countries have high levels of expertise and constant monitoring through digital networks, whereas Southern and Eastern European countries exhibit bureaucratic fragmentation. The Directive's "check and balance" regulatory measures, such as the European Network of Supervisory Authorities, are unable to overcome capacity and cultural differences, creating a "variable geometry" and uneven enforcement in different jurisdictions and industries. Risk-based prioritization exacerbates this, allowing companies to engage in forum shopping and the "variable geometry" of sustainability within and between jurisdictions and sectors (Ciacchi et al., 2026).

5.2 Civil Liability and Procedural Constraints

The Directive, alongside regulatory enforcement, includes civil liability. Companies can be liable for harm caused by non-compliance with due diligence. This offers, at least in theory, a way for interested parties to seek redress and hold companies to account for poor practices (Mercier, 2026). It also enhances accountability, albeit in a more complex way. One difficulty is establishing causation between the company's conduct and the harm. This will be structurally complicated in a fragmented global value chain. Environmental damage has a domino effect, with multiple actors across different jurisdictions. It can be hard to determine the impact of a particular firm's failure to conduct due diligence (Scheper, 2026). The standard of proof is also high, and it is not always clear how the court will approach questions of causation in such cases.

The information can be accessed, which poses a similar limitation. Much of the information about supply chain arrangements and corporate decision-making is held by companies. Plaintiffs, particularly Union citizens, could potentially face obstacles in accessing the information they need to substantiate their claims (Parroco, 2026). Without some means to redress these imbalances, the availability of liability in theory cannot necessarily be translated into practice. Earlier cases under national regimes - such as the French duty of vigilance law - suggest that the evidential hurdles are not only theoretical but are affecting the practical application of claims (Koos, 2024, p. 165). This is compounded by the fact that it is mixed with domestic law. While the Directive sets the ground for liability, procedural law is largely national. There are differences between Member States in terms of the burden of proof, disclosure, and remedies. As a result, there can be different approaches to frequent claims (Bueno et al., 2024, p. 298). The link between administrative and civil law proceedings is also not symmetrical. Supervisory decisions can be used to facilitate civil claims in certain systems by establishing evidence of non-compliance. In others, the two mechanisms operate more independently(Koos, 2024, p. 157). Collectively, these imply that civil liability is not a comprehensive enforcement tool, though it is important. It can be really significant in certain situations, particularly when the damage can be traced to a more identifiable malfunction. In a more general sense, though, it is limited in its efficacy by the systemic attributes of transnational production and by the circumstances of litigation(Friesenbichler et al., 2024, p. 8). The fact that there is a liability does not necessarily imply that due diligence requirements will yield substantive results. However, it is a constituent part of a larger enforcement environment that shapes how firms respond to regulatory expectations.

6. DUE DILIGENCE AS CONDITIONAL REGULATION

6.1 Essential Variables and Analysis Framework

The above discussion indicates that the impact of due diligence cannot be deduced directly based on its legal framework. The same obligations can yield different results depending on how they are interpreted and applied. This observation, although not new per se, needs a more specific analytical formulation. Rather than due diligence as a uniform regulatory instrument, it is better to consider it as a conditional regulation whose effect depends on the interaction of three factors: the capacity to enforce, the value chain structure, and corporate response (Felbermayr et al., 2024).

The capacity to enforce can be seen as the supervisory authorities' ability to understand and guide corporate conduct. This is not just a question of formal (supervisory) powers, such as the power to investigate and impose sanctions, but also of resources and capabilities and of the institutional readiness to understand complex organizational relations (Dill, 2025, p. 22). In cases of high capacity, the authorities might probe beyond form and into the actual content of due diligence systems. In areas where it is constrained, the control will be more inclined towards observable evidence, such as documentation and reporting (Uvarova & Bernaz, 2025, p. 14). The difference is consequential, as it determines the incentives firms have to invest in substantive risk management or a more formal compliance structure.

Value chain structure captures the organizational environment in which due diligence operates. The global production process is generally decentralized across multiple tiers of suppliers, with varying degrees of contractual stability and visibility. In more integrated or stable chains, firms can exert some influence, enabling them to track and shape practices beyond their direct operations. In less cohesive environments, such influence is diluted(Billedeau & Wilson, 2025). Visibility with first-tier suppliers is reduced, and the relationship can change as market conditions evolve. In this scenario, firms' capability to identify and manage risks is constrained, irrespective of their legal obligations (Sinnig & Zetzsche, 2025, p. 18).

The third element is firms' response: Due diligence is not practiced uniformly, as some adopt systems more oriented towards formal compliance, focusing on policies, code of conduct, and auditing practices. Others are more integrated in their practices, with due diligence impacting procurement, supplier management, and strategic planning (Lotfi & Walker, 2026). Enforcement expectations, financial limitations, and internal governance frameworks influence these reactions. In this regard, a firm can engage in procedural and more substantive activity in certain areas. These factors are interconnected, with corporate response relying on enforcement capacity, which sets expectations and incentives. Value chain structures affect strategies and the ability to implement. Corporate response, in turn, defines the translation of legal obligations into practice (Alexander & Orlando, 2025). Variation is created through their interaction. With credible enforcement, value chains are relatively stable, and firms undertake organizational change; as a result, due diligence is more likely to have a substantive impact. Without these, there is a greater likelihood of a formal impact.

6.2 From Formal Compliance to Substantive Impact

Substantive impact and formal compliance are effective ways of capturing the impact of due diligence. Formal compliance is the adoption of structures that meet legal requirements, such as policies and reporting, contractual arrangements, and audit procedures. These elements are typically mandatory and can be used to enable further engagement. However, they do not, in and of themselves, ensure a change in practices. Substantial impact, however, implies a change in the way the company handles risks, in terms of sourcing, supplier relations, resource allocation, etc. (Sinnig & Zetzsche, 2025, p. 2).

Between these two extremes is a third type, organizational adaptation. Here, firms go beyond minimal compliance, but do not transform their operations. They may construct supply chain maps, create more sophisticated risk management tools, or engage more with some suppliers. These changes can improve risk management, but may not eliminate it. The boundary between adaptation and impact is not always distinct, and impacts can evolve (Pítzen et al., 2025, p. 10). The shift between categories is driven by the interaction of the factors mentioned above. In particular, enforcement capacity is an important factor. Where enforcement agencies can assess the substance of due diligence processes, it incentivizes companies to show that their processes are effective. This may lead firms to invest in higher forms of organizational change (Rezarta & Richard, 2025). In places where supervisory authorities cannot evaluate the content of due diligence procedures, the incentives change. In this situation, it makes sense that companies will emphasize low-cost, highly visible actions with low impact (Sunny et al., 2025). Value chain structure creates even more restrictions. If firms operate in a fragmented industry with short-term contracting, they may be less visible or influential in changing practices beyond their direct suppliers (Gugkang et al., 2025). Even in areas where they want to enhance their capacity, their ability to exercise oversight is constrained. This does not preclude substantive influence, but makes it harder to achieve. The scope of influence is broader, and the potential for substantive influence is greater in integrated and stable chains (Kingston & Carnovale, 2026). These pressures are mediated by corporate response, as companies understand and apply due diligence in the framework of their priorities and limitations. Others tailor their systems to the most challenging regulatory frameworks and apply comparable criteria across jurisdictions. At the same time, others are more selective, targeting areas with the highest exposure or risk. Such decisions are not made solely on the basis of legal policies. They represent more general factors, such as cost, competition, and organizational capacity(Majekodunmi et al., 2025).

The connotation of this framework is that due diligence does not yield a single predictable outcome. It produces a variety of potential answers, both formal compliance and substantive change, in different circumstances in which it works. This conditionality is not a weakness of the regime but a defining feature of its operation (Sinnig & Zetzsche, 2025, p. 2). It is crucial to understand it in terms of both the design of the legal regime and the interplay between law, institutions, and organizations.

7. COMPARATIVE AND SECTORAL ANALYSIS

7.1 France and Germany: Dissident Enforcement Logics

The above analytical framework can be applied to understand the roles of different institutional arrangements in enforcing due diligence obligations. France and Germany are a good comparison, not because they reflect all possible approaches, but because they represent different enforcement approaches. While companies face comparable due diligence requirements in both systems, they have very different oversight and enforcement structures.

The French approach, based on the Duty of Vigilance Law, obligates firms to establish and implement a vigilance plan to manage risks arising from their activities, subsidiaries, and certain third parties. Litigation is central to enforcement, which, in theory, allows interested parties to contest shortcomings in corporate conduct and claim compensation for damages. Civil liability, therefore, creates an ex post accountability system focused on specific harms (Mu, 2025).

In practice, however, procedural hurdles severely limit its effectiveness. Proving standing, overcoming the lack of information, and establishing causality between corporate activities and environmental damage are all difficult, especially when production is transnational (Salih et al., 2025)1. The handful of cases litigated under the French vigilance regime already reveal that evidentiary and procedural challenges are inherent to the regime. It is litigated on a case-by-case basis for "flagship" cases with litigable damages. While it may put pressure on companies in some cases, it does not provide for ongoing monitoring of corporate activities(Bounfour & Chatelain, 2025). By contrast, the German approach is based on administrative monitoring, requiring companies to maintain risk-monitoring systems that are subject to scrutiny by competent authorities. This enables continuous oversight of corporate operations. Authorities can periodically assess firms' risks and their risk mitigation measures, if they have the resources to do so. This is not about particular instances of harm, but about organizational structures (Oijen et al., 2020, p. 1). However, this method has drawbacks, as administrative professionals have limited resources, and they cannot supervise all entities or areas. They may use signs (such as reporting requirements) as an entry point, but the signs may not be linked to risk management. So, administrative supervision is not just about powers but also the intensity of capacity (Black & Baldwin, 2012, p. 15).

This is reflected in the model in Section 6. In France, liability is the main issue, with the victim suing for compensation. In Germany, public servants enforce via bureaucratic processes of limited capacity and competency. These settings give rise to different incentives for firms: in France, potential liability creates incentives for compliance (where possible); in Germany, monitoring creates incentives for compliance through formal due diligence processes that depend on enforcement. Neither regime is entirely successful, which suggests that enforcement is part of due diligence (Schilling-Vacaflor & Lenschow, 2021, p. 689*). The Directive superimposes this divergence and, in effect, attempts to reconcile the two logics: it keeps the French model of ex post civil liability of harm, but overlays it with the German model of ex ante monitoring of risk-management systems by the administration. It remains an open question whether this hybrid will harmonize or merely replicate the shortcomings of both regimes at the level of the Union. A company may comply with the monitoring arm by documenting its procedures, but still be liable only when a claimant overcomes the evidential obstacles that already dull the French regime. Instead of merging the two enforcement cultures into a single standard, the CSDDD might simply leave multinationals operating in a patchy patchwork, where the level of scrutiny continues to depend on which national authority is in control and which national court.*

7.2 The Textile Industry: Value Chain Constraints

The textile sector is a good example where the structural features of global value chains are prominent. Manufacturing is broken down into several stages: processing raw materials, spinning, dyeing, finishing, and assembly. These are often located in different legal jurisdictions and are linked by contractual arrangements that may not be long-lived. The structure is fragmented and, at times, dynamic(Milnitz & Marchi, 2025). This is relevant for the implementation of the due diligence obligations. These environmental risks are concentrated at specific stages of this structure, particularly during dyeing and finishing operations that involve large volumes of water and chemicals. These are often seen in places with inconsistent or inadequate regulation. Consequently, the environmental effects of textile manufacturing are usually physically borne by companies that decide on sourcing and production policies. This segregation makes it hard to identify and address risks in the context of due diligence (Both & Wilhelm, 2025).

The visibility of lead firms is likely to fade out after the first layer of suppliers. The nature of relationships can change in response to shifts in cost, demand, or production capacity, making them less stable to rely on for long-term risk management. The ability to exercise oversight is limited, even when firms seek to expand it, due to the way the supply chain is organized (Járvás et al., 2025). Audits and certification schemes are among the monitoring mechanisms that can provide insight, although they offer only a partial, usually momentary picture of conditions. They are more inclined to show compliance at a particular point in time rather than practices (Rashid, 2025). The industry's response is mixed: Some companies try to map their supply chains more accurately to better understand and engage with suppliers. This can help address environmental risks, but it takes time and money and may come at the expense of flexibility. Other companies are more dependent on contractual arrangements that include due diligence provisions in supplier contracts, without any substantial change in sourcing policies. The majority of companies operate between these positions, balancing formal compliance with more specific forms of organizational adaptation (Rezarta & Richard, 2025).

The distribution of responsibility within the supply chain remains uneven. Whereas lead firms are legally liable parties, suppliers have to bear the brunt of compliance, which can be extremely strained when their economic capacity is limited. Changes in commercial terms do not always align with expectations regarding environmental standards, which limits suppliers' ability to make significant improvements. In these circumstances, there is a danger that compliance becomes more or less formal, whereby documentation hides actual limitations (Maiju, 2025). The process of this occurrence is more or less contractual. Lead companies usually offload their liabilities by imposing standardized codes of conduct and due diligence warranties into supplier contracts, transferring the compliance risk and much of the legal risk down the line to the least capable bearers. An initial-level supplier who must pass these terms on to its own subcontractors repeats the same trickle-down effect, thus shifting responsibility towards the weakest participants and leaving control and margin at the highest levels. The outcome is a mismatch that the Directive fails to resolve: the firm that can change practices due to leverage contracts out of liability, while the firm that incurs the liability cannot change anything.

At a more micro level, there is the issue of supplier tiers. While lead firms may have longer-term supplier relationships with first-tier suppliers, information tends to fade down the supply chain. Lower-level suppliers may have different contractual arrangements and be less closely tied to the lead firm. Environmental risks, such as those associated with dyeing and processing, are likely to be found in these more distant and less controlled lower echelons(Jian et al., 2025, p. 4). *This results in a mismatch between liability and control. Even if firms adopt due diligence processes, these have less impact on environmental outcomes at the lower end. So, the success of due diligence in the textile industry depends not only on the design of due diligence processes, but also on firms' ability to establish consistent, transparent relationships with suppliers further down the supply chain(OECD, 2025).*

8. EXTERNAL DIMENSION AND CORPORATE RESPONSE

8.1 External Reach Through Market Access

The Directive is extra-territorial in nature, given that non-compliance will deny firms access to the internal market. A company incorporated outside the EU may be subject to EU provisions if it meets a certain turnover threshold in the EU. This system does not operate on concepts of jurisdictional sovereignty. Rather, it relies on economic presence. Businesses that want to operate in the market must align their operations with due diligence requirements that extend throughout their operations(Sinnig & Zetzsche, 2025, p. 9).

On the face of it, this is similar to previous EU regulations, particularly regarding product standards. In these cases, compliance is checked at the border, and non-compliant products may be barred from the market. Due diligence does not operate this way, as it involves organizational processes that unfold over time and across many jurisdictions. It is more than a one-off test. This difference is not merely technical. It has implications for the Directive's impact outside the Union (Feld, 2026). The effect is therefore indirect. The Directive incentivizes firms to align their practices with EU law, but does not guarantee consistency. Companies differ in their exposure to the EU. Some, especially large multinationals with significant EU presence, may find it most efficient to align. Maintaining different systems for different jurisdictions can be costly and difficult to manage. Selectivity may be a possibility for others, especially with lower exposure(Mangiuc & Kagitci, 2025, p. 1204).

Selectivity may take many forms. A company may prioritize its due diligence obligations for EU markets but adopt different processes for other markets. Alternatively, it may selectively follow a standardized worldwide approach. The Directive does not prescribe how. It prescribes standards, but not how they may be implemented(Ciacchi et al., 2026). Patterns reflect the interactions of enforcement, value chains, and firms. Where enforcement is effective and supply chain structures are less complex, there is more pressure to adopt more formalized systems. Where access to the EU is limited and/or value chains are dynamic, there is more nuanced convergence (Partzsch, 2025). The Directive's extra-territoriality is thus contingent. Local conditions play an important role here, as value chains can extend into jurisdictions with less stringent regulation or enforcement. In these countries, companies may be constrained in terms of risk assessment and compliance. In addition, Information could be scarce, and monitoring could be constrained by limited access. When firms undertake due diligence, these factors influence their ability to do so. The Directive impacts corporate activities, but it does not remove external factors(Rezarta & Richard, 2025). *Such an external reach is not a neutral read by the people on the receiving end. Such process requirements are often perceived by producer states and lower-tier suppliers in the Global South as green protectionism: standards as human-rights and environmental regulation, but, in practice, as terms of entry to the EU that it sets for itself and others must meet. Since the requirements are based on economic presence rather than territorial jurisdiction, they increase the Article 114 tension previously described by imposing the costs of compliance on firms and governments that were not involved in the rulemaking process (Koos, 2024, p. 162). The danger is that those suppliers who are least able to shoulder the administrative burden are either forced to exit or to engage in shallow compliance, in such a way that the coverage of the Directive is greatest where its developmental legitimacy is weakest.*

8.2 Corporate Responses and Adaptation

If the Directive's external reach is mediated, the same can be said of corporate response. Legal requirements do not necessarily translate into corporate action. They are mediated through firms, which are influenced by their priorities, capabilities, and internal processes. To understand the workings of due diligence, it is therefore important to consider how firms respond to these obligations(Rezarta & Richard, 2025).

One form of response is oriented towards formal compliance. Companies draft policies, issue codes of conduct, and include contractual provisions requiring suppliers to meet certain standards. Oversight is usually based on audits or certification. These arrangements are relatively easy to put in place and offer external visibility of due diligence regimes. They also correspond to the information readily available to supervisory agencies(Sinnig & Zetzsche, 2025, p. 4). However, it is uncertain how they might actually affect practices. Policies may not be fully integrated into decision-making processes, and audits may not reflect conditions across the supply chain (Partzsch, 2025). Another type of reaction is institutionalized adaptation. In this case, firms institutionalize due diligence across procurement and supplier management, as well as in longer-term strategies. They might, for instance, expand mapping of the supply chain, establish mechanisms to manage risk across the production chain, and commit to monitoring. These steps can shift firms' environmental risk-management practices, but they may also come at higher costs and less flexibility (Chu et al., 2025). Most firms operate between these two poles. They combine compliance with adaptation, targeting areas of higher risk or vulnerability. This reflects both strategic calculation and practical constraint. Firms have to deal with regulatory and economic considerations (such as costs, competition, and suppliers).

A second constraint is corporate governance structures. Due diligence systems are usually managed by compliance and risk management units, which are not linked with commercial units. This may lead to the recognition of environmental risks that do not affect procurement strategies, prices, or supplier selection (Both & Wilhelm, 2025). If the sustainability (or risk) function is not connected with the financial and operational functions, due diligence can be "siloed" within the organization. Siloing reduces the likelihood that risks will translate into changes in business practices, despite compliance arrangements (Eller, 2025). *This is not merely a matter of weak coordination but of competing internal mandates. The legal or compliance role is charged with identifying and avoiding adverse effects, and the procurement role is rewarded with low prices, short lead times, and flexible sourcing. These goals can often work at cross-purposes: the supplier that compliance has identified as high risk can be the very supplier that procurement depends on to save money and time. In cases where the commercial interests of procurement are at play, due diligence results are observed but not implemented, and corporate formal obligations are drained internally. It is not, however, merely the presence or absence of the detection of the risks, but the power of the detecting function to override the function that has a reason to disregard them.* Hence, there is a link between the external and internal due diligence. Whether EU standards are likely to influence corporate conduct outside the EU is determined not only by market forces but also by the same structural and institutional factors that shape the effects of due diligence within the EU. This supports the article's main argument: that due diligence is a conditional regulatory mechanism whose impact depends on the regulatory environment (Kiewisch, 2025).

9. CONCLUSION

The Corporate Sustainability Due Diligence Directive represents a change in the European Union's strategy for regulating corporate behavior within global value chains (Sinnig & Zetzsche, 2025, p. 1). In doing so, it moves beyond disclosure to processes to address a problem that previous regulatory approaches struggled to tackle. Environmental damage is frequently caused in places institutionally distant from the companies that make production decisions. The Directive seeks to address this gap without taking direct control, but by requiring firms to manage risk through their organizational practices. However, it is not clear that this will have predictable and significant effects(Zumbansen, 2025).

The analysis also sheds light on the constitutional position of process-based regulation in EU law. Although Article 114 TFEU has been subject to a flexible interpretation by the Court of Justice, its application has traditionally been premised on a sufficiently close link to market functioning(Marušić, 2025). The Directive challenges this proposition by seeking to regulate corporate processes with environmental impacts. Consequently, the legal basis not only enables harmonization; it is part of a regulatory strategy that involves "organizing". Process-based harmonization has the potential to transform Article 114 from a market-correcting competence to a general competence for regulatory governance. This poses a structural challenge for future EU laws: can process-based obligations be accommodated within the internal market regime, or do they mark a redefinition of the scope of Union competence(Zöllmer, 2026)?

This view helps explain how the same legal provisions may result in different types of engagement across jurisdictions and industries. Where regulators have the capacity to evaluate the substance of due diligence practices, there are greater incentives for firms to embed risk assessment in their activities. Where this is not the case, compliance may be more formal. Value chain organization adds complexities. With decentralized, dynamic production, the firm may not have access to or control over the supply chain beyond first-level suppliers. Even if they try to extend their reach, they do so within economic and contractual constraints(Tathavadekar, 2025).

Corporate response mediates these pressures. Corporate response is not a uniform application of the law. They adopt and apply due diligence in line with their priorities and limitations. They establish mechanisms that comply with legal requirements without changing established practice, or embed due diligence more fully in procurement and governance processes. These actions are shaped by legal, market, cost, and capacity factors (OECD, 2025). Taken together, these factors suggest that the Directive should not be regarded as a system that delivers consistent results. Its importance is in creating a framework that makes a corporate organization subject to environmental regulation. At the same time, its effectiveness is dependent on factors outside the Directive's terms. It facilitates the potential for change but does not ensure predictable change (Sørensen, 2025).

This has implications for the evaluation of the Directive in EU law. It cannot be evaluated solely based on its normative content and consistency. It must be considered in light of its interpretation and enforcement, the development of enforcement across the Member States, and firms' responses to the incentives it creates. Otherwise, we risk assuming that the impacts of the Directive are contingent on institutional and organizational factors(Rezarta & Richard, 2025). More broadly, the research suggests that regulatory responses that focus on corporate processes rather than economic structures pose challenges. Due diligence may affect corporate perceptions of environmental risk, but it occurs within value chains that differ in power and responsibility allocation. Due diligence is unlikely to achieve substantive change without these changes. This means that harmonization of the process, although based on Article 114 TFEU, might not, on its own, be enough to tackle the structural drivers of environmental problems (Sinnig & Zetzsche, 2025, p. 5).

This does not mean that the Directive is unimportant. Rather, it is a step forward in regulatory technique in recognizing the need to regulate the organizational structure of global supply chains. It has played a pivotal role in redefining the interaction between law and business conduct, albeit to varying degrees. So, the challenge is not simply to sharpen the requirements for due diligence, but also to clarify how they interact with other enforcement and economic considerations. Without it, there is a risk that due diligence will become a procedural rather than a transformative concept (Rezarta & Richard, 2025). *A number of design decisions might help bridge the gap between formal compliance and substantive accountability. First, the regime itself would be effective as a matter of supervisory capacity, thus providing a plausible floor: minimum standards of resourcing and technical capability of national authorities would restrain the variable geometry and forum shopping described above. Second, safe-harbor protection might depend on evidence of substantive engagement, including risk-based supplier remediation and stakeholder consultation, rather than on the presence of policies and audit trails, which would create an incentive to invest in visible paperwork rather than actual change. Third, a targeted disclosure requirement and a more gradual redistribution of the burden of evidence when a prima facie failure is proved would help overcome the causation and information asymmetries that have dulled civil liability in regimes like the French Duty of Vigilance Law. These measures do not eliminate the structural causes of harm in global value chains, but they help reduce the gap between what the Directive says on paper and its actual implementation.*

REFERENCES

Alexander, A., & Orlando, E. (2025). The Triple Alignment Model for Effective Sustainable Supply Chain Regulations. *Academy of Management Proceedings, 2025*(1). DOI: 10.5465/AMPROC.2025.19814abstract

Assche, A. V., & Narula, R. (2022). Internalization strikes back? Global value chains and the rising costs of effective cascading compliance. *Economia e Politica Industriale, 50*(1), 161–173. DOI: 10.1007/s40812-022-00237-x

Barbieri, N., Consoli, D., Napolitano, L., Perruchas, F., Pugliese, E., & Sbardella, A. (2022). Regional technological capabilities and green opportunities in Europe. *The Journal of Technology Transfer, 48*(2), 749–778. DOI: 10.1007/s10961-022-09952-y

Bartl, M. (2022). Towards the imaginary of collective prosperity in the European Union (EU): Reorienting the corporation. *European Law Open, 1*(4), 957–986. DOI: 10.1017/elo.2023.5

Billedeau, D. B., & Wilson, J. (2025). Misalignment in the automotive supply Chain: Sustainability commitments of automotive manufacturing firms and their suppliers. *Journal of Cleaner Production, 520*, 146122. DOI: 10.1016/j.jclepro.2025.146122

Black, J. (2001). Decentring regulation: Understanding the role of regulation and self-regulation in a 'post-regulatory' world. *Current Legal Problems, 54*(1), 103–146. DOI: 10.1093/clp/54.1.103

Black, J., & Baldwin, R. (2012). When risk-based regulation aims low: Approaches and challenges. *Regulation & Governance, 6*(1), 2–22. DOI: 10.1111/j.1748-5991.2011.01124.x

Blažo, O. (2025). The Mirage of Territoriality: EU Law's Extraterritorial Reach Between Protectionism and Altruism. *Comenius Bratislava Legal Forum, 1*, 545. DOI: 10.46282/bpf.2025.33

Boström, M., Jönsson, A. M., Lockie, S., Mol, A. P. J., & Oosterveer, P. (2014). Sustainable and responsible supply chain governance: Challenges and opportunities. *Journal of Cleaner Production, 107*, 1–7. DOI: 10.1016/j.jclepro.2014.11.050

Both, C. M., & Wilhelm, M. (2025). Supply chain sustainability risk management in the era of mandatory due diligence: A literature review. *Journal of Purchasing and Supply Management, 32*(2), 101083. DOI: 10.1016/j.pursup.2025.101083

Bounfour, T., & Chatelain, L. (2025). A (So Far Unlikely) Day in Court: An Overview of the First Judicial Decisions under the French Duty of Vigilance Law. *Business and Human Rights Journal*, *10*(2-3), 434–439. DOI: 10.1017/bhj.2025.22

Bueno, N., Bernaz, N., Holly, G., & Martín-Ortega, O. (2024). The EU Directive on Corporate Sustainability Due Diligence (CSDDD): The Final Political Compromise. *Business and Human Rights Journal*, *9*(2), 294–300. DOI: 10.1017/bhj.2024.10

Cherednychenko, O. O. (2023). Private enforcement of EU law. In *Edward Elgar Publishing eBooks* (p. 19). Edward Elgar Publishing., DOI: 10.4337/9781802208030.00010

Chu, H., Liu, N., Tsang, A., & Zeng, X. (2025). Environmental Risk, Environmental Action, and Firm Value: Evidence From Climate-Linked Contracts. *Australian Accounting Review*, *35*(3), 215–240. Advance online publication. DOI: 10.1111/auar.70010

Ciacchi, S., Faure, M. G., & Oded, S. (2026). Navigating the EU's Corporate Sustainability Due Diligence Directive: Strategic Implications for US Companies. *Michigan Business & Entrepreneurial Law Review*, *34*, 34. Advance online publication. DOI: 10.36639/mbelr.15.1.navigating

Cohn, M. (2001). Fuzzy Legality in Regulation: The Legislative Mandate Revisited. *Law & Policy*, *23*(4), 469–497. DOI: 10.1111/1467-9930.00121

Court of Justice of the European Union. (2004). Omega Spielhallen- und Automatenaufstellungs-GmbH v Oberbürgermeisterin der Bundesstadt Bonn (Case C-36/02), ECLI:EU:C:2004:614. https://eur-lex.europa.eu/legal-content/EN/TXT/?uri=CELEX:62002CJ0036

Court of Justice of the European Union. (2000). Germany v European Parliament and Council (Tobacco Advertising I) (Case C-376/98), ECLI:EU:C:2000:544. https://eur-lex.europa.eu/legal-content/EN/TXT/?uri=CELEX:61998CJ0376.

Court of Justice of the European Union. (2000). Germany v European Parliament and Council (Tobacco Advertising I) (Case C-376/98), ECLI:EU:C:2000:544. https://eur-lex.europa.eu/legal-content/EN/TXT/?uri=CELEX:61998CJ0376.

Court of Justice of the European Union. (2006). Germany v European Parliament and Council (Tobacco Advertising II) (Case C-380/03), ECLI:EU:C:2006:772. https://eur-lex.europa.eu/legal-content/EN/TXT/?uri=CELEX:62003CJ0380.

Court of Justice of the European Union. (2010). Vodafone Ltd and Others v Secretary of State for Business, Enterprise and Regulatory Reform (Case C-58/08), ECLI:EU:C:2010:321. https://eur-lex.europa.eu/legal-content/EN/TXT/?uri=CELEX:62008CJ0058.

Crovini, C., Højrup, E. M., Jensen, M., Sørensen, K., Jensen, S., Castellano, N. G., D'Onza, G., Melis, A., Luca, F. D., & Maffei, M. (2025). *The Integration of Sustainability into the Risk Management Process and the Impact of the Energy Crisis and Green Transition: Insights from a Danish Energy Trading Company. In the SIDREA series in accounting and business administration.* Springer International Publishing., DOI: 10.1007/978-3-031-76618-3_16

Cuomo, F., Mallin, C., & Zattoni, A. (2015). Corporate Governance Codes: A Review and Research Agenda [Review of *Corporate Governance Codes: A Review and Research Agenda*]. *Corporate Governance An International Review, 24*(3), 222. Wiley. DOI: 10.1111/corg.12148

de Freitas, J. T. (2024). Prioritization under Value Chain Due Diligence. *European Journal of Risk Regulation, 1*(4), 985–998. Advance online publication. DOI: 10.1017/err.2024.19

Dill, A. (2025). The role of corporate governance in macro-prudential regulation of systemic risk. SSRN *Electronic Journal.* DOI: 10.2139/ssrn.5072748

Elbel, J., Böse-O'Reilly, S., & Hrzic, R. (2023). A European Union corporate due diligence act for whom? Considerations about the impact of a European Union due diligence act on artisanal and small-scale cobalt miners in the Democratic Republic of Congo. *Resources Policy, 81*, 103241. DOI: 10.1016/j.resourpol.2022.103241

Eller, K. H. (2025). Corporate Environmental Due Diligence and Value Chains. In *Edward Elgar Publishing eBooks* (p. 143). Edward Elgar Publishing., DOI: 10.4337/9781035324408.00018

European Parliament and Council of the European Union. (2022). Directive (EU) 2022/2464 of the European Parliament and of the Council of 14 December 2022, as regards corporate sustainability reporting. *Official Journal of the European Union, L, 322*, 15–80. https://eur-lex.europa.eu/eli/dir/2022/2464/oj

European Parliament and of the Council Directive (EU) 2022/2464 of 14 December 2022. OJ L 322, 16.12.2022, p. 15–80. https://eur-lex.europa.eu/eli/dir/2022/2464/oj

Evans, J., & Silbey, S. S. (2021). Co-Opting Regulation: Professional Control Through Discretionary Mobilization of Legal Prescriptions and Expert Knowledge. *Organization Science, 33*(5), 2041–2064. DOI: 10.1287/orsc.2021.1525

Felbermayr, G., Friesenbichler, K. S., Gerschberger, M., Klimek, P., & Meyer, B. (2024). Designing EU Supply Chain Regulation. *Inter Economics, 59*(1), 28–34. DOI: 10.2478/ie-2024-0007

Feld, L. (2026). Leading the Way or Crossing the Line? The Extraterritorial Dimension of the EU Directive on Corporate Sustainability Due Diligence. *Business and Human Rights Journal*, *1*(1), 23–45. Advance online publication. DOI: 10.1017/bhj.2025.10038

Foerster, A., Laundau, I., & House, M. (2025). Sustainable Finance and Corporate Sustainability Due Diligence: Interactions, Complementarities and Accountability Logics. *The University of New South Wales Law Journal*, *48*(3). Advance online publication. DOI: 10.53637/GUXD7806

Ford, J., & Nolan, J. (2020). Regulating transparency on human rights and modern slavery in corporate supply chains: The discrepancy between human rights due diligence and the social audit. *Australian Journal of Human Rights*, *26*(1), 27–45. DOI: 10.1080/1323238X.2020.1761633

Friesenbichler, K. S., Felbermayr, G., Gerschberger, M., Klimek, P., & Meyer, B. (2024). Designing EU Supply Chain Regulation. SSRN *Electronic Journal*. DOI: 10.2139/ssrn.4917535

Gugkang, A. S., Tanakinjal, G. H., & Hamid, M. R. A. (2025). The Challenges of Implementing Socially Sustainable Supply Chain Management to Enhance Fair Labor Practices for Small-Scale Farmers in Malaysia. *Global Conference on Business and Social Sciences Proceedings*, *17*(1), 56. DOI: 10.35609/gcbssproceeding.2025.1(56)

Guye, V., Meyfroidt, P., & zu Ermgassen, E. (2025). The need for improved public transparency in the era of due diligence regulations. *SPIRE - Sciences Po Institutional Repository*. https://hal.science/hal-05420553

Járvás, T., Illés, B., & Tóth, Á. B. (2025). Uncertainty factors in supply chains. *Advanced Logistic Systems -. Theory into Practice*, *19*(1), 53–60. DOI: 10.32971/als.2025.005

Jian, Z., Xu, G., & Li, X. (2025). Intersection of ESG performance and corporate innovation: A perspective of vertical linkages within the industrial chain. *Frontiers in Sustainability*, *6*, 1642317. Advance online publication. DOI: 10.3389/frsus.2025.1642317

Kalimo, H., Ville, F. D., & Happersberger, S. (2023). The Unilateral Turn in EU Trade Policy? The Origins and Characteristics of the EU's New Trade Instruments. *European Foreign Affairs Review*, *28*(Special Issue), 15–34. DOI: 10.54648/EERR2023012

Kiewisch, M. (2025). Navigating the EU Corporate Sustainability Due Diligence Directive: How Multinational Enterprises Approach Regulatory Familiarization in the Chocolate Sector. *Regulation & Governance*, rego.70042. Advance online publication. DOI: 10.1111/rego.70042

Kingston, E., & Carnovale, M. (2026). Social Upgrading in Global Supply Chains: Reframing Corporate Responsibilities. *Business Ethics Quarterly*, *1*, 1–28. Advance online publication. DOI: 10.1017/beq.2025.10103

Koos, S. (2024). Civil Law, Conflict of Laws, and Extraterritoriality in the European Supply Chain Due Diligence Law. *Hasanuddin Law Review*, *10*(2), 144. DOI: 10.20956/halrev.v10i2.5535

Kübek, G., & Mancini, I. (2023). EU Trade Policy between Constitutional Openness and Strategic Autonomy. *European Constitutional Law Review*, *19*(3), 518–547. DOI: 10.1017/S1574019623000226

Lotfi, M., & Walker, H. W. (2026). *Beyond paper compliance: why supply chains struggle to confront forced labor. ORCA Online Research @Cardiff*. Cardiff University.

Maiju, V. (2025). Drivers and barriers for lower-tier suppliers' sustainability compliance in emerging countries : a narrative review. In *Theseus (Ammattikorkeakoulujen)*. https://www.theseus.fi/handle/10024/906481

Majekodunmi, A. O., Edohen, A., & Conteh, J. Y. (2025). Regulatory Divergence and Security Implementation: Compliance-Driven Security Architecture in Multi-Jurisdictional Financial Organizations. *Research Journal in Civil Industrial and Mechanical Engineering*, *2*(2), 53–71. DOI: 10.61424/rjcime.v2i2.336

Mangiuc, M.-D., & Kagitci, M. (2025). CSRD as a Paradigm Shift in Sustainability Reporting: From Voluntary Practice to Legal Requirement. *Amfiteatru Economic*, *27*(70), 1193. DOI: 10.24818/EA/2025/70/1193

Marco, D. D., & Cagliano, R. (2026). Toward Sustainable Organizations: Routines to Embed Sustainability in Corporate Culture. *Creativity and Innovation Management*, *35*(3), 590–611. Advance online publication. DOI: 10.1111/caim.70047

Marušić, B. (2025). The role of Article 114 TFEU in balancing fundamental rights and product safety in the AI Act. *Papers from the International Scientific Conference of the European Studies Department Jean Monnet Center of Excellence, Faculty of Philosophy at Sofia University "St Kliment Ohridski," 12*, 277. DOI: 10.60054/PEU.2025.12.277-289

Mercier, V. (2026). The Protection of Human Dignity Through the European Directive on Corporate Sustainability Due Diligence. *Bulletin of the Transilvania University of Braşov Series VII Social Sciences • Law*, 129. DOI: 10.31926/but.ssl.2025.18.67.3.14

Milnitz, D., & Marchi, J. J. (2025). TEXTILE AND APPAREL SUPPLY CHAIN: STRUCTURE, COMPLEXITY, AND GLOBAL TRANSFORMATION TRENDS. In *Seven Editora eBooks*. DOI: 10.56238/sevened2025.036-005

Mu, Y. (2025). Divergence in HRDD Legislation: A Comparative Study of France, Germany, and the United States. *Lecture Notes in Education. Psychology and Public Media, 107*(1), 27. DOI: 10.54254/2753-7048/2025.LD25578

OECD. (2025). Measuring the uptake and impact of due diligence for responsible supply chains. In *OECD Publishing eBooks*. DOI: 10.1787/52ef761f-en

Parroco, V. (2026). The right to an effective remedy and due diligence legislation in the EU. *Archivio Istituzionale Della Ricerca (Alma Mater Studiorum Università Di Bologna)*. https://hdl.handle.net/11585/1038050

Partzsch, L. (2025). Hardening sustainability: Supply chain laws complement cocoa and coffee certifications. *Forest Policy and Economics*, *181*, 103661. DOI: 10.1016/j.forpol.2025.103661

Peters, A., Gleß, S., Thomale, C., & Weller, M.-P. (2023). Business and Human Rights: Towards a 'Smart Mix' of Regulation and Enforcement. *Zeitschrift für ausländisches öffentliches Recht Und Völkerrecht / Heidelberg Journal of International Law, 83*(3), 415. DOI: 10.17104/0044-2348-2023-3-415

Pítzen, S., Kivimaa, P., Hildén, M., & Mosoni, C. (2025). From informal coordination to formalized policy integration: Options for strengthening climate change adaptation in EU finance and trade policy. *Frontiers in Climate*, *7*, 1543732. Advance online publication. DOI: 10.3389/fclim.2025.1543732

Rashid, H. (2025). Evaluating the efficacy of multi-tier supplier audits in enhancing supply chain reliability. *International Journal of Multidisciplinary Research and Growth Evaluation*, *6*(1), 1918–1923. DOI: 10.54660/.IJMRGE.2025.6.1-1918-1923

Rezarta, T., & Richard, T. (2025). Navigating the Future: Assessing the Impact of Due Diligence Rules in the European Union on Business, Environment, and Workers. *Università Degli Studi Di Bari*. DOI: 10.15162/2612-6583/2390

Rodríguez-Gómez, S., Sánchez, I., Arco-Castro, M. L., & Pérez, M. V. L. (2025). The Effect of Institutional Pressures and Business Incentives on the Level of Corporate Compliance System Implementation. *Corporate Social Responsibility and Environmental Management*, *33*(1), 18–38. DOI: 10.1002/csr.70155

Salih, R. H., Khinger, I. K., Ubaid, H. R., Aldabagh, D., & Alani, S. S. (2025). Legal Perspectives on Corporate Accountability in Environmental Sustainability. *Environment and Social Psychology*, *10*(11). Advance online publication. DOI: 10.59429/esp.v10i11.3992

Salminen, J., Sobel-Read, K., Viljanen, M., & Eller, K. H. (2022). Digital Platforms as Second-Order Lead Firms: Beyond the Industrial/Digital Divide in Regulating Value Chains. *European Review of Private Law/Revue Européenne de Droit Privé/ Europäische Zeitschrift für Privatrecht, 30*, 1059. DOI: 10.54648/ERPL2022049

Scheper, C. (2025). Corporate due diligence in global value chains. In *Edward Elgar Publishing eBooks* (p. 370). Edward Elgar Publishing., DOI: 10.4337/9781035300907.00039

Scheper, C. (2026). Due Diligence Legislation and the Politics of Implementation: The Case of the German Supply Chain Law. *Regulation & Governance*, rego.70134. Advance online publication. DOI: 10.1111/rego.70134

Schilling-Vacaflor, A., & Lenschow, A. (2021). Hardening foreign corporate accountability through mandatory due diligence in the European Union? New trends and persisting challenges. *Regulation & Governance*, *17*(3), 677–693. DOI: 10.1111/ rego.12402

Schilling-Vacaflor, A., & Lenschow, A. (2021). Hardening foreign corporate accountability through mandatory due diligence in the European Union? New trends and persisting challenges. *Regulation & Governance*, *17*(3), 677–693. DOI: 10.1111/ rego.12402

Schmidt, S. K. (2018). The European Court of Justice and the Policy Process. In *Oxford University Press eBooks*. Oxford University Press., DOI: 10.1093/ oso/9780198717775.001.0001

Sellare, J., Börner, J., Brugger, F., Garrett, R., Günther, I., Meemken, E., Pelli, E. M., Steinhübel, L., & Wuepper, D. (2022). Six research priorities to support corporate due diligence policies. *Nature*, *606*(7916), 861–863. DOI: 10.1038/d41586-022-01718-8 PMID: 35764799

Shulman, S. (2024). Corporate Sustainability Due Diligence. *Columbia Journal of Environmental Law*, *49*(2), 479–523. DOI: 10.52214/cjel.v49i2.12633

Sinnig, J., & Zetzsche, D. A. (2025). The EU's Corporate Sustainability Due Diligence Directive: From Disclosure to Mandatory Prevention of Adverse Sustainability Impacts in Supply Chains. *European Journal of Risk Regulation, 1*(2), 628–652. Advance online publication. DOI: 10.1017/err.2024.100

Sørensen, K. E. (2025). Applying CSDDD in Corporate Groups. *European Company and Financial Law Review*, *22*(2), 129–164. DOI: 10.1515/ecfr-2025-0006

Sunny, S. A., Cottle, G. W., & Coffman, C. D. (2025). Moving the Needle: Evaluating Firm Responsiveness to the Issue of Global Climate Change. *Academy of Management Proceedings, 2025*(1). DOI: 10.5465/AMPROC.2025.14769abstract

Tathavadekar, V. P. (2025). Optimizing Supply Chain Sustainability: Leveraging the Circular Economy for Effective Resource Management and Reduced Environmental Impact. *Advances in Consumer Research. Association for Consumer Research (U. S.)*, *2*(4), 602. DOI: 10.61336/acr/25-04-01

Taylor, M. B. (2020). Human rights due diligence in theory and practice. In *Edward Elgar Publishing eBooks*. Edward Elgar Publishing., DOI: 10.4337/9781786436405.00011

Uvarova, O., & Bernaz, N. (2025). Heightened Human-Rights Due Diligence as Risk Governance: Food-Retail Supply Chains in Wartime Ukraine. *European Journal of Risk Regulation*, *1*. Advance online publication. DOI: 10.1017/err.2025.10049

van Oijen, J. C. F., Wallenburg, I., Bal, R., & Grit, K. (2020). Institutional work to maintain, repair, and improve the regulatory regime: How actors respond to external challenges in the public supervision of ongoing clinical trials in the Netherlands. *PLoS One*, *15*(7), e0236545. Advance online publication. DOI: 10.1371/journal.pone.0236545 PMID: 32735568

Villiers, C. (2022). New Directions in the European Union's Regulatory Framework for Corporate Reporting, Due Diligence and Accountability: The Challenge of Complexity. *European Journal of Risk Regulation*, *13*(4), 548–566. DOI: 10.1017/err.2022.25

Wilhelm, M. (2024). Mandatory due diligence legislation: A paradigm shift for the governance of sustainability in global value chains? *Journal of International Business Policy*, *7*(4), 459–465. DOI: 10.1057/s42214-024-00193-4

Yuan, L., & Mähönen, J. (2024). Can Integrate a Sustainable Business Model and Global Value Chains Revive the Value Chain's Sustainable Growth? *Circular Economy and Sustainability*, *4*(4), 2957–2980. Advance online publication. DOI: 10.1007/s43615-024-00352-y

Zöllmer, J. (2026). The end of negative integration within the internal market? Reassessing the EU's rigidity in the free movement of goods. *Figshare*. DOI: 10.6084/m9.figshare.31119056

Zumbansen, P. (2025). Devaluing Sustainability: Financialized Disclosure Governance and Transparency in Modern Slavery and Climate Change. *Journal of Law and Political Economy*, *5*(2). Advance online publication. DOI: 10.5070/LP60065156

Chapter 5
Climate Advocacy, Corporate Speech, and the Limits of Greenwashing:
A Human Rights-Based Accountability Framework

Mahmoud Masud
http://orcid.org/0009-0003-8210-3416
Prince Mohammed Bin Fahd University, Saudi Arabia

Reema Yaghmour
http://orcid.org/0009-0000-9174-5371
Prince Mohammad Bin Fahd University, Saudi Arabia

ABSTRACT

This chapter examines the limits of existing legal frameworks in addressing greenwashing, arguing that its current characterisation as a form of consumer deception or failure to comply with regulations is insufficient. It argues that greenwashing possesses a hybrid nature, combining both commercial and political elements, thereby challenging traditional doctrinal classifications. Relying on Article 19 of the International Covenant on Civil and Political Rights, Article 10 of the European Convention on Human Rights, and UNECE Aarhus convention, this chapter situates greenwashing within the broader freedom of expression framework, while analysing the conditions under which its restriction may be justified. It proposes that greenwashing should instead be viewed as a form of expression capable of producing

DOI: 10.4018/979-8-2600-3678-5.ch005

structural harm within a democratic society and environmental governance as it undermines rights by distorting the informational foundations on which several rights are depended on, not merely by deceiving consumers, investors and the regulator.

I. INTRODUCTION

Multinational companies make net-zero promises, claim eco-friendly products, sustainability disclosures and transition narratives that influence regulatory policies, investor behaviour, and the public's opinion. The growth of corporate environmental expression is often discussed alongside the systematic trend of greenwashing, which is referred to as misleading the public to believe that a company or entity is doing more to protect the environment through claims that are strategically used to exaggerate, conceal, or distort facts about environmental performance (United Nations, n.d.). Despite the longstanding issue of greenwashing, current frameworks characterise it as either a form of consumer deception or failure to comply with the disclosure requirements prescribed by law. Therefore, this chapter challenges this characterisation as analytically insufficient, and argues that greenwashing must be viewed as a form of corporate expression that can cause structural harm within a democratic environmental regime (Debnath et al., 2023, pp. 3-7,10-12). The main question this chapter addresses is not whether corporate environmental expression is considered a form of protected expression, but rather determining the foundational grounds where is necessary and justifiable to restrict it.

This Chapter positions this issue under the freedom of expression framework, Article 19 of the International Covenant on Civil and Political Rights (ICCPR) in particular, further supported by Article 10 of the European Convention on Human Right (ECHR) (Council of Europe, 2017, pp. 21–29, 52–57). While proposing limitations that the law prescribes where necessary and proportional in a democratic society, for the protection of the rights of others, including their right to be informed, these provisions provide the general normative framework where the communicative accountability should be assessed. This assessment is based on the protection of several elements guaranteed under the previous mentioned Articles ranging from facts and extends to guaranteeing informational pluralism and the right to individuals to participate in a democratic society (Green & Temper, 2026, pp. E176–E180).

To this effect, the chapter provides a balanced human rights-based framework, informed by regional jurisprudence, such article 19 of the ICCPR (International Covenant on Civil and Political Rights, 1966, art. 19(3)), and reflected in other regional instruments, such as Article 10 of the European Convention on Human Rights (ECHR hereinafter). This approach is further supported by the UNECE's Aarhus convention, which similar to Article 19, grants individuals the right to receive truthful and sufficient information relating to the environment (UNECE, 1998, arts. 4–5, 6, 9). The proposed framework is based on three interdependent criteria: intentionality, materiality, and structural impact. This strategic framework focuses on analysing the deception of markets in isolation and the systemic democratic implications. It places environmental responsibility into the well-established freedom of expression dogma and narrows it down to the climate environment. The chapter shows that instead of building greenwashing as a special regulation form, the current human rights approach can be used in response to the communicative aspects of environmental degradation in a structural perspective. The goal is founded on dignity, equality and participatory legitimacy (Hill et al., 2025, pp. 4–9, 15–18).

The significance of this chapter lies in the analysis of Article 19 of ICCPR, combined with proposed framework, providing a consistent normative framework that can be applied in courts, regulatory bodies, and to further policy-makers in various jurisdictions.

This chapter proceeds as follows. Section II develops the conceptual foundations such as corporate expression, democratic legitimacy, and structural harm. Section III explicates Article 19 of ICCPR and Article 10 of ECHR while relating it to UNECE's Aarhus convention. Section IV demonstrates the hybrid nature of corporate environmental speech, demonstrating the limits of characterising it as either commercial or political. Section V introduces and analyses the proposed three-part framework based on intentionality, materiality, and structural impact. Finally, section VI applies this framework in a judicial context.

II. CONCEPTUAL FOUNDATIONS: CORPORATE EXPRESSION, DEMOCRATIC LEGITIMACY, AND STRUCTURAL HARM

a. Corporate Expression Within Article 19 ICCPR

Under articles relative to human rights, there is a threshold between exercising the right i.e granting the individual the protection of exercising it, and between narrowing the scope of protection. Under Article 19 of the International Covenant on Civil and Political Rights the ratio between protectiveness of exercising the right and narrowing the individual's exercise deliberately has a broad protective scope. Human Rights Committee has made it evidently clear that Article 19(2) encompasses all types of expression, which includes commercial advertising and corporate communication, regardless of who the speaker is. The drafters of the Article did not distinguish the merits to which the protection extends, whether to natural or juridical persons. Rather, the protection extends on the nature and social role the expression plays not based on the ontology of the speaker (Human Rights

Committee, 2011, paras. 11–16, 21–27, 34–36). The condition of narrowing the scope of expression is by testing the effects and gravity the speech holds on the society against the individual's exercise; granted this limitation is prescribed by law, necessary and proportional in a democratic society.

As a result, when applying Article 19 on corporate environmental communication, it can be seen that, prima facie, corporate environmental communication comes under Article 19. Sustainability reporting, net-zero pledges, product environmental assertions, ESG (Environmental, Social, Governance) reporting, and climate-transition narratives are communicative actions that are directed to the common space. Such expression helps in the spread of information that is pertinent to collective decision making even when it is motivated by economic imperatives. However, the normative complexity of greenwashing under the Covenant is not automatically relieved by the fact that it is recognized as being protective. Article 19 is a human rights framework that is based on principles of dignity, equality, and democratic participation. When the communicative infrastructures of corporate actors are dominating, formal equality of speech protection is confronted with substantive inequalities of discursive power (Post, 2012, pp. 23–41). The assumption of equal communicative standing between corporations and the public is fragmented by the discrepancy between the power and dominance of the corporation's effect to sway the public's opinion and view.

This tension is also structurally important as pluralism and free discussion are guaranteed by the ICCPR, but the concentrated communicative power can pervert the very principles to which pluralism functions. The speech of corporate climate is not like normal advertising, it interferes with regulatory schedules, investment patterns and the sense of urgency of the population. Therefore, being safeguarded, its democratic weight must be questioned and analysed when evaluating what restrictions can be allowed (Sunstein, 1993, pp. 17–38). The Human Rights Committee has recognized that restrictions can be made where there is need to safeguard the rights of other people including but not limited to the rights to life, health, and private life which are all directly involved in environmental degradation. Indirectly, therefore, informational distortion that slows down mitigation measures or mis-calibrates regulatory settings can jeopardize rights guaranteed under other parts of the Covenant (Human Rights Committee, 2018, paras. 62–64).

Therefore, the question of doctrine that is raised, is whether communicative acts that carry material distorting effect on the epistemic basis of democratic environmental governance is covered by the right of others limitation clause. Regulating this issue under the lens of a "pure market" namely such as consumer deception, competition, or misleading advertisement, is not adequate as it situates the intent being purely economic as opposed to greenwashing containing a more complex and hybrid intent base. As a result, the Covenant needs to be understood in accordance with its object and purpose: to achieve the conditions under which human dignity can thrive. Accordingly, it is necessary to demonstrate how does current freedom of expression articles situates greenwashing which will be demonstrated in the following section.

b. Commercial, Political, and Hybrid Speech

The traditional separation of commercial and political speech has traditionally framed the jurisprudence of freedom-of-expression. Political speech is normally given greater protection because it directly contributes to democratic deliberation, compared to commercial speech which has been considered instrumental and therefore more controllable. This dichotomy is destabilised by corporate climate communication (Debnath et al., 2023, pp. 3–12). The net-zero ad by a fossil-fuel company is not just a promotion, but it also suggests implicitly the adequacy of regulations, the technical viability of technology, and the need to reform the system. In this way, it helps to form a larger story about the course of the energy transition, thus participating in the discussion with society. However, its economic motive is still paramount. This hybridity makes judicial review difficult. It should not be given the full political-speech deference because it would protect the distortions of strategy; it should not be treated as commercial because it would undervalue its normative influence. Article 10 of the ECHR on comparative jurisprudence exemplifies this tension. The European Court of Human Rights acknowledges that commercial expression should be protected but with broader regulatory discretion in cases where the considerations of the public interest are high. The proportionality analysis that is structured by the Court is contextually sensitive, as opposed to being categorical. Corporate environmental speech must then be a measured doctrinal speech. Its hybrid nature requires a threshold structure that is sensitive to expressive freedom and systemic risk (Human Rights Committee, 2011, paras. 11–16, 21–36). The question of characterisation is not the sole question that needs to be addressed; but the harms posed by greenwashing also needs to be examined, to which the following section will demonstrate.

c. Deliberative Democracy and Informational Integrity

Deliberative democratic theory provides the normative skeleton on which the rhetoric of corporate climate should be questioned. Democratic form of governance is legitimate because the citizens are assumed to be able to develop opinions and preferences based on information that is at least reasonably reliable. Contemporary regulatory institutions have similarly recognised the systemic implications of misleading sustainability communication. ESMA's 2024 Final Report on Greenwashing observed that sustainability-related misleading claims may undermine investor confidence, distort sustainable investment markets, and weaken the informational reliability upon which market actors rely. Informational pluralism does not require that the truth be told, but it depends on the critical condition that the systemic distortion should not exist. The Habermasian discourse theory anticipates the concept of the better argument that is not enforced (Habermas, 1996, pp. 359–387). In the case of asymmetric resource distribution and manipulative narrative production, the very notion of deliberative equality is undermined. Even without any direct falsehoods, corporate sustainability efforts, however, can still use selective disclosure and aspirational framing which, when combined in large quantities, changes the expectations of the population about the sufficiency of transition pathways. The resulting damage is structural and not accidental. In addition, it is pertinent to note that greenwashing extends beyond isolated misleading advertisements and increasingly operates through broader ESG narratives, net-zero commitments, and future-oriented sustainability claims (Montgomery et al., 2024, pp. 223–224, 232–234).At the personal level, actionable misdirection may not be felt in any cognisable legal sense. However, the aggregate impact of such practices may diffuse regulatory drive, influence investor capital flows and remake the policy debate tone. International policy guidance has similarly recognised that deficiencies in transparency, comparability, and ESG methodologies may undermine market integrity, investor confidence, and the effectiveness of climate-transition frameworks (OECD, 2022, pp. 6–7). Such systemic ramifications cannot be silent in the presence of any doctrine of freedom of expression. Article 19 provides the right to information, which is not only satisfied by the existence of speech, but also by the delivery of conditions that allow evaluative discernment. In the absence of these circumstances, the right is a far-fetched one. This concern has also been recognised internationally. In the 2024 Joint Declaration on the Climate Crisis and Freedom of Expression, international freedom of expression mandate holders warned that misinformation and disinformation may undermine public trust and hinder informed participation and decision-making concerning climate change. The Declaration further emphasised that corporations should provide detailed disclosures regarding the environmental and social impacts of their operations in accordance with international standards (UN Human Rights Office of the High Commissioner, 2024). Moreover, existing legal instruments tend to relate greenwashing to falsehood, intent and/or evidence of consumer reliance, thereby regulating greenwashing through the mandate of providing adequate product information, such instruments include Rio declaration and UNECE (Rio Declaration on Environment and Development, 1992, Principle 10). However it is pertinent to note that there is yet no direct incitement or explicit illegality does not neutralise its normative value in regional or international instruments (UN Human Rights Office of the High Commissioner, 2024, pp. 3–6(.In this regard, greenwashing has structural similarities with other types of communicative distortion that undermine democratic integrity without necessarily going beyond standard incitement limits. While such criteria (falsehood, intent and/or evidence of consumer reliance) are relevant and applicable to the proposed argument, it however does not reflect deeper normative concerns. Deception about environmental performance does not simply mislead consumers but also distorts the informational prerequisites upon which decisions made about the collective climate is based. Moreover, climate governance presumes informational integrity, regulatory calibration, investor allocation, civic mobilisation and intergenerational policy design, all rely on sufficiently true environmental disclosures. By corporately, strategically, managing sustainability discourses with selective disclosure, aspirational overstatement, or reputational positioning out of material transition, corporations re-establish an epistemic space upon which climate policy is debated and constructed. The harm created by greenwashing, therefore, is structural and not merely isolated. Its danger lies in the ability of such expression to subtly reshape the conditions of deliberation, which can be observed in the background, weaken the participatory legitimacy without the need to promote illegal actions. It undermines democratic equality by allowing resource-dominant actors to influence environmental discourse by establishing, perpetuating, and reinforcing asymmetries of power, slowing responsiveness to regulation, and disproportionately over-burdening communities already at risk of climate change. This concern is further reflected within international human rights realm. In its 2025 report, the United Nations Special Rapporteur on Human Rights and Climate Change described climate misinformation, greenwashing, and "climate obstruction" as practices capable of distorting public understanding of climate realities, delaying effective climate action, and undermining democratic participation and access to information. The report further recognised that fossil fuel actors have strategically relied on misinformation campaigns, selective communication, and greenwashing practices to influence public perception and regulatory responses to climate change. Such observations reinforce the argument that greenwashing extends beyond isolated consumer deception and may instead operate as a structural communicative harm capable of undermining the informational foundations upon which democratic environmental governance depends. (United Nations Human Rights Council, 2025, paras. 37–41)

Notwithstanding the gravity and severe harm associated with greenwashing, a substantial regulatory reaction may risk creating a chilling negative effect on the legitimate participation of corporations in environmental dialogue. Another factor to consider is the fact that corporations transcend being the subject of regulation; they are central participants in the technological innovation, capital change and intercultural discussion. Therefore, a wide-ranging suppression of environmental claims can discourage transparency, diminish aspirational commitment and encourage tactical silence. Instead, an ethical principled system, needs to avoid being either laissez-faire permissive or reactionary suppressive. Thus, the challenge faced is establishing a threshold that will maintain freedom of expression, while focusing on mitigating systemic informational distortion. Greenwashing, therefore, should not only be seen as a deceptive act but also as the possible perversion of the deliberative infrastructure on which informed civic participation is based.

d. Structural Harm Theory in Communicative Contexts

Structural harm is not the familiar category of direct incitement or immediate harm; rather it refers to the gradual, usually insidious changes that run throughout institutional and epistemic milieus. Greenhouse-gas emissions are the material agents of harmful consequences in the specific field of climate change, but the communicative practices of the actors that drive mitigation policies are what makes the difference between implementing it in a timely and healthy way. Corporate environmental misrepresentation works by creating the illusion that the necessary transition has already been undertaken at a sufficient rate and thus slows down the necessary regulatory actions. These distortions create an atmosphere of complacency between investors and policymakers and disintegrate civic mobilisation by spreading the sense of crisis urgency. The damage in question, not meeting the requirement that would make it incitement, still takes place within the boundaries of the expression that is formally legal. Article 19(3) of the applicable legal framework explains that a restriction is not only justified on the grounds of illegality of the content, but the need to protect the rights of other persons. The use of a structural-harm lens is consistent with proportionality reasoning. Any limitation should be strictly specific and supported by empirical evidence; however, justification can be made where cumulative distortion proves to undermine participatory equality and violate environmental rights. This theoretical background, thus, leads to the operationalisation of the three-part framework which is expounded in the later sections of this chapter.

III. NORMATIVE ARCHITECTURE: ARTICLE 19 ICCPR, PROPORTIONALITY, AND INFORMATIONAL RIGHTS

a. Article 19 ICCPR: Scope, Protective Reach, and Corporate Speakers

Article 19(2) of ICCPR protects the freedom to pursue, obtain and impart information and ideas of any nature without reference to borders. The Human Rights Committee has long given this provision a liberal interpretation and General Comment No. 34 states categorically that Article 19 applies to commercial advertising and other types of business communication. The Covenant does not draw the distinction between speakers in terms of corporate status; instead, protection is applied to expression itself. This inclusivity is based on the principle on which the free flow of information is the basis of democratic participation. Corporate actors are at the centre of the modern market democracies as prolific information producers. The disclosure of climate, sustainability, and environmental social governance (ESG) communications are part of the new informational ecosystem, which affects markets, directs investment flows, and regulative expectations. However, the Committee has emphasized that Article 19 is not a right in absolute form. The limitations can be applied in accordance with Article 19(3) and prescribed by law and required to comply with the rights or reputations of others, or to safeguard national security, law and order, health, or morals. This means that the interpretive question is not whether the speech of corporations on environmental issues is safeguarded, a question that is answered in the affirmative, but whether limitations to prevent greenwashing may meet the structured test of Article 19(3). There are two doctrinal clarifications that are required to understand it subtly. To begin with, the right of others provision is a dynamic one that should be interpreted in the context of the changing concept of environmental harm and interdependence between civil-political and socio-environmental rights. According to the Human Rights Committee, which is mentioned in General Comment No. 36, the right to life is endangered by environmental degradation and climate change. The protected rights may thus be involved, albeit indirectly, but substantially, in informational distortion that weakens effective climate mitigation (Human Rights Committee, 2018, paras. 62–64). Second, Article 19 defends the right to disseminate information as well as the right to access information. Where corporate communication systematically misleads the environmental performance or transition preparedness, it may possibly intrude on the right of the people to obtain correct environmental information to enable informed participation. Article 19, therefore, extends protection and incorporates normative processes that make accountability easier.

b. Permissible Restrictions Under Article 19(3): Legality, Legitimate Aim, Necessity

Article 19(3) of the tripartite test states that any limitation must be provided by law (legality), pursue a legitimate aim and necessary and proportionate. Each part of the test must be fulfilled cumulative manner.

• Legality

The legality requirement requires accessibility, transparency and predictability. The corporate players should be able to know the norms of environmental claims.

This condition is gradually met by the modern regulatory developments. Climate-related financial disclosure frameworks, anti-greenwashing principles by financial regulators and consumer protection laws stipulate evidentiary standards of environmental representations.

As an illustration, sustainable finance disclosure laws demand supporting ESG assertions and outlaw deceptive environmental labelling. Such regimes give the normative clarity that is required to satisfy the legality requirement of Article 19. Limitations based on well-defined criteria of verifiability, substantiation and consistency are not likely to infringe foreseeability principles. This can be observed in frameworks that require corporations advertising environmentally friendly products/services to provide sufficient evidence to prove the credibility of their claim advertised. The requirement served does not conflict with the principles under freedom of expression; rather reinforces the principles enshrined such as foreseeability, certainty, and accessibility.

• Legitimate Aim

The regulation of greenwashing may reasonably address a number of valid objectives identified in Article 19(3), such as guarantee of the rights of others (such as the right to life, right to health, and the right to participation), protection of public health, and social functioning based on environmental stability where the protection of the public order is central. The Human Rights Committee has affirmed that the States should take positive steps to deal with environmental harms that pose threats to Covenant rights (Portillo Cáceres v. Paraguay, 2019, paras. 7.3–7.5). The linkage to safeguarded rights is tangible as opposed to intangible where deceptive corporate climatic communication postpones mitigation policies or undermines regulatory determination. Notably, this does not mean that corporate speech should be widely suppressed. Instead, one of the arguments raised is that targeted restrictions to material misleading environmental representations are aimed at pursuing recognised legitimate ends under the Covenant framework.

• Necessity and Proportionality

The most demanding limb under the tripartite test is the necessity requirement. The Committee has reiterated a number of times that limitations imposed should be the least restrictive method to attain the protective role and that they should not compromise the nature of the right. The requirement is very much in line with proportionality reasoning (Human Rights Committee, 2018, paras. 62–64). The State must demonstrate, first, suitability, that the restriction can meet the legitimate aim, secondly, necessity, that there would be no less restrictive alternative which would afford similar protection and thirdly, proportionality stricto sensu — that the good is greater than the evil to expressive freedom. In the case of greenwashing, proportionality denies laissez-faire permissiveness and categorical prohibitions (Human Rights Committee, 2011, paras. 21–36). A ban on aspirational climate commitments would probably not pass necessity analysis. Proportionality, however, may be met by specific demands of evidentiary substantiation, the ban on materially misleading statements, and injunction disclosure orders, when narrowly designed. There is also proportionality analysis that needs to be evaluated contextually. In cases where the corporations have superior communicative power, there might be need to have limited corrective regulation in order to maintain democratic equality.

c. Proportionality as Structured Democratic Reasoning

Proportionality is not just a balancing game; it is a methodical way of constitutional argument that checks judicial discretion. It is an analytical field that enables courts to strike a balance between interventions without falling into the trap of suppression or permissiveness (Council of Europe, 2017). In Article 19, it provides that the limitation of speech is not arbitrary, over-broad or punitive. This doctrine has been further developed in Article 10 jurisprudence by the European Court of human Rights. In its analyses it takes into consideration the character of the speech, the degree to which the speech is contributing to the debate in the society, the speaker and the severity of the sanction. Commercial expression is also open to protection but often falls within a wider margin of appreciation in which the consumer protection or health issues comes into play. In *Markt Intern Verlag GmbH v Markt*, the Court recognised that commercial speech may be regulated in Germany to avoid unfair competition and misleading the consumer *(Markt Intern Verlag GmbH v. Germany*, 1989, paras. 26–35). *Verein against Tierfabriken Schweiz (VgT) v Switzerland (No 2)*, the Court emphasized the significance of the discourse of the public interest but retained the proportionality test on restrictions (*Verein gegen Tierfabriken Schweiz (VgT) v. Switzerland (No. 2)*, 2009, paras. 92–95).

Proportionality applied to the climate communication should consider:

- The corporate environmental speech has a hybrid nature,
- Its systematic impact on public opinion and the policy,
- The threat of freezing valid sustainability discourse.

d. Article 10 ECHR and the Aarhus Convention: Comparative Sensibilities and the Right to Environmental Information

Article 10 of the European Convention on Human Rights is consistent with Article 19 of the ICCPR, but it is characterised by a rich body of jurisprudence. The European Court has continued to maintain that freedom of expression is one of the fundamental pillars of a democratic society, and at the same time has recognized that commercial discourse can be held to a higher level of scrutiny under some circumstances. In *Animal Defenders International v United Kingdom*, the Grand Chamber upheld bans on paid political advertisement, highlighting the need to prevent the subversion of the discourse by financially strong forces. Though the case does not directly address the issue of corporate greenwashing, the rationale behind it is enlightening (*Animal Defenders International v. United Kingdom*, 2013, paras. 100–108). The Court upheld that a material imbalance of resources may justify the regulation intervention aimed at maintaining democratic equality. Equally, the European approach to tobacco advertising can bear restraints based on the public-health necessity, thus providing a helpful analogy to the larger public-interest interests posed by climate change. Even though the doctrine of the margin of appreciation allows some flexibility in the contexts of different jurisdictions, the fundamental principle is the same: the expressions that contribute to the democratic discourse are strongly safeguarded but cannot be deemed immune to the proportional regulation in the case of structural risks.

The Aarhus Convention on the other hand, is a major improvement of the normative framework as it introduces the access to environmental information into the participatory governance. The Convention possesses similar elements to Article 19's guaranteed right to be informed (Aarhus Convention, 1998, arts. 4–5, 6, 9). Its core objective, is to ensure the public's participation in making environmental decisions based on transparency; enforcing principles of environmental law, including access to information, informed public participation, and accountability (UNECE, 2014, pp. 50–68, 96–103). This can be seen under article such as: Article 4, this Article is categorical in ensuring that the public has access to the environmental information possessed by the public authorities but Article 6 provides the obligation of the public to be involved in decision making processes concerning the environment. Although the Convention mostly deals with the State requirements, the logic behind it assumes the existence of an informational integrity environment. In case the corporate bodies take over the climate communication with false representations, the rights of participation provided in the Aarhus Convention can be in danger of being gutted. This concept is also supported under the Rio Declaration, and states that the best way to tackle environmental problems is by involving all the interested citizens, which is based on the proper access to information (Rio Declaration on Environment and

Development, 1992, Principle 10). The limitations on greenwashing can therefore be theorised not as a means of speech repression, but as an implementation of the informational preconditions of participatory environmental governance.

e. Reconciling Protection and Accountability

The normative architecture that has developed based on Article 19 of the ICCPR, Article 10 of the ECHR, and the Aarhus Convention does not consider greenwashing as a sui generis category that is not subject to freedom of expression. Instead, it places the corporate environmental communication directly on the doctrinal frameworks, which have been already established, thus, providing it with the cover of the well-established legal principles.

Three implications follow:

- The doctrine of free expression provides the protection of corporate environmental speech presumptively, as it contributes to the public debate and democratic process.
- Any such speech must meet the high dual standards of legality and proportionality such that the restriction must be law based, have a legitimate purpose and be proportioned to the interests at risk.
- It is only in cases where structural informational distortion is invasive of the rights of others that it may be considered as a legitimate basis of targeted intervention to establish a threshold that the substantive interest in truthful communication is not compromised.

The architecture, in its turn, preconditions the development of the operational framework that is elaborated later, as it offers a consistent legal framework on the basis of which the policy and jurisprudence can be applied consistently.

IV. CORPORATE ENVIRONMENTAL SPEECH AS HYBRID EXPRESSION

a. The Presumption of Protection: Pluralism, Tolerance and Democratic Society

Human rights are vital to environment degradation. They both support one another and act like two sides of the same coin. The first guarantees substantive rights, such as the right to health, property, speech. On the other hand, the other guarantees participation in democratic debates and access to information. Therefore, environmental participation does not operate in isolation, it affects relative rights such as freedom of expression (Jesenko, 2018, pp. 296–301).

Articles 19 of ICCPR and Article 10 of ECHR does not protect a specific type of speech, but protects speech that may even offend, shock and disturb others. As established in the *Handyside v UK* before the European court, the essence of freedom of expression is to guarantee pluralism and democratic debate (*Handyside v. United Kingdom*, 1976, para. 49). This concept is further supported in ECHR's *Muller v Switzerland* where the court mentioned Article 10 does not distinguish between the various forms of expression (*Müller and Others v. Switzerland*, 1988, para. 27). When shedding the light of the previous two cases on corporate environmental speech, it can be assumed that it is a guaranteed and protected right that falls under Article 10 of the Court and Article 19 of ICCPR. The challenge however is considering the test to adopt when narrowing the scope of corporation's expression.

There has been an ongoing debate and instability classifying commercial speech as "commercial" when the expression carries acute political nature. In ECtHR *Verein against Tierfabriken Schweiz (VgT) v Switzerland (No 2)*, this instability was apparent. VgT, wanted to broadcast a commercial for the purpose of telling people to eat less meat. They took videos of pigs living calmly in a forest, and then cut the scene to pigs crammed nervously and squealing, end ended the advertisement with "Eat less meat, for the sake of your health, the animals and the environment!". The broadcasting company refused to air the commercial because it possessed clear political character, and conflicts with the state's law which prohibits advertisement from financially powerful groups from obtaining a competitive political advantage. The court therefore had to clarify whether VgT's advertisement is considered commercial or political. The Court acknowledged that a commercial speech may reflect political controversial opinions making them fall outside the nature of commercial speech and into political and thereby narrowing the scope of protection (*VgT Verein gegen Tierfabriken v. Switzerland*, 2001, para. 57) Furthermore, it is the nature of the speech, not the form, which determined the appropriate evaluation. In this case the Court did not evaluate the speech as a matter of pure commercial but as a speech involving the general interest (*VgT Verein gegen Tierfabriken v. Switzerland*, 2001, para. 66). Accordingly what matters is not how the message is written, but the message that lies behind it. Even if the expression did not contain people or topic of politics, but conveyed or expressed a controversial message, it can be classified as political. This factor, which will be further explained in the following section, is an essential element when dealing with greenwashing. The intent behind the expression is vital to be assessed.

This case is one of many examples where the Court evaluated the speech and characterised it against political and commercial speech nature, rather than leaving it in the grey areas. However, in the case of greenwashing, it cannot not simply be classified into political or commercial speech under freedom of expression. For this reason, this chapter proceeds on the basis that corporate environmental speech is neither purely political nor commercial speech, but a hybrid speech that needs an extra layer of protection imposed by its own framework which will be illustrated as follows.

b. Complex Composition of Greenwashing: Commercial and Political

Greenwashing lies in an area between marketing and politics. It also seeks to achieve economic benefit even as it seeks to influence the public perception of the environmental issues, making it hard to evaluate it under Article 10. Therefore, seeing as it possesses qualities of both commercial and political, a distinction is necessary to illustrate that it does not fall squarely within one sole form of speech.

• Commercial Nature

Several cases under ECtHR have discussed the question of commercial speech under Article 10. Commercial speech is often characterised as intending to profit and attract consumer's attention. Regardless of the goal of commercial speech, whether solely monetary gain or attention grabbing, it is not deprived of protection under Article 10. This is illustrated by judges Gölcüklü, Pettiti, Russo, Spielmann, De Meyer, Carrillo Salcedo and Valticos's joint dissenting opinion where they mentioned in ECtHR's *Markt Intern Verlag Gmbh and Klaus Beermann v Germany,* "it is just as important to guarantee the freedom of expression in relation to the practices of a commercial undertaking as it is in relation to the conduct of a head of government" (*Markt Intern Verlag GmbH and Klaus Beermann v. Germany*, 1989, joint dissenting opinion, sec. 1). The judges also noted freedom of expression serves above all interests, the general interest. Also, ECtHR's case further supports this reasoning, in *Case of Casado Coca v Spain*, the Court delved into the question of the inclusion of commercial speech under Article 10 and concluded that Article 10 does not apply solely to certain types of information or ideas or forms of expression, in particular those of a political nature; it also encompasses artistic expression information of a commercial nature. While states have a margin of appreciation to broaden or narrow the scope of the expression that power is still under the supervision of the Court, and such a margin of appreciation is particularly essential in the complex and fluctuating area of unfair competition and advertising (*Casado Coca v. Spain*, 1994, paras. 35–37, 51–53). Even in the case of freedom of expression, while applying the tripartite test, matters can get complex and difficult to resolve.

In the case of greenwashing the matter is difficult to attribute because it is not only used for commercial purposes.

• Political Nature

While greenwashing can serve a commercial purpose, it is important to stress on the political effects it imposes. There are key principles that articles relevant to freedom of expression carry. They do not only function as means to convey a message but also as a means for society to participate in debates (Human Rights Committee, 2011, paras. 2, 13). To be able to participate in debates and share their opinion, the information imparted must be truthful and sufficient for them to act upon the information. As previously established political speech shapes the nature of the society.

Applying greenwashing onto political speech, it can cause a few problems when applying the very broad protective scope that the ECtHR has held, namely in *Lingens v. Austria* that freedom of political debate is at the very core of the concept of a democratic society, and that the public has the right to receive the ideas and expressions (*Lingens v. Austria*, 1989, paras. 41–53). While this case revolves around persons of politics and the balance between privacy and expression of people of the public the same principles outlined can still be imposed in greenwashing. As the Court highlighted, the public has the right to receive information and participate in a debate and express their opinions. However, in the case of greenwashing, the public cannot participate in a meaningful debate if the information that was imparted was factually incorrect. Not only that, but even when the public does participate and advocate, it does so on the basis of distorted facts, which ultimately leads to detrimental effects on the environment.

c. Corporate Environmental Speech as a Hybrid Speech

Based on the previous mentioned reasons, corporate environmental speech cannot be regulated under current frameworks because they possess traits of both commercial and political. For instance, in the United States, the Federal Trade Commission's Green Guides aim to ensure that environmental claims made by corporations remain truthful and non-deceptive. Pursuant to section 260.1 and 260.2, the claims such as "eco-friendly" are deemed prohibited not because they directly affect the environment by misleading consumers and the market, but because it affects consumers' perception deceiving them into buying the product, as well as there are no ways for the corporation to substantiate the claim. While the guide addressed the topic of advertising environmentally friendly products under section 260.4, but that was mirrored to market deception and not the effects on the environment (Federal Trade

Commission, 2012, pp. 3–6). This is reflected in Europe's Green guide (European Commission, n.d.); While this guide focused on both consumer perception and the environment. However, it is important to note that the reference to the environment is to make sure that consumers are provided with adequate and trustworthy information, and not for the effect of greenwashing on policies and on the society.

The issue with current frameworks is that they treat greenwashing only as a commercial problem or transparency problem. By regulating this side of greenwashing, its other harms are left unregulated. Greenwashing possesses traits of both commercial and political. It has the power to influence society for the goal if gaining profit, and at the same time it has to power to influence public opinion.

Another issue that is crucial to address, the guidelines and frameworks relating to greenwashing require substantiation. That can risk causing corporations to cease expressing all together which creates a chilling effect on freedom of expression. As the European court noted in *Barthold v. Germany*, a veterinary surgeon gave a newspaper interview discussing the shortage of emergency veterinary services available at night in the area. He went further and mentioned that his services provide 24-hour service. The domestic court sanctioned hum under unfair competition law, as this interview may provide Barthold with an advantage, i.e advertising himself and diminishing other veterinary services and therefore considered the newspaper interview as prohibited commercial advertising. The European court highlighted that the restriction imposed is secondary to the relevant problem at hand, which is the ability of Barthold to form an opinion and share it with the public, the court further acknowledged the issue, but highlighted it was not necessary in a democratic society as there was an apparent problem with shortages in vets at night in Germany (*Barthold v. Germany*, 1985, para. 58).

Parallel to this case, requiring substantiation is secondary to the original problem at hand, the original problem is not one-sided, it concerns both consumer deception and the safety of the environment. Thus, requiring corporations to substantiate will only solve one part of the equation- consumer deception- but can risk becoming too demanding and impose a chilling effect on freedom of expression. The goal is not to silence corporation but implement a framework which will balance all rights: the corporation, the public, and the safety and welfare of the environment. The following sections builds on this line of reasoning and proposes a well construction framework relying on intentionality, materiality, and structural impact.

V. THE THREE-PART FRAMEWORK: INTENTIONALITY, MATERIALITY, STRUCTURAL IMPACT

While the tripartite test related to Article 10 of ECHR and Article 19 of ICCPR is highly applicable in the case of greenwashing. It does not cover all necessary sides. The distinctive factor of the two Articles, they treat speech as either commercial or political. However, as it has been established, the nature of greenwashing possesses both, making it difficult to classify the speech as purely commercial or purely political. Therefore, this framework serves to respond to greenwashing's attributes, while focusing on all key factors and not only consumer deception. Also, similar to the tripartite test under Article 19 of ICCPR which mandates the satisfaction of all three grounds, this too adopts the same approach requiring the satisfaction of all grounds.

a. Intentionality

The first criterion of determining whether the type of speech falls under the category of greenwashing is intentionality. Similar to how Article 10 of ECHR distinguishes between commercial and political speech based on certain factors, intentionality functions similarly. While intentionality is defined as "the characteristic of being conscious of intending an object" (Merriam-Webster, n.d.). The definition in the proposed framework adopts a similar scope but aims to differentiate the intention behind the claim advertised as carrying elements of greenwashing- consumer deception, the use of false information, and supressing public participation in environmental issues. Therefore, this criterion assesses whether the representation of the environment is deliberate and attentive or a mere careless mischaracterisation, as opposed to the good-faith projection or adaptive sustainability policy.

It is crucial to question the intention which lies behind the claims advertised because not all claims made by corporations to support the environment is an automatic case of greenwashing. A deeper assessment needs to be taken. A prime example illustrating this criterion is *Deutsche Umwelthilfe v Lufthansa* (Landgericht Köln, 84 O 29/24, 21 March 2025). Lufthansa advertised that passengers could offset flight-related CO_2 emissions through contributions to climate protection projects and through the purchase of sustainable aviation fuel (SAF). Deutsche Umwelthilfe (DUH), a German environmental organisation, challenged these claims on the basis that they created a misleading impression that consumers could effectively compensate or neutralise the climate impact of their flights. The Cologne Regional Court held that the advertising was misleading because consumers were not adequately informed about the extent and operation of the alleged compensation mechanisms.* This case provides primal questions relative to these criteria:

- Is the claim made by the corporation highly achievable?

When, for example, corporations advertise their future project to go net zero by 2040, while providing a concrete and practical illustration on how this can be achieved, it demonstrates the true intention of their advertisement. In the above-mentioned case, the airline made a misleading claim regarding achieving CO2 emissions, thereby clearly showing the intention behind the advertisement, as contradictory to the claim advertised.

- Can the corporation substantiate their claim advertised?

Another crucial question the corporation must answer is whether they are able to prove their claim. The corporation must rely on data dating back to the beginning of the date of the advertisement, not merely recent data to silence greenwashing allegations. They must also not have documents or data contradicting their public claim. As crucial as it is to prove their claim, it is equally important not to have files which negate their claim. The EU Green directive will impose a structured and ex-ante obligation of substantiation, rather than allowing corporations to justify claims post-ante (European Commission, 2023, arts. 3–5). While EU's approach is justified on grounds of consumer protection and environmental integrity, it imposes a burden on expression, potentially restricting corporate speech before it enters the public eye. By contrast, the proposed framework adopts a more freedom of expression-friendly approach by allowing claims to be made subject to subsequent verification, thereby avoiding overly rigid pre-approval constraints on speech.

b. Materiality

The second criterion assessed after the satisfaction of the intentionality ground is materiality. This criterion assesses whether the claim is capable of influencing regulatory design, investor behaviour, or consumer choice in environmentally relevant manners. This ground performs a function quite similar to two established legal principles. First, it mirrors the causation requirement under Tort law; just as liability arises does not rise from the wrongful conduct alone but requires a link between the wrongful conduct and the harm suffered, materiality requires a demonstrable connection between the misleading claim and a tangible effect on those who encountered it. secondly, it mirrors the legitimate aim requirement under Article 19(3) ICCPR and Article 10(2) ECHR where the restriction on the expression must pursue a legitimate aim; in this ground (materiality) the intervention must be grounded in a demonstrated material harm. Without materiality, limitation lacks legitimate foundation regardless of how misleading the claim may be.

Therefore, materiality asks a two-part question: whether the claim was capable of causing material harm, and whether a reasonable person encountering the claim could have made a different decision had they known the truth behind the claim. The "reasonable person standard" is similar to what is depended on across different fields of law which also serves as a determining anchor to assess harm.

In this ground, the assessment is weighed against three aspects simultaneously, as opposed to current frameworks which assesses the damage against solely the consumer, the assessment is made against the individual consumer, the regulator, and the market. All three groups play a role in causing material harm, a company's reputation is a valuable intangible asset that influences its relationships with stakeholders, including customers, investors, employees, and partners (Lin et al., 2025, p. 79). The following analysis will set to demonstrate how each group plays a role in detail:

The first audience is the individual consumer. Similar to the concept of anti-deception law, this ground also recognises that the consumer must choose a product based on truthful information; and this decision has the potential to cause environmental harm due to the factually distorted claim. When the consumer buys a product convinced it is environmentally friendly, this instance is not a standalone issue but will eventually cause detrimental effects on the environment. Therefore, materiality is satisfied where misleading claims distort consumer purchasing decisions, thereby sustaining demand for environmentally harmful products.

The second audience is the investor. In an effort to regulate and prevent greenwashing, the European Union issued a Sustainable Finance Disclosure regulation that require financial players to formally declare their degree of compliance to ESG disclosure and reporting obligations (Cremasco & Boni, 2024, p. 1019). Therefore, a materially misleading environmental claim has the potential to divert capital towards companies that operate contrary to what they advertise themselves to be; in turn leading to a misallocation of funding and reinforcing such companies, rather than contributing to putting an end to the deceptive chain of advertising.

In addition, greenwashing does not only materially affect consumers and investors but also the regulator. This audience is what current frameworks neglected to consider but plays a principal role in regulating greenwashing. The reason being, if companies submit misleading data and advertisement, regulators do not see the need to refine current laws pertaining to advertising and consumer deception which in turn will risk causing a huge gap between current laws and reality. Furthermore, misleading environmental advertising contributes to a false perception that adequate action is already being taken. This is achieved through creating doubt, distraction, or an illusion of progress (Tinger et al., 2024, p. 69). This problem also conflicts against principle of legality, which stipulates that the law must be written in a matter to ensure foreseeability, clarity and accessibility and remain adapted to the reality of the society to serve its purpose. If current frameworks operate based on the illusion created by corporations, this is not only problematic in one dimension, namely greenwashing, but opens a pandora box of problems all going back to taming greenwashing through the sufficient and adequate regulation.

Furthermore, it must be highlighted that corporate environmental speech and corporate responsibility do not stand on opposite ends of the spectrum, but rather they both complement each other. They must be reconciled rather than ranked. Similar to the principle pertaining that freedom of expression is protected as long as it does not undermine relevant rights such as the right to privacy, both corporate speech and responsibility must be balanced and ensure that no one overrides the other without necessary and proportionate justification. Corporate environmental speech cannot override its democratic responsibility and accountability without a material basis for intervention. Where no material effect imposed on the previously established areas, the speech is protected. This balance test ensures against over regulation which risks causing a chilling effect on freedom of expression.

c. Structural Impact

While verifying the corporation's intention behind the claim and assessing the material harm on the previously mentioned audience are crucial steps in regulating greenwashing, an additional step is necessary to ensure that all areas are tackled and regulated. This third and last criterion assesses greenwashing against the line of greenwashing in the long run, not only as isolated incidents. Therefore, this criterion assesses whether the expression is capable of contributing to systemic informational distortion with respect to climate regulation.

The United Nations Special Rapporteur on the Promotion and Protection of the Right to Freedom of Opinion and Expression considered disinformation to carry series consequences namely, having the effect to destroys people's trust in democratic institutions (Khan, 2021). While disinformation and greenwashing are not identical, but they carry similar traits such as disseminating factually incorrect information. Therefore, greenwashing if unregulated sufficiently can lead to dangerous consequences in the society.

In order to establish the presence of structural impact, there are key grounds that needs to be assessed, which will be formed into the following questions: first, recurrency; is the material harm of greenwashing recurrent, or a standalone case? If the issue is common among a number of corporations, then this ground is satisfied. Second, persistence, it essential to consider the time period in which the issue is persisting in. if the issue has been persistent for period of time, in a matter without corrective action. Thus, the idea behind this is not merely with the time period by itself, but if the action was repeated persistently over this period of time without corrective action. Only then will this sub-ground be satisfied.

With greenwashing, individuals may presume that companies are acting in compliance with environmental standards and obligations, thereby reducing the urgency of the problem. the danger lies in the fact that in reality such misperception contributes to such corporations unknowingly enforcing the very practice that the goes against individual's beliefs with supporting the environment (Tinger et al., 2024, p. 68).

Third, corrective capacity must also be assessed against the risk of irreversibility. Similar to the consequences of prolonged conflict become increasingly difficult to reverse the longer they persist unchecked, the epistemic damage caused by sustained corporate environmental misrepresentation intensifies over time. Non-compliance, misdirected investment, and weakened public urgency do not automatically correct themselves once the false narrative of greenwashing is lifted; the effects they created and caused persists. This line of reasoning is reflected in Rio declaration Principle 15, also recognised as the precautionary principle. It recognises where there are threats of serious or irreversible damage, lack of full scientific certainty shall not be used as a reason for postponing cost-effective measures (Rio Declaration on Environment and Development, 1992, Principle 15). For this reason, the corrective capacity asks not only whether self-correction has occurred, but whether it can still occur in time to be meaningful. Where the window for effective correction is narrowing, the justification for external intervention is strongest.

The three-ground test ensures no over-restriction. This is influenced by the current frameworks of Article 10(2) ECHR and 19(3) of ICCPR. In order to illustrate the application of this framework before the court, the following section will provide an illustrative example.

VI. JUDICIAL AND REGULATORY APPLICATION

Before illustrating how the proposed framework can be applied, it is essential to note that this framework is not confined to specific regions but can be applied consistently across different jurisdictions. The reason behind this lies in the mirage between freedom of expression and consumer deception both of which are protected and provided for worldwide. An example of applying this framework will be demonstrated

a. Summary of the Case

A case to illustrate the application of this framework is the case of *Greenpeace France and Others v TotalEnergies* before the Paris Court (*Greenpeace France et autres v. TotalEnergies SE et TotalEnergies Électricité et Gaz France*, 2025). The proceedings were initiated in March 2022 by three French environmental associations: Greenpeace France, Les Amis de la Terre France, and Notre Affaire à Tous, hereinafter referred to as the applicant; against TotalEnergies SE (which is the holding company of the TotalEnergies Group, specializing in the energy industry, whose subsidiary), and its retail subsidiary TotalEnergies Électricité et Gaz France hereinafter referred to as the defendant. The claim was brought on the basis of Articles L 121-1 et seq. of the French Consumer Code (Code de la consommation, 2016), transposing Directive 2005/29/EC (European Parliament & Council, 2005) on unfair commercial practices. When the defendant changed its name from "Total" to "TotalEnergies," it launched a large communication campaign across websites, advertisements, and social media (*Greenpeace France et autres v. TotalEnergies SE et TotalEnergies Électricité et Gaz France*, 2025, para. 6). In that campaign, it presented itself as moving toward carbon neutrality and as being a major actor in the energy transition, mentioning claims such as:

- "Our ambition is to be a major player in the energy transition while continuing to meet the energy needs of populations" "We place sustainable development at the heart of our strategy, our projects and our operations to contribute to the well-being of populations, in line with the Sustainable Development Goals defined by the United Nations. "
- "We place sustainable development at the heart of our strategy, our projects and our operations to contribute to the well-being of populations, in line with the Sustainable Development Goals defined by the United Nations."
- "More energy, less emissions: this is the double challenge that we must take up alongside our stakeholders in order to contribute to the sustainable development of the planet and face the climate challenge. Our ambition is to contribute to achieving carbon neutrality by 2050 together with society [...] By offering our customers energy products that are less and less carbon-intensive" (*Greenpeace France et autres v. TotalEnergies SE et TotalEnergies Électricité et Gaz France*, 2025, paras. 3–4, 79).

The applicants alleged that defendant's 2021 rebranding campaign emitted a misleading environmental claim concerning its ambition to achieve carbon neutrality by 2050 and its aspiration to become a major player in the energy transition.

More specifically, the applicants contended that the company's communications did not merely promote individual products but constructed an environmental narrative capable of influencing how consumers perceive the company's role in addressing climate change. They argued that such representations amounted to greenwashing, as the defendant conveyed a level of environmental responsibility and transition commitment that was not supported by the company's actual operations and strategic direction (*Greenpeace France et autres v. TotalEnergies SE et TotalEnergies Électricité et Gaz France*, 2025, para. 5). More specifically, they contended that the only way to achieve carbon neutrality by 2050, in line with the Paris Agreement's 1.5°C objective, requires a drastic and immediate reduction in greenhouse gas emissions and the cessation of new fossil fuel development was to drastically and rapidly reduce greenhouse gas emissions and to immediately cease the exploration and exploitation of new oil deposits; whereas the defendant's strategy planned to develop numerous fossil fuel projects and increase its fossil fuel production(*Greenpeace France et autres v. TotalEnergies SE et TotalEnergies Électricité et Gaz France*, 2025, para. 15). The Tribunal examined 44 communications in total across three categories: claims relating to carbon neutrality and the energy transition; claims relating to fossil gas; and claims relating to agrofuels(*Greenpeace France et autres v. TotalEnergies SE et TotalEnergies Électricité et Gaz France*, 2025, paras. 61, 136, 155). Of these, only three communications drawn from the commercial website of the defendant were found to constitute misleading commercial practices under Articles L 121-2 and L 121-3 of the Consumer Code *Greenpeace France et autres v. TotalEnergies SE et TotalEnergies Électricité et Gaz France*, 2025, paras. 97, 135). The remaining communications, including the defendant's communications such as its rebranding and social media strategy posts, were dismissed as falling outside the material scope of the Directive on the ground that they lacked a direct connection with the promotion, sale, or supply of a product to consumers *Greenpeace France et autres v. TotalEnergies SE et TotalEnergies Électricité et Gaz France*, 2025, paras. 62, 67–80, 144–153, 155–162). The Tribunal ordered the cessation of the offending communication, its removal from the commercial website within one month under a daily penalty of 10,000 euros, and publication of the operative part of the judgment on the homepage of the French consumer website for 180 consecutive days, together with moral damages of 8,000 euros per applicant association consumers *Greenpeace France et autres v. TotalEnergies SE et TotalEnergies Électricité et Gaz France*, 2025, paras.3-8, 12) . The following sections will analyse the Court's reasoning against the proposed framework this chapter proposes.

b. Applying the First Criterion: Intentionality

The intentionality criterion requires examining whether the environmental claims made at issue were made deliberately, with awareness of their potential to distort public understanding of transition progress, or whether they reflected a good-faith projection of genuinely held sustainability commitments. The defendant's own Board of Directors report of 28 May 2021 (the very document on the basis of which the campaign was launched) explicitly confirmed that its carbon neutrality claim referred to the global scientific concept under the Paris Agreement, requiring net zero emissions by 2050 in line with Intergovernmental Panel on Climate Change (IPCC) recommendations consumers *Greenpeace France et autres v. TotalEnergies SE et TotalEnergies Électricité et Gaz France*, 2025, para 118). At the same time, the company's operative strategy planned a 4% annual growth in energy production globally between 2023 and 2030 and continued investment in new fossil fuel projects consumers *Greenpeace France et autres v. TotalEnergies SE et TotalEnergies Électricité et Gaz France*, 2025, paras.127-128); which posed as a trajectory the company's own internal documents acknowledged was incompatible with Paris-aligned pathways.

The Tribunal found that defendant had deliberately made an environmental claim likely to mislead consumers by presenting a carbon neutrality ambition without disclosing that its own internal scenario ran counter to the scientific recommendations it was invoking consumers *Greenpeace France et autres v. TotalEnergies SE et TotalEnergies Électricité et Gaz France*, 2025, paras.130). This is precisely what the intentionality criterion is designed to capture. The criterion does not require the presence of criminal fraud but rather it requires a conscious choice to present an environmental commitment while omitting material information that contradicts it. in respect to paragraph 130 it is apparent that the defendant knew the gap existed but chose not to disclose it, the Tribunal stated "the TotalEnergies group deliberately made an environmental claim likely to mislead consumers, by leading them to believe that by buying its products or services, they were participating in the emergence of a low-carbon economy, by following the recommendations of the scientific community, based on the Paris Agreement." consumers *Greenpeace France et autres v. TotalEnergies SE et TotalEnergies Électricité et Gaz France*, 2025, para 130)

The Tribunal also highlighted that by stating: "Faced with the challenge of climate change, the TotalEnergies Group is transforming itself with the ambition of becoming a major player in the energy transition, in particular by accelerating its development in renewable energies and aiming for carbon neutrality by 2050 together with society", the TE Group clearly expresses that it is committed to a positive approach to the environment, irrespective of the fact that the term "ambition" rather than "commitment" was chosen for this communication, both terms covering the same reality from the consumer's point of view consumers *Greenpeace France et autres v. TotalEnergies SE et TotalEnergies Électricité et Gaz France*, 2025, para. 104). Therefore, the first criterion is satisfied, now the question of materiality must be assessed.

c. Applying the Second Criterion: Materiality

The materiality criterion is satisfied where it can be demonstrated that the misleading representation was capable of producing a tangible distorting effect on the decisions of identifiable audiences. The Tribunal noted that in respect to the claims/ communications raised by the defendant, the question is not to investigate whether there is a contradiction between the group's stated ambitions of carbon neutrality and its activity from the point of view of the company, but of determining whether, from the consumer's point of view, the communications contain false or misleading environmental claims, and if so, whether they may have had an influence on his business decision consumers *Greenpeace France et autres v. TotalEnergies SE et TotalEnergies Électricité et Gaz France*, 2025, para. 108). In respect to the communications disseminated by the defendant, the Tribunal highlighted that those claims have manifestly substantially altered the economic behaviour of a normally observant and circumspect consumer whose choice, while still guided by price, increasingly incorporates the environmental qualities of the product or service consumers *Greenpeace France et autres v. TotalEnergies SE et TotalEnergies Électricité et Gaz France*, 2025, para. 134); thereby applying the reasonable person standard proposed in the preceding section.

With regards to the material effect on the investor, because one of the legal instruments applied on this case was the Directive, the investor was excluded from this scope pursuant to recital 7 in the preamble to the Directive consumers *Greenpeace France et autres v. TotalEnergies SE et TotalEnergies Électricité et Gaz France*, 2025, para. 60). However, if this case were to have been seen while applying this proposed framework, the investor would not have been excluded, seeing as they play a crucial role in mitigating the issue of greenwashing.

Furthermore, regarding the regulator the Tribunal's found that defendant omitted to disclose to consumers that its internal climate scenario ran counter to Paris' Agreement scientific recommendations carries direct regulatory implications consumers *Greenpeace France et autres v. TotalEnergies SE et TotalEnergies Électricité et Gaz France*, 2025, paras. 131-131). Where regulators rely on corporate environmental disclosures to assess whether existing legal frameworks are adequate, systematically misleading representations create a false picture of compliance and transition progress. If companies present themselves as Paris-aligned actors while continuing to expand fossil fuel production, regulators receive a distorted signal that current law is working thereby reducing the perceived need for more stringent intervention.

d. Applying the Third Criterion: Structural Impact

The structural impact criterion requires demonstrating prevalence of the misleading conduct across a pattern rather than an isolated incident, persistence over time without corrective action, and diminishing corrective capacity. The case at hand satisfies each indicator. First, persistence is established by the sheer breadth of the campaign. The Tribunal examined 44 separate communications deployed across Facebook, Instagram, X, LinkedIn, YouTube, press, and television in a coordinated manner following the May 2021 rebranding.

Second, persistence without corrective action is demonstrated by the fact that Communication No 3 (the claim that defendant's ambition was to be a major player in the energy transition while continuing to meet the energy needs of the population) remained accessible on the commercial website as of 4 October 2024, more than three years after the initial summons was filed in March 2022 consumers *Greenpeace France et autres v. TotalEnergies SE et TotalEnergies Électricité et Gaz France*, 2025, paras. 83,94). It was accessible through the tabs "Why choose us", then "About TotalEnergies", then "TotalEnergies is committed", alongside a red button entitled "Our commitments", embedded, in other words, in the heart of the commercial interface through which consumers choose their energy supplier consumers *Greenpeace France et autres v. TotalEnergies SE et TotalEnergies Électricité et Gaz France*, 2025, para. 94) . The defendant had yet not removed it. The court found it necessary to order its cessation under a daily penalty of 10,000 euros precisely because no voluntary correction had occurred consumers *Greenpeace France et autres v. TotalEnergies SE et TotalEnergies Électricité et Gaz France*, 2025, dispositif, point 5).

Third, diminishing corrective capacity follows from the combined effect of the preceding two sub-grounds. The longer a misleading environmental narrative is maintained, which in this case was across 44 communications, over three or more years, embedded in a company's commercial identity and name, the more it shapes background expectations among consumers, investors, and regulators. These expectations do not automatically reset once a judgment orders their removal. The repositioning achieved by the defendant's rebranding, and the consumer and regulatory perceptions it generated, persist beyond the corrected communication.

VII. CONCLUSION

This chapter revealed that the argument of proposing a new framework is based on three propositions. First, environmental responsibility is in part communicative and climatic harm is not only worsened by emissions but also driven by distortions that undermine informed public participation making collective response difficult. Secondly, corporate environmental speech possesses a multi-dimensional traits, falling somewhere between economic, political, and public normativity, which requires a fine threshold analysis as opposed to traditional expression, which blankets deference or suppression. Thirdly, pluralistic societies involved in democratic environmental governance must have and maintain a legal structure that protects both expressive freedom and safeguards against systemic informational distortion. With the increasing climate crisis, companies' involvement in community dialogues has in turn parallelly increased. Therefore, maintaining a healthy discussion and epistemic integrity determines the credibility of environmental management. Thus, the proposed framework situates, reflects and compliments these issues all the while acting as the suitable applicable body. Furthermore, this manuscript contributes to a principled synthesis that can balance freedom of expression with environmental responsibility on an international level by re-conceptualising greenwashing as structurally harmful corporate expression, addressed by the regional jurisprudence as an extension of Article 19 ICCPR.

This chapter raised the question: on what grounds can greenwashing by corporations be legitimately regulated under a human rights framework, without silencing the corporate voice in environmental debate? The answer this chapter has developed is that the existing human rights framework rooted in Article 19 of the ICCPR and Article 10 of the ECHR already contains the normative tools to address greenwashing in a principled and proportionate way however misses targeting greenwashing in all its dimensions (consumer deception, market and regulator effect, as well as freedom of expression.

Current regulatory frameworks approach greenwashing as a market problem; a form of consumer deception to be addressed through substantiation and competition law. This chapter has argued that this characterisation is analytically incomplete. Greenwashing is not only a misleading product claim. It is a form of corporate expression that distorts the informational environment within which citizens, investors, and regulators make decisions about climate change. When a corporation systematically represents itself as complaint with the law, namely green initiatives, while expanding fossil fuel production, it does not merely mislead the consumer at the point of purchase, but contributes to a wider epistemic distortion that slows regulatory action, misdirects investment, and weakens civic mobilisation.

To address this gap, the chapter proposed a three-part framework built on intentionality, materiality, and structural impact. The three criteria must be satisfied cumulatively. Together, the three criteria protect freedom of expression while providing a principled basis for accountability where corporate environmental speech crosses into structural harm.

The implications of this chapter extend across different levels. At the doctrinal level, it offers a re-characterisation of greenwashing that is consistent with established human rights principles rather than dependent on the creation of a new regulatory category. Greenwashing does not need to be considered as a case outside the freedom of expression framework; but rather needs to be understood as a form of expression that, when it satisfies the three criteria, falls within the legitimate scope of proportionate restriction under Article 19(3) ICCPR and Article 10(2) ECHR.

This chapter also contributes to an emerging understanding that environmental responsibility is partly communicative (i.e resolved under relative articles related to freedom of expression). In addition, the way in which corporations present their environmental performance plays a direct role in shaping how environmental issues is perceived, debated, and regulated. when communication creates a misleading impression of progress or compliance, it does not only affect individual decisions but weakens the conditions under which informed participation and effective regulation can take place. For this reason, regulating greenwashing is not a contradiction to freedom of expression, but a necessary extension of it. The aim is not to silence corporate speech, but to ensure that it does not distort the informational environment on which collective climate action depends on.

REFERENCES

Animal Defenders International v. United Kingdom (2013). European Court of Human Rights.

Casado Coca v. Spain (1994). European Court of Human Rights.

Code de la consommation. (2016). *Article L121-1.* https://www.legifrance.gouv.fr/codes/article_lc/LEGIARTI000032227301/2016-07-01

Convention on access to information, public participation in decision-making and access to justice in environmental matters (Aarhus Convention). (1998). United Nations Economic Commission for Europe

Council of Europe. (2017). *Protecting the right to freedom of expression under the European Convention on Human Rights: A handbook for legal practitioners.* Council of Europe.

Cremasco, C., & Boni, L. (2024). Is the European Union (EU) sustainable finance disclosure regulation (SFDR) effective in shaping sustainability objectives? An analysis of investment funds' behaviour. *Journal of Sustainable Finance & Investment, 14*(4), 1018–1019. DOI: 10.1080/20430795.2022.2124838

Debnath, R., Ebanks, D., Mohaddes, K., Roulet, T., & Alvarez, R. M. (2023). Do fossil fuel firms reframe online climate and sustainability communication? A data-driven analysis. *npj. Climate Action, 2*(1), 47. DOI: 10.1038/s44168-023-00086-x PMID: 38694952

European Commission. (2023, March 22). *Proposal for a directive on substantiation and communication of explicit environmental claims (Green Claims Directive), COM(2023) 166 final.*

European Commission. (n.d.). *Green claims.* https://environment.ec.europa.eu/topics/circular-economy-topics/green-claims_en

European Parliament and Council. (2005). Directive 2005/29/EC concerning unfair business-to-consumer commercial practices. *Official Journal of the European Union, L, 149*, 22–39.

Federal Trade Commission. (2012). *Guides for the use of environmental marketing claims* (16 C.F.R. Part 260). https://www.ftc.gov/sites/default/files/attachments/press-releases/ftc-issues-revised-green-guides/greenguides.pdf

Green, S., & Temper, L. (2026). The health case for a fossil fuel advertising ban. *Canadian Medical Association Journal, 198*(5), E176–E180. DOI: 10.1503/cmaj.250022 PMID: 41663133

Habermas, J. (1996). *Between facts and norms.* Polity Press. DOI: 10.7551/mitpress/1564.001.0001

Handyside v. United Kingdom (1976). European Court of Human Rights.

Hill, S., Nemes, N., Montgomery, A., Scanlan, S., McNally, B., Tubiello, F., Aronczyk, M., Wood, T., Smith, T., & Kaupa, C. (2025). Testing the greenwashing assessment framework. *Ecology and Society*, *30*(2), 31. DOI: 10.5751/ES-16106-300231

Human Rights Committee. (2011). *General comment No. 34: Article 19: Freedoms of opinion and expression* (CCPR/C/GC/34). United Nations.

Human Rights Committee. (2018). *General comment No. 36: Article 6 (Right to life).* United Nations.

International Covenant on Civil and Political Rights. (1966). United Nations.

Jesenko, K. (2018). Freedom of expression in environmental cases before the European Court of Human Rights. *ERA Forum*, 19, 295–301.

Khan, I. (2021, April 13). *Disinformation and freedom of opinion and expression: Report of the Special Rapporteur on the promotion and protection of the right to freedom of opinion and expression.* United Nations. https://digitallibrary.un.org/record/3925306

Landgericht Köln. (2025, March 21). *84 O 29/24* (ECLI:DE:LGK:2025:0321.84O29.24.00) [Judgment]. Justiz NRW. https://nrwe.justiz.nrw.de/lgs/koeln/lg_koeln/j2025/84_O_29_24_Urteil_20250321.html

Lin, W. L.. (2025). The impact of greenwashing: Risks and implications for corporate performance and stakeholder trust. In *Proceedings of the International Conference on Environmental, Social, and Governance (ICESG 2024)* (p. 79). DOI: 10.2991/978-2-38476-358-0_8

Lingens v. Austria (1989). European Court of Human Rights.

Markt Intern Verlag GmbH v. Germany (1989). European Court of Human Rights.

Merriam-Webster. (n.d.). *Intentionality.* In *Merriam-Webster.com dictionary.* https://www.merriam-webster.com/dictionary/intentionality

Montgomery, A. W., Lyon, T. P., & Barg, J. (2024). No end in sight? A greenwash review and research agenda. *Organization & Environment, 37*(2), 221–256. DOI: 10.1177/10860266231168905 PMID: 39091586

Müller and Others v. Switzerland (1988). European Court of Human Rights.

OECD. (2022). *Policy guidance on market practices to strengthen ESG investing and finance a climate transition* (OECD Business and Finance Policy Papers). OECD Publishing. https://doi.org/DOI: 10.1787/bf84ff64-en

Paris Judicial Court. (2025, October 23). *Greenpeace France and Others v. TotalEnergies SE and TotalEnergies Électricité et Gaz France* (machine-translated English version). ClientEarth. https://www.clientearth.org/media/1dmfkuvo/gpf-vs-totalenergies_english_machine-translated.pdf

Portillo Cáceres v. Paraguay (2019). Human Rights Committee, Communication No. 2751/2016.

Post, R. (2012). *Democracy, expertise, and academic freedom.* Yale University Press.

Rio Declaration on Environment and Development. (1992). United Nations.

Sunstein, C. R. (1993). *Democracy and the problem of free speech.* Free Press.

Tinger, M., Walton, S., & O'Kane, P. (2024). The climate impact of misinformation, disinformation, and greenwashing in advertising. In Vandenberg, K. M., & Tinger, M. (Eds.), *Rethinking advertising: Ethics and effectiveness* (pp. 68–69). Palgrave Macmillan.

UN Human Rights Office of the High Commissioner. (2024, May 3). *Joint declaration on the climate crisis and freedom of expression.* United Nations.

UN Human Rights Office of the High Commissioner. (2024, May 3). *Freedom of expression and media crucial to tackling climate crisis: International freedom of expression rapporteurs.* United Nations. https://www.ohchr.org/en/statements-and-speeches/2024/05/freedom-expression-and-media-crucial-tackling-climate

United Nations. (n.d.). *Greenwashing – the deceptive tactics behind environmental claims | United Nations.* https://www.un.org/en/climatechange/science/climate-issues/greenwashing

United Nations Economic Commission for Europe. (1998). *Convention on access to information, public participation in decision-making and access to justice in environmental matters.* Aarhus Convention.

United Nations Economic Commission for Europe. (2014). *The Aarhus Convention: An implementation guide* (2nd ed.). United Nations Economic Commission for Europe.

United Nations Human Rights Council. (2025). *The imperative of defossilizing our economies: Report of the Special Rapporteur on the promotion and protection of human rights in the context of climate change, Elisa Morgera* (A/HRC/59/42). United Nations. https://docs.un.org/en/A/HRC/59/42

Verein gegen Tierfabriken Schweiz (VgT) v. Switzerland (No. 2) (2009). European Court of Human Rights.

VgT Verein gegen Tierfabriken v. Switzerland (2001). European Court of Human Rights.

ENDNOTE

* The original judgment was issued in German. Any English translations provided in this chapter are unofficial translations generated through machine-assisted translation for research purposes.

Chapter 6
A Comparative Study of Environmental Accountability Frameworks in India, the European Union, and International Climate Litigation:
From Soft Law to Hard Liability

S. Abraham
http://orcid.org/0009-0001-7672-7607
Saveetha School of Law, Saveetha Institute of Medical and Technical Sciences (SIMATS), India

G. Aswathy Prakash
http://orcid.org/0000-0003-0806-4562
Saveetha School of Law, Saveetha Institute of Medical and Technical Sciences (SIMATS), India

ABSTRACT

Environmental governance is shifting from voluntary principles to enforceable legal obligations. This paper compares frameworks in India, the EU, and international climate litigation, showing a clear move toward hard legal liability. It highlights India's activist judiciary, the EU's advanced corporate accountability rules, and

DOI: 10.4018/979-8-2600-3678-5.ch006

new international legal concepts like ecocide. The study recommends stronger global enforcement, greater corporate liability, and institutionalized climate justice mechanisms.

INTRODUCTION

Over much of the past century, the organization of environmental governance at the global level has revolved around a paradox. Recognizing the seriousness of ecological degradation, making international agreements, adopting sustainability principles, and articulating the concepts of intergenerational justice were all efforts made by states. The legal basis of such efforts, however, had a structure which did not facilitate compliance. Indeed, as illustrated by the declarations of Stockholm (1972) and Rio (1992), the Convention on Biological Diversity, the Kyoto Protocol, and finally, the Paris Agreement of 2015, there have been efforts aimed at creating frameworks that had a significant normative value but little in terms of operation: they defined obligations of endeavor rather than result, relied on self-reporting by the states, and had no enforcement mechanism. Such an approach to environmental governance may be referred to as soft law, or the use of normative frameworks without the ability to coerce.

The results of that governance approach are now apparent in rapid climate change, biodiversity destruction, and acidification of the oceans, deforestation, and transboundary pollution to the degree where human and natural existence itself becomes threatened. As of September 2023, atmospheric carbon dioxide concentration exceeded 420 parts per million, which means that it is higher than it has been in the past three million years (Intergovernmental Panel on Climate Change [IPCC], 2023). Average global temperature increased by more than 1.1 degrees centigrade above pre-industrial levels, with the warmest year ever recorded being 2023 (IPCC, 2023). The fact that even the strictest nationally determined contributions are far from sufficient to prevent catastrophic climate change proves the inefficiency of voluntary compliance.

Consequently, the transformation of environmental law is now taking place in several directions simultaneously. The courts of various countries delivered landmark decisions that redefine environmental regulation in terms of constitutional obligation, fundamental right, and corporate legal responsibility. In addition, the European Union adopted some of the most advanced pieces of legislation concerning mandatory corporate sustainability and environmental due diligence. The supreme and high courts of India elaborated a doctrine of environmental constitutionalism forcing state and non-state actors to consider ecological consequences. Finally, the International Court of Justice prepared to deliver advisory opinions on climate responsibilities, while coalitions of environmental activists pushed for the criminalization of ecocide as the fifth crime under the Rome Statute. All this amounts to a move from soft law to hard liability in environmental governance, as described by this paper.

In order to examine the nature and scope of that transformation, it is necessary to compare the developments taking place in three specifically selected focal points. For one thing, India offers some of the best examples in which constitutional courts enforce environmental obligation despite the lack of legislation on this matter. Another aspect of such evolution can be found within the European Union that has been setting precedents through adopting or planning to adopt new instruments regulating environmental responsibility and due diligence of corporations. Finally, international climate litigation represents the frontier of global agreement about environmental law, with novel doctrines emerging along with transnational disputes and international tribunal processes.

LITERATURE REVIEW

According to (Ram Mohan et. al., 2024), the history of climate litigation in India shows the shift of climate cases from peripheral environmental matters to a distinct constitutional category. The typology of Indian cases provided by the authors, which includes suits on government accountability, sectoral emission disputes, and rights cases, demonstrates that litigation is an evolving practice of translating international climate commitments into domestic ones through the mechanism of constitutional obligations. At the same time, Ram Mohan, Reynaers Kini, and Prasad warn that recognizing the right not to suffer from climate effects is only the beginning of the process since further action on legislation and administration is necessary. This work is relevant in explaining how the judiciary of India is filling in the regulatory gap that exists due to the lack of climate laws. The article published in the Indian Journal of Legal Review (2025) on the topic of corporate environmental accountability in India shows how constitutional mandates, environmental statutes, tort law, and PIL influence each other in shaping new liability standards in the country. In the process, the study identifies persistent problems in applying the law in practice because of the inability to prove causation, obtain standing, and deal with greenwashing defenses. However, at the same time, Indian courts are using their experience and comparing it to the international experience in dealing with similar problems, thus making reasonable use of comparison despite their fragmented statutes. This article is directly relevant to the goal of the chapter that deals with corporate liability in India.

In its discussion of the case M.K. Ranjitsinh of the Supreme Court of India, the Vidhi Centre for Legal Policy (2024) explains how the court found a way to recognize a new right from climate changes in the framework of life and equal rights. It differs from the public trust doctrine used in the earlier ruling Ridhima Pandey. To make the discussion of the case more productive, the commentary makes a parallel between the Indian Supreme Court and the European Court of Human Rights judgment in Verein KlimaSeniorinnen, where similar questions of causation and victimhood were discussed. Shodhsagar's Indian Journal of Law (2025) investigates the question of climate law and environmental governance in India and shows that without an omnibus climate legislation, the court became the main institution where climate factors are taken into account during environmental rulings. Using the methodology of doctrinal analysis of Supreme Court and High Court judgments, the research demonstrates a trend from indirect consideration of environmental factors to direct use of constitutional arguments in climate disputes. Among other findings of the study, the author emphasizes the accountability gap of institutions, particularly weak implementation of pollution control measures as an intrinsic feature of ineffective functioning of courts. This article is directly connected with the goals of the chapter as it is based on enforcement aspects.

Reporting by the American Society of International Law (2024) about Verein KlimaSeniorinnen Schweiz and Others v. Switzerland describes the landmark decision of the European Court of Human Rights on the positive obligation of states under the Convention to mitigate climate change. According to the ruling, the deficient regulation of Switzerland violated this obligation. The value of the decision consists in its transformation of the soft political climate ambition expectation into a legally justiciable standard related to Articles 8 and 6 of the Convention. This source serves as a basis for the analysis in the chapter of how the regional human rights adjudication makes soft climate commitment obligations legally binding. The European Constitutional Law Review (2025) presents the doctrinal analysis of the consequences of the KlimaSeniorinnen judgment, emphasizing the control powers of the Committee of Ministers under Article 46 ECHR. As the authors note, the continuous use of the "Action Report" instead of the correcting "Action Plan" by the country under scrutiny demonstrates the limitations of monitoring mechanisms even after the landmark finding of violation. This source helps analyze the enforcement mechanisms in the chapter. Firstly, Johannsen (2025) in the Review of European, Comparative and International Environmental Law critically reviews The Hague Court of Appeal's 2024 reversal of the earlier decision to impose a 45 percent reduction in Shell's emissions. While accepting a non-binding social obligation of care linked with human rights protection and horizontal application of the doctrine to private persons, the court refused to set any quantitative target because of the absence of scientific and legal consensus on corporate carbon budgeting.

According to Johannsen, this is an example of an "empty victory" when recognition of the ethical principle has not been combined with the enforceable measure. Herein lies the problem of the dichotomy between the soft law and hard liability in transnational climate litigations. Next, The Cambridge Journal of Business and Human Rights (2024) describes the Corporate Sustainability Due Diligence Directive (CSDDD) adopted in Europe as the first regional mechanism requiring mandatory human rights and environmental due diligence obligations across global supply chains of both EU and non-EU businesses above the required turnover level. The text considers this directive as the result of prolonged negotiations on liability issues within the Brussels Effect a wider framework when EU regulations have a global impact. This research serves as one of the foundations for comparative discussion of the EU transition from voluntary ESG disclosure to mandatory liability for human and ecological risks management within the chapter.

Finally, the European Journal of Risk Regulation (2025) discusses the complex CSDDD enforcement regime that combines civil liability and administrative supervision. It is argued in the article that this approach can cause proportionality issues for smaller EU companies which compete with larger foreign businesses that fall outside the scope of the directive. The text will help evaluate the efficiency of CSDDD enforcement design to achieve effective accountability versus the mere burden for EU enterprises within the context of the chapter's discussion about ESG accountability tools.

RESEARCH OBJECTIVES

The objectives which this chapter seeks to achieve are as follows:

- ✓ To chart the journey of environmental accountability from soft law obligations to legally binding commitments.
- ✓ To carry out a comparative study of the environmental accountability framework in India, the EU and the mechanism of international climate litigations.
- ✓ To analyze the role played by courts, human rights and the constitutional safeguard in environmental enforcement.
- ✓ To examine the expanding legal liability of multinationals in relation to environmental destruction and climate commitments.
- ✓ To establish emerging legal principles relating to ecocide, environmental due diligence, ESG accountability and transnational liability.
- ✓ To formulate recommendations for improving global environmental accountability and enforcement.

RESEARCH METHODOLOGY

The research adopts a comparative and doctrinal/qualitative methodology. It is based on secondary sources like international treaties, constitutions, environmental laws, case laws, policies, academic studies and instruments of international environmental law.

THE EVOLUTION OF GLOBAL ENVIRONMENTAL ACCOUNTABILITY

From Stockholm to Paris: The Architecture of Soft Law

The modern legal order on the international level is rooted back in the 1972 UN Conference on the Human Environment held in Stockholm. This resulted in the creation of the Stockholm Declaration and the establishment of the United Nations Environment Programme (UNEP). Twenty-six principles of the declaration include the principles of state responsibility for environmental damages, conservation of natural resources, and the interconnectedness of environment protection with people's lives. At the same time, it is important to note that none of those principles had a binding nature; they merely reflected the aspirations (Sands et al., 2018).

During the following two decades, there has been much activity in the development of international environmental law through the adoption of numerous conventions, protocols, and treaties covering specific environmental problems like ozone depletion, marine pollution, biodiversity, climate change, desertification, and toxic waste. In 1992 at the Rio Earth Summit three landmark agreements namely the United Nations Framework Convention on Climate Change (UNFCCC), the Convention on Biological Diversity (CBD), and the Rio Declaration on Environment and Development were adopted, thereby forming the basis of global environmental governance. The precautionary principle mentioned in Principle 15 and the polluter pays principle mentioned in Principle 16 of the Rio Declaration are now regarded as key principles in environmental legislation around the world, although still not in their legally binding version.

The first international treaty imposing binding quantitative limits on emissions by developed states was the Kyoto Protocol signed in 1997. It did not, however, achieve success due to several reasons, among which the US's withdrawal from the agreement in 2001 and the lack of participation of major developing country emitters must be emphasized. In 2015, the Paris Agreement introduced an entirely new approach to climate policy based on the bottom-up approach and NDCs. (Bodansky, 2016) notes that it is essentially a new way of designing climate governance that replaced binding targets with self-determined contributions under the supervision of the international community.

Limits of Voluntary Compliance and the Pressure for Enforcement

It is clear from experience that the inadequacy of the soft-law approach to environmental governance can hardly be disputed. Greenhouse gas emissions around the world continue to increase despite the climate commitments that countries have made throughout the decades. Despite the creation of the framework of the Convention on Biological Diversity, the rate of biodiversity decline continues. Deforestation, ocean pollution, and soil degradation proceed at such a speed that is completely unacceptable considering the goals of sustainability expressed in international declarations. A key limitation inherent in international environmental soft law relates to the requirement for the coincidence of political willingness, economic motivations, and norms; however, this cannot always happen.

There is an emerging pressure for more sophisticated accountability mechanisms. Environmentalists and other non-governmental organizations are using judicial institutions to force the authorities to address environmental challenges through means of legislation and international law when politics fails. The establishment of human rights frameworks provides an efficient tool for linking the environmental problems to human rights issues which allows for creating legal obligations. Improvements in attribution science have made it possible to trace the role of various parties in climate change and its consequences providing evidence that could not be gathered earlier. Increasing discussion about the need for ESG accountability of corporations creates pressure on them.

Erosion of Traditional Sovereignty Doctrines

The doctrine of absolute sovereignty of the state in international law has undergone consistent weakening when applied to environmental contexts. The Trail Smelter Arbitration case (1941) in which Canada had to accept liability for causing air pollution to the United States is considered to have introduced the principle according to which states are required to take measures preventing activities in their jurisdiction from harming other states. This principle is enshrined in the Stockholm Declaration Principle 21 and further elaborated by the jurisprudence of the International Court of Justice in such cases as the Gabcíkovo-Nagymaros Project case (1997) and Pulp Mills on the River Uruguay case (2010).

Additionally, erga omnes obligations, which are obligations owed to the international community as a whole, and not just a particular state, continue to undermine the protection provided by sovereignty in terms of environmental issues. The idea of the common concern of humankind, upon which climate change has been conceptualized in the UNFCCC preamble, means that there is no way for states to hide behind their sovereignty in relation to the way their environmental actions have performed. The work carried out by the International Law Commission regarding the prevention of transboundary harm and the responsibility of states for internationally wrongful acts forms the legal doctrine for pursuing accountability in this area (Crawford, 2019).

ENVIRONMENTAL ACCOUNTABILITY IN INDIA

Constitutional Environmentalism

The Indian constitution can be described as one of those countries with strong foundations for environmental accountability, but this cannot be traced to the 1950 Constitution since it came into existence through the means of constitutional amendments and interpretation. The 42nd Constitutional Amendment in 1976 saw the introduction of two clauses: Article 48A (which directs that it is the duty of the state to protect and improve the environment) and 51A(g) (it states that the citizens must protect the environment). Neither of these can be termed as a fundamental right, but the Indian Supreme Court has used both of them in determining the existence of environment in the scope of Article 21's right to life.

Article 21 as an environmental right forms the foundation of India's constitutional environmentalism. For instance, Subhash Kumar v. State of Bihar (1991) declared that Article 21's right to life includes the right to enjoy pollution free water and air for complete enjoyment of life. Later on in jurisprudence, in Virender Gaur v. State of Haryana (1995), it was ruled that pollution of environment was a violation of the right to life. Also, in M.C. Mehta v. Union of India (Oleum Gas Leak Case, 1987) the absolute liability of hazardous enterprises was determined through the use of Rylands v. Fletcher (England), but with the elimination of previous defenses.

Public Interest Litigation as Environmental Enforcement

Environmental accountability in India has primarily been achieved by way of public interest litigation (PIL). With the Supreme Court lowering the standards for locus standi in the early 1980s, allowing for any public-spirited individual or organization to bring a matter before the Supreme Court on behalf of those who were unable to speak for themselves, constitutional litigation has become an accessible tool for ensuring environmental accountability. This jurisdiction of PILs has been used in several cases dealing with environmental matters, such as pollution from industry, illegal mining, deforestation, wetlands conservation, air quality, solid waste management, and protection of ecosystems.

Perhaps the most well-known example of a series of judicial efforts at environmental accountability anywhere in the world is the series of M.C. Mehta cases spanning more than three decades beginning in the 1980s. Among other orders, the Supreme Court ordered the relocation of industries causing pollution in Delhi, conversion of public transport buses to run on compressed natural gas, establishment of compliance committees, and has developed many important principles for the enforcement of environmental law that have had a wide-ranging impact in the field of environmental law across the globe (Rajamani & Ghosh, 2020). Another landmark case is that of Vellore Citizens Welfare Forum v. Union of India (1996), where the Supreme Court expressly adopted the principles of precaution and polluter pays as part of Indian environmental law under Article 21.

National Green Tribunal and Institutional Architecture

The establishment of the National Green Tribunal under the provisions of National Green Tribunal Act 2010 represents one of the major developments in India's enforcement mechanism for environmental regulation. The NGT was established as a judicial body with the power to hear and decide civil matters relating to environmental protection, conservation of forests and other natural resources and enforcement of environmental rights. The NGT utilizes multi-bench systems whereby a bench is made up of judicial members and expert members who are highly trained scientifically and technologically. This enhances the NGT's ability to handle complex environmental disputes in ways that cannot be done by generalist courts.

The NGT has achieved a lot in terms of the number of orders issued in thousands of cases on matters like discharge of pollutants, solid wastes, air pollution due to crop burning, construction activities in sensitive areas, and rehabilitation of rivers. It has developed a jurisprudence relating to precautionary principle, polluter pays principle and sustainable development principle and delivered orders with huge financial implications for states, municipalities, and corporations (Gill, 2016). However, compliance with the orders of NGT is inconsistent, and NGT has struggled to enforce its directions where it involves continuous administrative actions by several governmental organizations.

Corporate Environmental Liability in India

There exist other mechanisms besides the Oleum Gas Leak case rule for the imposition of environmental liability on corporations in India. These mechanisms are provided by the Environment (Protection) Act of 1986. This act enables the Central government to take actions for pollution control. There are also the Water (Prevention and Control of Pollution) Act of 1974 and the Air (Prevention and Control of Pollution) Act of 1981, whereby there is an environmental protection mechanism based on the consent system. Criminal sanctions are provided under these Acts. Lastly, there is the Companies Act, 2013. It provides for CSR expenditures for qualifying companies. This has been directed towards environmental initiatives by the Ministry of Corporate Affairs.

While improvements continue to be made in the Indian environmental liability regime, implementation has proven challenging due to several issues. One is that enforcement by state pollution control boards has not been consistent due to lack of resources, political pressure, and inadequate technical expertise. Also, prosecution of cases of environmental violations through criminal laws has proven difficult in India. The gap between the progressive environmental jurisprudence of India and the challenges faced in its implementation remains.

EUROPEAN UNION ENVIRONMENTAL GOVERNANCE

European Green Deal and Regulatory Architecture

There has been a paradigm shift in the EU's approach to environmental governance after its decision to adopt the European Green Deal in December 2019, which envisions the making of Europe as the first continent with zero-net emissions through a comprehensive legislative package to support such aspirations. It should be noted that this legislation goes far beyond an environmental statement as it represents a full-fledged economic and governance strategy designed to ensure coherence between the EU's economic regulation and the idea of sustainability. The legislative proposal under consideration covers different areas including energy, transport, agriculture, industry, finance, trade, and land-use, thus creating a complex web of environmental obligations for various entities engaged in diverse activities.

Specifically, the European Climate Law (Regulation (EU) 2021/1119) enshrines the vision of Europe becoming a continent with zero-net emission in 2050 and sets a mandatory target of achieving at least 55 percent net reductions in emissions of greenhouse gases as compared with 1990. The document obligates all the EU bodies and member states to implement policies and measures to reach the objectives set, while establishing a European Scientific Advisory Board on Climate Change to oversee the process. The revised Emissions Trading System (EU ETS), Carbon Border Adjustment Mechanism (CBAM), the Renewable Energy Directive, and Effort Sharing Regulation represent another set of obligations.

Corporate Sustainability Due Diligence 'Directive' (CS3D)

The introduction of the Corporate Sustainability Due Diligence Directive (CS3D) in April 2024 by the European Parliament is the most significant event in the evolution of EU's environmental jurisprudence on the responsibility of corporations to date. The Directive obliges the large companies that conduct business inside the EU territory to perform due diligence with respect to both the human rights and the environment. This means that such businesses are supposed to prevent, address and remediate any adverse effects related to the human rights and the environmental degradation. The concept of due diligence becomes the standard of assessing whether a company complies with its responsibility.

According to the CS3D, the duty of care implies that a business is expected to carry out due diligence procedures if it has more than 1,000 employees as well as the worldwide annual net sales of at least EUR 450 million. For the businesses outside the EU area that have considerable sales inside the European Union, the same criteria should be met. A corporation is obliged to incorporate such practices into its policy, to recognize any adverse effects, take measures, establish complaint handling system, monitor and report on its practices. Thereby, the directors of such enterprises must consider the sustainability issues related to their decisions (European Commission, 2022).

ESG Disclosure and the Corporate Sustainability Reporting Directive

The Corporate Sustainability Reporting Directive (CSRD) has been in effect since January 2023 and applies to large corporations from the reporting year 2024. It signifies the most significant enhancement in terms of non-financial disclosure requirements within the European Union. In comparison to the Non-Financial Reporting Directive, the CSRD introduces a much more robust framework where disclosure must be made on a wide range of sustainability topics, according to the European Sustainability Reporting Standards (ESRS) created by EFRAG. Most importantly, the CSRD framework requires that sustainability disclosure is subject to independent audit, effectively transforming sustainability communication from an optional narrative practice into an audit procedure. The ESRS framework necessitates disclosure on environmental aspects, such as mitigation and adaptation against climate change, pollution, water and sea resources, biodiversity and ecosystems, and resource use and circular economy. ESRS employs the approach of double materiality, which considers the materiality of sustainability factors in terms of their impact on the financial position of the organization but also considers the organization's impact on humans and the environment. Double materiality is a major development conceptually, expanding the scope of environmental responsibility of a business far beyond the traditional concept of financial materiality.

EU Taxonomy and Sustainable Finance

Under the EU Taxonomy Regulation (Regulation (EU) 2020/852), there is a system for categorizing sustainable economic activities in such a way that it indicates what activities can be considered environmentally sustainable. Under the CSRD, firms have to report the proportion of activities they undertake which fit into the taxonomy. On the other hand, financial firms have to disclose the percentage of their portfolio holdings that are linked to Taxonomy activities. The Taxonomy creates robust incentives in the marketplace for improving environmental performance through the provision of funds for activities meeting strict environmental criteria such as the Do-No Significant Harm criterion and minimum social safeguards.

Together with the SFDR, Low Carbon Benchmarks Regulation, and the draft Green Bond Standard, the Taxonomy contributes to creating an essential part of the EU sustainable finance regime. Through this, private capital will be diverted towards sustainable activities through the creation of standardized and reliable information on environmental performance. The EU sustainable finance regime is arguably one of the best worldwide and is currently influencing sustainable finance regimes in other countries, including the UK, Japan, and, increasingly, the US (Achterfeldt, S., & Traverso, M. 2026).

INTERNATIONAL CLIMATE LITIGATION AND TRANSNATIONAL ACCOUNTABILITY

Climate Litigation Landscape

There have been increasing numbers of climate lawsuits in the last decade, with climate litigation expanding geographically and in its sheer numbers. According to the Sabin Center for Climate Change Law at Columbia University, there have been 2,500+ climate-related cases worldwide, an increase from more than two-fold increase from 2015 numbers (Setzer & Higham, 2023). The cases include various legal bases, including lawsuits filed against fossil fuel companies, environmental permit applications, constitutional actions filed against national governments, climate disclosures and greenwashing, and lawsuits before international courts/tribunals. The main categories under which climate litigation takes place are as follows. Human rights-based climate litigation argues that inadequate action on the climate crisis by the government contravenes the human or constitutional right of individuals to the protection of their lives and livelihoods, health, dignity, and the right to a healthy environment. Urgenda Foundation v. State of the Netherlands (2019) is a seminal case in this regard, where the Dutch Supreme Court upheld a ruling ordering the government of the Netherlands to reduce greenhouse emissions by a minimum of 25% of the 1990 levels by 2020. The Dutch state had breached its obligation of care as per the European human rights laws, which included the right to life and the right to privacy/family life.

Landmark Cases and Doctrinal Developments

It should be noted that the decision of the German Federal Constitutional Court in Neubauer et al. v. Germany (2021) is highly significant from the perspective of the development of human rights-based approaches to climate accountability. In this case, the Federal Constitutional Court considered that the target of reducing emissions set by the Federal Climate Action Act by 2030 was too modest and violated the basic rights of future generations, because it imposed an excessively high burden on reducing emissions in the post-2030 period. It is reasonable to view the doctrine of intertemporal freedom, i.e., the idea that current climate inaction limits the freedom of future generations, as the main innovative aspect of this decision. Milieudefensie et al. v. Royal Dutch Shell (2021) another milestone case decided by the Hague District Court – demonstrates how climate accountability duties have been expanded from states to companies based on principles of private law. Namely, Shell Company has been required to reduce its CO2 emissions by 45 percent by 2030 compared with 2019, including all Scope 3 emissions from the usage of its products by customers. The relevant obligation of the company has been derived from the unwritten standard of care in torts under Article 6:162 of the Dutch Civil Code, supplemented with human rights considerations, such as Articles 2 and 8 of the European Convention on Human Rights and the UN Guiding Principles on Business and Human Rights.

Another interesting case, where the concept of climate-related duties of care was introduced by the courts, is Sharma v. Minister for the Environment (2021) (Australia). In particular, this litigation has resulted in defining a duty of care for the Environment Minister to the Australian youth not to cause personal injuries to them associated with approving a coal mine extension project. However, while the Full Federal Court ruled against this claim, the case itself significantly contributed to discussion regarding governmental duties of care in climate-related cases (Preston, 2021). For example, a similar approach to the analysis of climate duties of care was employed in Billy v. Australia (2022) at the United Nations Human Rights Committee level.

Advisory Opinion Proceedings

One of the latest trends in international climate litigation is the start of advisory opinion procedure before several international judicial bodies. Thus, in May 2024, the ITLOS delivered an advisory opinion upon the request made by the Commission of Small Island States on Climate Change and International Law on whether greenhouse gas emissions can be considered a type of pollution regulated by the United Nations Convention on the Law of the Sea and what obligations exist under the UNCLOS for states concerning prevention, reduction, and control of pollution. An important consequence of this advisory opinion will be the identification of legal aspects of climate change regulation in the framework of existing treaty relations.

The ICJ is currently drafting an advisory opinion regarding the obligations of states regarding climate change issues at the request of the UN General Assembly, which was adopted in March 2023 through an initiative of Vanuatu. It can be expected that this opinion will be among the most significant decisions in the area of climate law because it will contain considerations about the obligations of states to protect the climate system under international law, consequences of non-compliance with such obligations by states, and consequences for states, which acts and omission have caused substantial damage to small island developing states and other particularly affected regions.

Corporate Climate Liability

The developing theory of corporate liability for climate change is based on the integration of tort law, human rights law, securities law, and consumer protection law from different jurisdictions. Scientific attribution has become increasingly sophisticated at identifying the extent to which specific fossil fuel companies contribute to greenhouse gas levels in the Earth's atmosphere and their consequences, providing the evidentiary basis for establishing a causal link in tort litigation. The Carbon Disclosure Project's Carbon Majors Database shows that only 100 corporations have been responsible for about 71% of industrial greenhouse gases emitted into the atmosphere since 1988.

In the USA, several states and municipalities have instituted tort proceedings against fossil fuel corporations by claiming public and private nuisance, strict product liability, negligence, and consumer protection. Among the plaintiffs are the city of New York, the state of Hawaii, the state of Rhode Island, and others. Although these actions were met with a number of procedural difficulties, including the claim of federal pre-emption and remittal of certain claims to federal courts, some managed to pass these barriers. In the case of BP plc. v. Mayor and City Council of Baltimore decided in 2021, the Supreme Court of the USA made some determinations regarding procedure.

COMPARATIVE LEGAL ANALYSIS

Convergence Toward Enforceable Obligations

A comparative analysis of the frameworks for environmental accountability in India, the European Union, and international climate litigation highlights a trend of convergence towards the imposition of legal obligations despite the significant legal, institutional, and economic differences that exist between the three spheres. In each of the three domains, legal developments follow a trajectory of moving from voluntarism to obligations, from declaration to enforcement, and from aspirations to accountability. This is not likely to be a mere coincidence, since the ecological emergency has reached such a level that voluntary legal governance has proven insufficient and legal actors have shown their capacity to utilize established legal doctrines to incorporate environmental concerns into them.

There is a set of common doctrinal trends among the three domains. The association of environmental harm with human rights such as rights to life, health, and dignity serves as the fundamental doctrine in constitutional law in India, human rights court decisions in the EU, and climate litigation at international tribunals. What used to be the polluter pays principle enshrined in various soft law provisions in environmental declarations has come to be included in constitutional laws in India, directive laws in the EU, and increasingly in international tribunal decisions. Similarly, what was once a soft law doctrine has now become hard law in all three jurisdiction.

Enforcement Divergences and Institutional Gaps

Despite the similarities in the normative direction, there is a marked difference between the three legal systems considered regarding enforcement capacity, institutional mechanisms, and efficiency. Firstly, it should be noted that the enforcement mechanisms in the European Union are fairly sophisticated and advanced. Specifically, the enforcement mechanisms include both supranational legislation as such and the obligations of member states, as well as Commission infringement procedures and the competence of the European Court of Justice. Legally binding directives and regulations in the EU, as well as the principle of supremacy, ensure an extremely efficient enforcement mechanism that does not exist in international environmental law.

Secondly, despite having made considerable strides forward, which was mostly the result of judicial creativity, there is much room for improvement when it comes to enforcement mechanisms in India. In other words, despite having a fairly advanced judicial stance towards environmental issues in India, the actual enforcement practices are far from ideal. Regulatory agencies in India are chronically underfunded and politically influenced; the use of criminal sanctions is sporadic and slow; most importantly, there are often no adequate monitoring and compliance mechanisms that would allow translating the judicial decision-making into action. Nevertheless, the NGT has had a positive impact on India's enforcement practices, although some problems persist.

Thirdly, there are several enforcement challenges in international climate litigation because, by definition, international climate litigation uses national mechanisms. Namely, the use of national law and national courts is inherent to international climate litigation, and therefore international mechanisms are rather limited in this sphere. Notwithstanding the successful cases such as Urgenda, there still remains a question as to how effectively such decisions will be implemented due to the need for governments to make the required efforts. Advisory opinions in international law may have substantial normative force; however, they cannot be regarded as legally binding decisions.

Role of Civil Society and Strategic Litigation

Among the various similar characteristics of the three examples, one is that they all involve the intervention and participation of civil society actors, the NGOs in environmental matters, and the local communities in the pursuit of environmental accountability using litigation. In the case of India, the role of organizations such as the Centre for Science and Environment and environmental law experts is crucial because they were able to initiate ground breaking litigation against the environment through PIL. In Europe, environmental groups such as Urgenda, Greenpeace, and Client Earth have been known to lead some of the most significant environmental lawsuits against corporations and states regarding climate accountability. Such cases have also included international cases involving small island states, the indigenous peoples in the Pacific Islands, and the youth climate activists. The increased involvement of civil society actors in the use of litigation as an approach to environmental matters results from the fact that there is no longer an elite monopoly on the use of the courts, which means that now people do not sue just for the sake of doing it. Some of the factors behind strategic litigation include the fact that the requirement for standing in the process of bringing about litigation has been relaxed.

EMERGING CONCEPTS IN ENVIRONMENTAL ACCOUNTABILITY

Ecocide as an International Crime

The initiative to define and declare ecocide as the fifth crime under the Rome Statute has become quite popular in recent years. A group of experts convened by Stop Ecocide International defined ecocide as an unlawful act committed "with knowledge that there is a substantial likelihood of severe and either widespread or long-term damage to the environment being caused" in 2021. The proposed definition was adopted by several nations that advocated for beginning the process of amending the Rome Statute with respect to the new crime – namely, Vanuatu, Fiji, Samoa, and Belgium. If ecocide is recognized as one of the international crimes, it will signify an important shift in the relationship between international criminal law and the law protecting the environment. According to the Rome Statute and Geneva Conventions, any environmental destruction during the period of an armed conflict is considered a war crime. However, there is no legal remedy provided for the mass destruction of the environment in peacetime, including acts committed by corporations that damage ecosystems significantly. As such, recognizing ecocide under the Rome Statute would mean imposing individual criminal responsibility on the corporation's heads responsible for authorizing destructive actions against the environment. This would serve as a major deterrent against damaging the environment by corporations since no economic or legal penalty can prevent their actions (Higgins et al., 2013).

Rights of Nature

The doctrine of rights of nature stands out as an extremely important and philosophical challenge to the anthropocentric basis on which environmental laws are built. According to the doctrine of rights of nature, nature, ecosystems, or living things have the same rights as humans, and these rights may be enforced independently of the interests of people. It was for the first time in 2008 that the Ecuadorian constitution recognized the right of nature (Pachamama) to exist, persist, regenerate its life, and maintain its vital processes. In 2017, the Whanganui River in New Zealand became the first body of water in the country to receive legal personhood status via the Te Awa Tupua act. From a legal perspective, recognizing nature as having rights is highly important for addressing the issue of accountability in relation to the environment because it means that we introduce a new category of interest that should be protected under any circumstances and may not be sacrificed for the benefit of others. If nature has the right to flow, ecosystems have the right to exist, forests have the right not to be destroyed, it follows that our actions as human beings should be significantly restrained. Rights of nature as a concept is relatively new, and it faces a number of legal challenges. Nevertheless, the emergence of the doctrine in several constitutions around the globe, including India, Colombia, and Uganda, indicates an important shift in legal philosophy (Kauffman & Martin, 2021).

Intergenerational Equity and Climate Justice

This idea of intergenerational equity, namely the responsibility to ensure that the ability of the natural environment to satisfy future needs is not destroyed, has strong roots within international environmental law since its definition by the Brundtland Commission in 1987. Its evolution into a binding principle of environmental law was propelled by climate litigation, especially in the light of intertemporal freedom used by the German Constitutional Court in Neubauer to define the obligations towards future generations. The term climate justice can be defined as the connection between distributive features of climate change namely, the uneven effects of global warming affecting people least responsible for CO2 emissions and liability for those who have contributed more significantly to the problem. The fact that climate change has disproportionate effects on women, children, indigenous peoples, and developing countries was acknowledged by numerous international institutions such as UN Human Rights Council, Committee on the Rights of the Child, and Committee on the Elimination of Discrimination against Women. Thus, climate justice can be applied not only within the context of environmental law but discrimination laws as well.

RECOMMENDATIONS FOR STRENGTHENING GLOBAL ENVIRONMENTAL ACCOUNTABILITY

Binding International Environmental Obligations

The primary lesson one can draw from this comparative analysis is that the creation of enforceable international environmental duties along with measures to guarantee their fulfillment becomes urgent. The current structure of international environmental law, relying on NDC-based voluntarism, transparency, and soft-law approaches, does not have the institutional capacity necessary to induce sufficient ambition and actions in terms of the ecological crisis. The creation of an international accountability regime for climate based on its capacity to monitor, report, and verify would provide the institutional underpinning for efficient governance in global environmental issues.

The advisory opinion procedures before the International Court of Justice, International Tribunal for the Law of the Sea, and Inter-American Court of Human Rights become an opportunity to elaborate on the existing international legal climate accountability regime. It would be wise for states to act constructively regarding these procedures and fulfill their obligations resulting from them. The recognition of obligations not only to cut emissions but also to make sure that climate policy corresponds with the temperature goals of the Paris Agreement, as well as the right of communities adversely affected to bring a challenge for such an approach, would signify hardening the international climate law.

Harmonised Corporate Environmental Due Diligence

The European Union Corporate Sustainability Due Diligence Directive provides an example of compulsory environmental responsibility of corporations that should be replicated elsewhere. The application of the directive to businesses not established within the borders of the EU, but doing their business in the EU market, creates pressure to improve standards elsewhere. An international effort to implement corporate due diligence in the environmental field, which could be achieved within the UN framework of business and human rights or by adopting a new multilateral instrument, would help address the 'race to the bottom' in environmental standards in the face of corporate globalization.

Compulsory corporate due diligence needs to go hand-in-hand with liability systems that allow communities affected by corporate activities to seek compensation for the resulting harm. The existing obstacles to bringing cross-border environmental actions, such as forum non conveniens, conflict of laws, and difficulties with recognition and enforcement of foreign judgments, need to be overcome by means of a multilateral treaty designed to streamline the process of adjudicating environmental claims.

Institutional Strengthening and Judicial Capacity

The success of any environmental accountability regime relies on the ability of the enforcing agencies, their expertise, and their capacity. In the Indian case, improving the capacities of the National Green Tribunal, ensuring proper funding for the pollution control boards, and maintaining the independence of environmental regulators from government intervention become crucial prerequisites for turning the judicial pronouncements of the Supreme Court in the field of environment into actual environmental achievements. It can thus be concluded that although normative legal frameworks play a pivotal role, they cannot achieve anything without the needed institutional structures behind them. On the international level, strengthening the mandate of the United Nations Environment Programme, creating an International Court for the Environment, and improving the monitoring and compliance procedures of multilateral environmental agreements could help enhance global environmental governance. The proposal of the creation of an International Environment Agency, similar to the International Energy Agency, but with regulatory competencies, has found its proponents in the academic community and deserves further consideration at the international stage.

Recognition of Ecocide and Rights of Nature

The inclusion of "ecocide" in the list of international crimes through formal legal recognition within the context of the Rome Statute will lead to a new paradigm of accountability in cases of massive environmental damage, placing the responsibility on individuals in the most egregious cases. Countries should assist with the existing procedure in order to consider this amendment to the Rome Statute in the Assembly of States Parties.

It is necessary that the gradual acceptance of rights of nature within domestic legal regimes be encouraged and facilitated through international normative structures that recognize the inherent value of ecosystems and biodiversity apart from any instrumental value that they may have for humanity. The formulation of international legal standards for the protection of vital ecosystems such as oceans, rainforests, wetlands, coral reefs, and glaciers as common heritage of mankind with legal personality apart from state sovereignty may form an approach to environmental responsibility beyond the state-centered international legal system.

CONCLUSION

The trajectory of environmental accountability in this essay has been mapped out from the soft law voluntariness seen at the Stockholm and Rio conferences to the hard law framework represented in India's constitutional laws, European Union environmental governance, and climate litigation at the international level. It is evident that, based on the examples provided above, international environmental law is currently going through a structural shift, whereby environmental protection is treated as a legal obligation rather than a voluntary action. Constitutional courts working within the frameworks of human rights and flexible procedural methods, including public interest litigation, can drive significant progress in environmental responsibility despite weak legal and regulatory frameworks. Absolute liability of hazardous activities, the concept of the right to a clean environment as an aspect of the right to life, and international soft-law principles adopted as constitutional norms represent the important achievements in developing environmental law.

The experience of the EU shows that sophisticated environmental regulation is feasible when implemented by region-wide authorities with sufficient powers to ensure the compliance of all member states with applicable legislation. The framework consisting of the European Climate Law's binding targets, the EU emissions trading system (ETS), carbon border adjustment mechanism (CBAM), corporate sustainability reporting directive (CSRD), CSRD 3.0 Directive, and taxonomy represent a highly developed example of environmental governance which has no parallels elsewhere. The willingness of EU to hold non-member corporations accountable for their environmental impact due to access to EU's markets resulted in the creation of de facto global standards of corporate environmental accountability.

Climate change cases represent an emerging trend in international environmental governance involving enforcement of commitments assumed in international environmental agreements using innovative application of constitutional rights, tort laws, and human rights treaties. Landmark decisions in Urgenda case, Neubauer v. Germany, Milieudefensie v. Shell, alongside with the process of issuing advisory opinions from the international courts, will have a profound effect on the way states and corporations operate. Comparison of three case studies shows significant differences in terms of institutions, resources available for enforcement, and the actual results but reveals a common normative commitment towards environmental accountability. The emergence of doctrines of ecocide, rights of nature, and intergenerational equity reflects the new perspective of the relationship between humanity and nature that transcends the anthropocentric approach to environmental governance. The transition of environmental governance from soft law to hard law and liability does not concern only technicalities of legal systems. Rather, it reflects a fundamental social and moral shift concerning the understanding of humanity's position in the natural world. Ecological destruction becomes recognised as legally cognisable damage against which governments bear legal duty to protect future generations. Corporations also acquire new responsibilities for ensuring the absence of environmental damages along the entire supply chain. Furthermore, the ecosystem becomes an interested party which deserves protection through legal measures.

REFERENCES

Achterfeldt, S., & Traverso, M. (2026). Classification of the EU taxonomy for sustainable activities among other sustainable finance taxonomies worldwide. *Clean Technologies and Environmental Policy, 28*(4), 88. Advance online publication. DOI: 10.1007/s10098-025-03384-6

Bodansky, D. (2016). The Paris Climate Change Agreement: A new hope? *The American Journal of International Law, 110*(2), 288–319. DOI: 10.5305/amerjintelaw.110.2.0288

Crawford, J. (2019). *Brownlie's principles of public international law* (9th ed.). Oxford University Press. DOI: 10.1093/he/9780198737445.001.0001

European Commission. (2022). *Proposal for a directive on corporate sustainability due diligence* (COM/2022/71 final). Publications Office of the European Union.

Gill, G. N. (2016). Environmental justice in India: The National Green Tribunal and expert members. *Transnational Environmental Law, 5*(1), 175–205. DOI: 10.1017/S2047102515000278

Higgins, P., Short, D., & South, N. (2013). Protecting the planet: A proposal for a law of ecocide. *Crime, Law, and Social Change, 59*(3), 251–266. DOI: 10.1007/s10611-013-9413-6

Intergovernmental Panel on Climate Change. (2023). *Synthesis report of the IPCC sixth assessment report (AR6)*. IPCC. https://www.ipcc.ch/report/ar6/syr/

International Court of Justice. (1997). *Case concerning the Gabcikovo-Nagymaros Project (Hungary v. Slovakia), judgment of 25 September 1997. ICJ Reports, 1997*, 7.

International Tribunal for the Law of the Sea. (2024). *Advisory opinion on climate change and international law obligations under the United Nations Convention on the Law of the Sea* (Case No. 31).

Johannsen, K. (2025). An empty victory? Shell v. Milieudefensie et al. 2024, the legal obligations of carbon majors, and the prospects for future climate litigation action. *Review of European, Comparative & International Environmental Law, 34*(1).

Kauffman, C. M., & Martin, P. L. (2021). *The rights of nature: A legal revolution that could save the world.* Yale University Press.

Kotzé, L. J. (2019). *Global environmental constitutionalism in the Anthropocene.* Hart Publishing.

Kyoto Protocol to the United Nations Framework Convention on Climate Change, December 11, 1997, 2303 U.N.T.S. 148.

M.C. Mehta v. Union of India, AIR 1987 SC 1086 (India).

Milieudefensie et al. v. Royal Dutch Shell plc., (2021). Case No. C/09/571932/HA ZA 19-379 (District Court of The Hague, 2021).

Ministry of Corporate Affairs. (2013). *The Companies Act, 2013* (Act No. 18 of 2013). Government of India.

Neubauer et al. v. Germany, (2021). 1 BvR 2656/18, Federal Constitutional Court of Germany, BVerfGE 157, 30

Paris Agreement to the United Nations Framework Convention on Climate Change, December 12, 2015, T.I.A.S. No. 16-1104.

Preston, B. J. (2021). The influence of the Paris Agreement on climate litigation: Causation, corporate governance, and costs. *Journal of Environmental Law*, *33*(1), 1–30. DOI: 10.1093/jel/eqaa020

Rajamani, L., & Ghosh, S. (2020). India. In Rajamani, L., & Peel, J. (Eds.), *The Oxford handbook of international environmental law* (2nd ed., pp. 977–994). Oxford University Press.

Ram Mohan, M. P., Reynaers Kini, E., & Prasad, S. (2024). A worthy roadmap for global climate change litigation. *Georgetown Journal of International Law*, •••, 55.

Regulation (EU) 2020/852 of the European Parliament and of the Council of 18 June 2020 on the establishment of a framework to facilitate sustainable investment. (2020). *Official Journal of the European Union, L 198*, 13–43.

Regulation (EU) 2021/1119 of the European Parliament and of the Council of 30 June 2021 establishing the framework for achieving climate neutrality. (2021). *Official Journal of the European Union, L 243*, 1–17.

Sands, P., Peel, J., Fabra, A., & MacKenzie, R. (2018). *Principles of international environmental law* (4th ed.). Cambridge University Press. DOI: 10.1017/9781108355728

Setzer, J., & Higham, C. (2023). *Global trends in climate change litigation: 2023 snapshot*. Grantham Research Institute on Climate Change and the Environment, London School of Economics and Political Science. https://www.lse.ac.uk/granthaminstitute/wp-content/uploads/2023/06/Global_trends_in_climate_change_litigation_2023_snapshot.pdf

Sharma v. Minister for the Environment, [2021] FCA 560 (Federal Court of Australia).

Stop Ecocide International. (2021). *Independent expert panel for the legal definition of ecocide: Commentary and core text*. Stop Ecocide Foundation. https://www.stopecocide.earth/legal-definition

Subhash Kumar v. State of Bihar, AIR 1991 SC 420 (India).

United Nations. (1992). *Rio Declaration on Environment and Development* (U.N. Doc. A/CONF.151/26, Vol. I).

United Nations Conference on the Human Environment. (1972). *Declaration of the United Nations Conference on the Human Environment (Stockholm Declaration)* (U.N. Doc. A/CONF.48/14/Rev.1).

United Nations Framework Convention on Climate Change, May 9, 1992, 1771 U.N.T.S. 107.

United Nations Human Rights Committee. (2022). *Billy et al. v. Australia: Views adopted by the Committee under Article 5(4) of the Optional Protocol* (CCPR/C/135/D/3624/2019).

Urgenda Foundation v. State of the Netherlands, ECLI:NL:HR:2019:2006 (Supreme Court of the Netherlands, 2019).

Vellore Citizens Welfare Forum v. Union of India, AIR 1996 SC 2715 (India).

Vidhi Centre for Legal Policy. (2024). *Holding the state accountable*. https://vidhilegalpolicy.in/

World Commission on Environment and Development. (1987). *Our common future*. Oxford University Press.

KEY TERMS AND DEFINITIONS

Climate Litigation: Lawsuits filed in either domestic, regional, or international courts in order to hold states and companies accountable for meeting their climate change obligations and reducing carbon emissions.

Corporate Environmental Liability: The legal liability placed on companies that cause environmental harms, which includes duties such as conducting environmental due diligence and reporting sustainability information.

Ecocide: A term being increasingly used in discussions about whether acts of extensive or serious environmental destruction should be treated as an international crime under international law.

Environmental Accountability: The means by which states, companies, and other entities are legally bound to prevent, minimize, and redress any damage to the environment.

Environmental Due Diligence: A process that involves companies in identifying and assessing, among other actions, any actual or possible environmental impacts associated with their activities.

Chapter 7
Recasting Responsibility at Sea:
Legal Accountability for Maritime Climate Harm in the Bay of Bengal

C. A. M. Al Wahi
http://orcid.org/0009-0009-7813-8820
Bangladesh University of Professionals (BUP), Bangladesh

ABSTRACT

This chapter examines the evolution of legal accountability for maritime climate harm and marine pollution in the Bay of Bengal. Drawing on doctrinal analysis of the UNCLOS, IMO frameworks, recent advisory opinions of the ITLOS and the ICJ, landmark domestic and regional climate litigation and emerging mandatory corporate sustainability obligations, the chapter argues that a coherent transition from soft law flexibility toward enforceable legal responsibility is now underway. Focusing on Bangladesh as a primary case study, the chapter analyzes progressive hardening of due diligence obligations for flag, coastal and port states through judicial innovation and rights-based litigation and the extension of enforceable environmental obligations to shipping corporations through mandatory due diligence laws. The chapter proposes a three-tier accountability model linking international adjudication, domestic constitutional litigation and corporate due diligence engagement and identifies strategic pathways for transforming persistent soft law commitments into enforceable legal responsibility.

DOI: 10.4018/979-8-2600-3678-5.ch007

INTRODUCTION

International shipping contributes approximately 2.89% of global greenhouse gas (GHG) emissions, a proportion projected to increase significantly absent decisive regulatory intervention (Tanhuanpää et al., 2025). Ship-source pollution encompassing oil discharges, ballast water transfers, sulphur emissions and plastic waste continues to degrade marine ecosystems at a rate outpacing the adaptive capacity of vulnerable coastal communities; the marine plastic burden in particular is now understood to be substantially trade-driven and inadequately addressed by prevailing legal and policy frameworks (Zhang et al., 2025b). Yet the legal architecture governing these harms has historically been characterized by what Birnie et al. (2009) aptly described as cooperative flexibility: a preference for voluntary technical standards, flag-state discretion and aspirational targets that has produced incremental environmental improvement while forestalling enforceable accountability.

The Bay of Bengal exemplifies this governance failure. Bordered by Bangladesh, India, Myanmar, Sri Lanka, Thailand and the Maldives and traversed by some of the world's most intensively navigated shipping lanes, the Bay combines strategic commercial significance with acute climate vulnerability. Bangladesh possesses over 700 kilometers of coastline encompassing the Sundarbans, the world's largest contiguous mangrove forest and a critically important carbon sink, and densely populated deltaic lowlands where tens of millions face existential threats from sea-level rise, intensifying cyclonic activity, saltwater intrusion and coastal erosion (IPCC, 2022). The ship-breaking industries concentrated around Chittagong, together with pollution profiles of the Chittagong and Mongla ports and the volume of transit traffic on major trade corridors, create overlapping sources of maritime environmental harm that existing legal frameworks have struggled to address with adequate specificity or rigour.

Figure 1. The Bay of Bengal and Its Littoral States

Source: Author

International law has not remained static. The United Nations Convention on the Law of the Sea (UNCLOS, 1982) provides the overarching constitutional framework for ocean governance, while the International Maritime Organization (IMO), through the International Convention for the Prevention of Pollution from Ships (MARPOL 73/78) and the 2023 Revised GHG Strategy (IMO, 2023), has progressively tightened technical standards. Most significantly, the International Tribunal for the Law of the Sea (ITLOS) delivered a landmark Advisory Opinion in May 2024 at the request of the Commission of Small Island States on Climate Change and International Law (COSIS), confirming that UNCLOS Articles 192 and 194 impose specific, enforceable due diligence obligations on states to prevent, reduce and control marine pollution from anthropogenic GHG emissions (ITLOS, 2024). The International Court of Justice (ICJ) reinforced this doctrinal trajectory in its Advisory Opinion on the Obligations of States in Respect of Climate Change (2025), confirming that climate inaction may constitute a breach of general international law.

Three research questions guide the analysis. First, to what extent do existing UNCLOS and IMO obligations require states to prevent climate-related and pollution harm from maritime activities significantly affecting vulnerable coastal states? Second, how are ITLOS and other judicial bodies interpreting these obligations in ways that harden previously soft environmental commitments at sea? Third, what pathways exist for Bay of Bengal states, particularly Bangladesh to leverage these developments through domestic courts, regional mechanisms, corporate accountability regimes and future international proceedings? The chapter proceeds as follows: section two maps the treaty framework and governance gaps; section three examines transformative judicial developments; section four surveys human rights and climate litigation; section five analyzes mandatory corporate due diligence; section six proposes a multi-level accountability model and the conclusion synthesizes findings and identifies future research priorities.

METHODOLOGY

This chapter adopts a doctrinal legal research methodology supplemented by comparative legal and institutional analysis. The doctrinal component proceeds through systematic analysis of primary legal materials, including UNCLOS, IMO conventions and guidelines, advisory opinions and judgements of ITLOS and the ICJ, EU due diligence legislation and Bangladesh's constitutional and statutory framework, to identify the content and trajectory of states' and corporations' maritime environmental obligations. The comparative component examines analogous regional governance frameworks (the Barcelona and Nairobi Conventions), comparable rights-based climate litigation across jurisdictions, and the institutional design of EU corporate due diligence regimes in order to assess which features of those frameworks are transferable to the Bay of Bengal context and which are contingent on institutional conditions that the region does not currently share.

Case and instrument selection followed three criteria. First, doctrinal relevance: only instruments and decisions that directly elaborate the content of due diligence, the no-harm rule or corporate environmental responsibility were included, rather than every instrument touching incidentally on maritime or environmental matters. Second, temporal significance: priority was given to developments since 2009, beginning with the Seabed Disputes Chamber's due diligence jurisprudence that has materially altered the legal standard governing flag, coastal and port state conduct, on the view that earlier materials are adequately synthesized in the existing treatise literature cited throughout. Third, extraterritorial reach: corporate due diligence instruments and litigation precedents were selected specifically for their capacity to create legal exposure for actors domiciled outside the Bay of Bengal whose conduct nonetheless generates harm within it, since this is the central accountability gap the chapter addresses.

The multi-level accountability model proposed in Section 6 is the product of comparative institutional analysis informed by legal feasibility assessment: for each of the three tiers, it asks what a Bangladeshi or Bay of Bengal state actor could plausibly invoke under existing law and institutional arrangements, rather than proposing new legal instruments or testing the model against quantitative cost-benefit criteria. The latter exercise is identified as a distinct and necessary further step in the Future Research Directions discussed in Section 6.

LITERATURE REVIEW

The chapter draws on four bodies of scholarship that are typically treated separately in the existing literature but are synthesised here to support its central argument. *Maritime environmental governance.* The treatise literature on the law of the sea (Rothwell & Stephens, 2016; Tanaka, 2023) and international environmental law (Birnie et al., 2009; Sands et al., 2018) establishes the baseline account of flag-state primacy, port-state control and the soft-law character of IMO standard setting on which Section 2's analysis builds. Harrison's (2017) characterization of port state control as a 'safety net' for an underlying flag-state system, rather than an independent regulatory regime, is adopted as the organizing premise for the governance gap analysis that follows.

Climate litigation and advisory opinions. Peel and Osofsky's (2015) account of the "regulatory mismatch" between cumulative, diffuse climate causation and the bilateral architecture of international dispute settlement, together with Mayer's (2018) treatment of the international law of climate change, frames this chapter's assessment of the practical limits of advisory opinions discussed in Section 3. Boyle's (2012) account of the "human-rightsization" of environmental law underpins the translation, in Section 4, of rights-based climate litigation to the maritime context.

Corporate due diligence and supply chains. The UN Guiding Principles on Business and Human Rights (Ruggie, 2011) supply the normative foundation for the corporate accountability turn examined in Section 5. The existing literature on mandatory due diligence legislation has not, however, been extended specifically to shipping and ship recycling, which is the gap Section 5's analysis seeks to address.

Bay of Bengal and Bangladesh maritime vulnerability. Existing scholarship on Bangladesh's coastal and maritime governance has documented the country's exposure to sea-level rise and cyclonic intensification (IPCC, 2022) and the persistent regulatory and occupational hazards of the Chittagong ship-breaking industry (NGO Shipbreaking Platform, 2026). It has not, however, connected these documented vulnerabilities to the post-2009 hardening of international due diligence obligations or to the mandatory corporate due diligence regimes examined in Section 5, a gap this chapter directly addresses. Notably, the existing literature has also generally overlooked the Bangladesh Supreme Court's own engagement with ship-breaking regulation: in *BELA v. Bangladesh (2009)*, the High Court Division ordered the closure of all Chittagong ship-breaking yards operating without the environmental clearance required under domestic law and in 2016 issued a contempt rule against both the government and yard owners for continued non-compliance with that order (NGO Shipbreaking Platform, 2024). This domestic precedent, discussed further in Section 4, supplies a directly relevant and previously uncited foundation for the public interest litigation pathway this chapter proposes and its absence from the existing literature on Bay of Bengal maritime governance is itself indicative of the regional gap this chapter seeks to close.

SOFT LAW AT SEA: UNCLOS, IMO FRAMEWORKS AND THE LIMITS OF DISCRETION

The international legal architecture governing maritime environmental protection embodies a constitutional paradox. UNCLOS Part XII (Arts. 192-237) constitutes the most comprehensive treaty codification of states marine environmental obligations yet achieved. Article 192 establishes a foundational duty to protect and preserve the marine environment. Article 194 requires states to take "all measures necessary to prevent, reduce and control pollution of the marine environment from any source," qualified by reference to states capabilities. Article 194(3)(b) specifically addresses vessel-source pollution; Article 194(3)(c) extends the obligation to atmospheric pollution, increasingly significant as the climatic consequences of ship-source GHG emissions become progressively better understood. Yet UNCLOS delegates the substantive elaboration of these obligations primarily to the IMO and Article 123's discretionary language that states in enclosed and semi-enclosed seas 'should endeavour' to coordinate reflects the broader pattern of establishing frameworks for cooperation without requiring substantive action. The result is a governance architecture that is formally ambitious but operationally permissive: one that delineates the parameters of legitimate environmental conduct without providing mechanisms ensuring those parameters are respected by the states and corporations most responsible for maritime climate harm.

The most consequential structural feature of the UNCLOS framework is its assignment of primary regulatory jurisdiction over vessels to the flag state. Under Articles 91-94, flag states exercise administrative, technical and social jurisdiction over registered vessels (Article 217). This requires flag states to ensure compliance with applicable international pollution standards. In practice, this system has generated open registries of so-called "flags of convenience" in which shipowners register vessels in states offering minimal regulatory oversight. The result is a systematic severing of the connection between formal legal obligation and enforcement capacity (Rothwell & Stephens, 2016). Vessels transiting the Bay of Bengal may be registered in Panama, Liberia or the Marshall Islands, states collectively accounting for a large share of the world fleet tonnage while exercising limited supervisory control over their fleets environmental conduct. A coastal state whose waters these vessels pollute has constrained authority under UNCLOS to impose its own standards on foreign-flagged vessels: within the exclusive economic zone (EEZ), a coastal state's power to regulate vessel-source pollution is confined to laws conforming to applicable international rules (Art. 211(5)), themselves set by the IMO. The coastal state thus becomes doubly dependent on the IMO to set adequate standards and on flag states to enforce them.

Port state control, the right of port states to inspect and detain non-compliant vessels under UNCLOS Articles 218-220 provides a partial corrective. The Indian Ocean Memorandum of Understanding on Port State Control (Indian Ocean MOU), to which Bangladesh, India, Sri Lanka and other Bay of Bengal states are parties, establishes a regional coordination mechanism. The scale of this mechanism, while administratively significant, remains modest relative to the volume of traffic transiting the region: the Indian Ocean MoU's 2024 Annual Report recorded 5,334 inspections across its seventeen member authorities, resulting in 225 detentions an overall detention rate of 4.22% (Indian Ocean MoU, 2025). Bulk carriers, the vessel type most heavily represented on Bay of Bengal commodity routes, accounted for the largest share of both inspections (2,700) and detentions (107), while fire-safety deficiencies, rather than environmental or GHG-related deficiencies, constituted the single largest category of recorded non-compliance, at 15.26% of all deficiencies recorded (Indian Ocean MoU, 2025). This distribution illustrates a structural limitation independent of inspection volume: existing port state control practice in the region remains calibrated overwhelmingly toward safety rather than environmental compliance, a gap the MARPOL Annex VI inspections proposed in Section 6 are specifically designed to close. However, its effectiveness is constrained by inconsistent inspection rates, limited technical capacity for MARPOL Annex VI assessments and the absence of sufficiently dissuasive sanctions. As Harrison (2017) observes, port state control is best understood as a safety net for the flag state system rather than an independent regulatory regime whose catching capacity diminishes precisely where harms are most diffuse and technically complex, as in the case of GHG emissions.

MARPOL 73/78 constitutes the primary multilateral instrument for regulating vessel-source pollution and through Annex VI, air pollution and GHG emissions from shipping. Through amendments adopted in 2011 and 2021, Annex VI introduced the Energy Efficiency Design Index (EEDI) for new ships; the Ship Energy Efficiency Management Plan (SEEMP) entering into force on January 1, 2023, the Energy Efficiency Existing Ship Index (EEXI) and the Carbon Intensity Indicator (CII), requiring ships to calculate and report annual carbon intensity subject to flag state review. These technical efficiency measures represent genuine regulatory progress. Yet a critical structural weakness runs through Annex VI's design: it addresses the carbon intensity of individual vessels without imposing any absolute cap on the total volume of sectoral GHG emissions. A fleet of individually compliant vessels may nonetheless generate increasing total GHG output proportional to trade growth. The CII's corrective action mechanism further illustrates MARPOL's soft-law character; persistent underperformance attracts primarily reputational rather than legal penal consequences.

The IMO's 2023 Revised GHG Strategy (MEPC.377(80)) adopted a net-zero GHG emissions target by or around 2050, with indicative checkpoints of 20-30% reduction by 2030 and 70-80% by 2040 relative to 2008 baselines (IMO, 2023). Notwithstanding its rhetorical ambition, the Strategy's legal character is that of a non-binding policy document. Its implementation depends on subsequent binding MARPOL amendments subject to consensus-based IMO governance dynamics. Mid-term market-based measures, including a proposed GHG levy, remain under active negotiation. As Sands et al. (2018) observe in a related context, this pattern of 'regulatory displacement' substitutes soft-law ambition for binding obligation: flag states pointing to the Strategy's existence as evidence of meaningful climate governance face no cognizable legal sanction when their fleets significantly exceed indicative benchmarks.

The governance gaps assume particular consequences in the Bay of Bengal. Bangladesh has ratified UNCLOS and core MARPOL annexes, but its domestic regulatory framework for vessel-source atmospheric pollution and GHG emissions remains underdeveloped relative to the scale of maritime activity in its coastal waters. The ship-breaking industry concentrated around Chittagong presents a further dimension: the Hong Kong International Convention for the Safe and Environmentally Sound Recycling of Ships (2009) only entered into force on 26 June 2025 (IMO, 2025) and in the interim, hazardous materials including asbestos, polychlorinated biphenyls and heavy fuel residues were routinely released into the intertidal zone of Chittagong's ship-breaking beaches, with documented ecological consequences that persist even as compliance improves. The NGO Shipbreaking Platform's 2025 data record eleven worker deaths and at least 62 serious injuries across South Asian yards in that year alone, alongside 321 vessels dismantled globally, 214 of which were broken down on three South Asian beaches, accounting for 85% of the gross tonnage scrapped worldwide (NGO Shipbreaking Platform, 2026). This is maritime environmental harm traceable to specific corporate actors whose end-of-life vessel disposal decisions generate foreseeable damage, yet for which the soft-law architecture has provided no direct liability mechanism. Understanding this structural governance gap is a prerequisite for analyzing how judicial innovation and new regulatory instruments might begin to close it.

JUDICIAL CATALYSTS: ITLOS, ICJ AND THE REINTERPRETATION OF MARITIME ENVIRONMENTAL OBLIGATIONS

The limitations of the soft-law framework have not gone judicially uncontested. Over the past two decades, international judicial bodies have subjected maritime environmental obligations to increasingly rigorous interpretive scrutiny, with this momentum accelerating in 2024 and 2025, substantially altering the landscape within which those obligations operate without rewriting the underlying treaties.

Doctrinal Antecedents

Cases predating contemporary climate concern articulated the foundational precepts governing transboundary environmental harm. The *Corfu Channel Case (ICJ, 1949)* established the no-harm rule, which states that states must not knowingly allow their territory to be used for acts contrary to the rights of other states. The Trail Smelter Arbitration *(United States v. Canada, 1941)* elaborated this rule in explicitly environmental terms: no state has the right to use or permit the use of its territory causing injury to another through pollution. The ICJ's 2010 judgement in Pulp Mills on the River Uruguay (*Argentina v. Uruguay*) elevated the obligation to conduct environmental impact assessments (EIAs) to a customary international law norm for activities with potential transboundary effects (ICJ, 2010, para. 204) with direct implications for flag states licensing new vessel classes and coastal states expanding port capacity in environmentally sensitive zones such as the Sundarbans.

The most doctrinally influential precursor to the 2024 ITLOS Advisory Opinion is the Seabed Disputes Chamber's Advisory Opinion on Responsibilities and Obligations of States Sponsoring Activities in the Area (Case No. 17, 2011). The Chamber characterized the relevant obligation not as one of guaranteeing no environmental harm but as one of conduct: to deploy adequate means and exercise the best possible efforts to prevent damage (ITLOS, 2011, para. 110). Critically, the Chamber confirmed that the content of due diligence is not fixed: "what counts as a reasonable measure of diligence may vary in the light of new knowledge" (ITLOS, 2011, para. 117). As the scientific evidence linking ship-source GHG emissions to marine environmental harm has strengthened through successive IPCC Assessment Reports, the required standard of diligence has correspondingly intensified. A flag state that could in 2005 claim genuine uncertainty about its fleet's climate implications cannot credibly advance the same argument in 2025 in the face of the IPCC Sixth Assessment Report (IPCC, 2022).

The COSIS Advisory Opinion (ITLOS, 2024): Hardening the Maritime Climate Obligation

The full ITLOS bench delivered the Advisory Opinion on May 21, 2024, in response to the COSIS request, marking the most significant development in international maritime environmental law in the contemporary period. The Tribunal confirmed that anthropogenic GHG emissions constitute 'pollution of the marine environment' within the definition provided by UNCLOS Article 1(1)(4), bringing them squarely within the mandatory obligations of Articles 192 and 194 (ITLOS, 2024, para. 175). The significance of this characterization cannot be overstated: it forecloses the argument that the IMO's technical standards framework constitutes an exclusive and exhaustive climate accountability regime for shipping.

The Tribunal expressly rejected the argument that compliance with IMO technical standards suffices to discharge the Article 194 due diligence obligation, holding instead that the adequacy of measures adopted must be assessed against the best available science (ITLOS, 2024, paras. 222-234). It confirmed that states must take all necessary measures to reduce GHG emissions from all sources within their jurisdiction or control including, critically, vessels operating under their flag and that the precautionary approach applies, meaning the absence of complete scientific certainty regarding harm in a specific locality does not relieve states of preventive obligations (ITLOS, 2024, para. 245). The Tribunal further confirmed that the cooperative obligation under Article 197, requiring active engagement in international standard setting, implies that systematic obstruction of adequate standard setting within IMO negotiations may itself constitute a failure of treaty obligations.

The ICJ Advisory Opinion on Climate Change (2025): Customary Law and Human Rights

On July 23, 2025, the ICJ delivered its Advisory Opinion on the Obligations of States in Respect of Climate Change, following a request by the United Nations General Assembly through Resolution 77/276. The Court confirmed three central holdings: that the customary international law no-harm rule requires states to take all necessary measures to prevent significant harm to other states from GHG emissions; that UNFCCC and Paris Agreement obligations including nationally determined contributions constitute legally binding procedural and substantive commitments and climate harm directly and foreseeably affects the rights to life, health and an adequate standard of living of affected populations, expressly connecting climate inaction to human rights obligations (ICJ, 2025, paras. 197, 388-412). This human rights connection opens pathways for the rights-based litigation strategies analyzed in section four.

Table 1. Judicial Developments in Maritime Environmental Obligation and Their Relevance to the Bay of Bengal

Case / Opinion	Year	Tribunal	Key Holding	Relevance to Bay of Bengal
Corfu Channel (UK v. Albania)	1949	ICJ	States must not knowingly allow their territory to be used for acts that violate the rights of other states (no-harm rule).	Foundational authority for coastal state claims against flag states whose vessels cause transboundary maritime harm.
Pulp Mills on the River Uruguay (Argentina v. Uruguay)	2010	ICJ	Obligation to conduct environmental impact assessments for activities with potential transboundary harm is a customary international law norm.	Supports the argument that flag and coastal states must conduct EIAs for major shipping operations and port developments in the Bay of Bengal.
Responsibilities and Obligations of States (Seabed Advisory Opinion, Case No. 17)	2011	ITLOS Seabed Disputes Chamber	Due diligence is an obligation of conduct, not results; its content evolves with scientific understanding.	Establishes that the due diligence standard applicable to maritime environmental obligations strengthens as scientific knowledge of GHG harm advances.
COSIS Advisory Opinion (Case No. 31)	2024	ITLOS (Full Tribunal)	GHG emissions constitute "pollution of the marine environment" under UNCLOS Art. 1(1)(4). States bear specific, science-aligned due diligence obligations under Arts. 192 and 194. IMO-standard compliance does not discharge the due diligence obligation.	Most significant development: directly frames ship-source GHG emissions as actionable marine pollution, hardening obligations on flag states operating in and through the Bay of Bengal.
Advisory Opinion on Climate Change Obligations	2025	ICJ	States bear obligations under customary international law to prevent, mitigate and remedy climate harm. The no-harm rule and human rights obligations apply to climate-related state conduct.	Confirms that climate inaction may constitute internationally wrongful conduct reinforcing claims by Bay of Bengal coastal states affected by shipping-related GHG emissions.

Note. GHG = greenhouse gas; EEZ = exclusive economic zone; ITLOS = International Tribunal for the Law of the Sea; ICJ = International Court of Justice. Source: Compiled by the author from primary judicial sources cited in the text.

Critical Assessment: The Gap Between Interpretive Innovation and Practical Accountability

A candid assessment must acknowledge significant limitations alongside genuine achievements. Advisory opinions are not formally binding on states not party to adjudicated disputes; they carry the authority of legal reasoning and moral weight without directly creating enforceable obligations. The diffuse, cumulative character of climate causation resulting from contributions by thousands of actors across extended timescales complicates the identification of specific respondent states whose conduct can be particularized as legal causes of harm to specific claimant states. Peel and Osofsky (2015) characterize the situation as a 'regulatory mismatch' between cumulative climate causation and the bilateral, state-centric architecture of international dispute settlement, an observation that applies with particular force in the maritime domain. These limitations contextualize rather than diminish the significance of the 2024-2025 judicial developments traced above, which have together created a materially more demanding interpretive environment. States maintaining permissive flag-state regimes now face growing reputational, political and prospective legal pressure. For Bay of Bengal states pursuing accountability through multiple simultaneous channels, this shift in the interpretive environment is a meaningful and exploitable resource.

HUMAN RIGHTS, CLIMATE LITIGATION AND THE RIGHT TO A HEALTHY MARINE ENVIRONMENT

International environmental law and international human rights law have historically developed along separate jurisprudential tracks. The past decade has witnessed their convergence through what Boyle (2012) terms the "human-rightsization" of environmental law, a structural shift generating important accountability opportunities for communities in vulnerable coastal regions such as the Bay of Bengal.

The right to a clean, healthy and sustainable environment has evolved from the precursor statement of the Stockholm Declaration of 1972, through the Human Rights Council's Resolution 48/13 (2021), to formal recognition as a universal human right by the United Nations General Assembly through Resolution 76/300 of July 28, 2022 (UNGA, 2022). Bangladesh's constitutional framework incorporates environmental protection through Article 18A, inserted in 2011, obliging the state to protect and improve the environment and preserve natural resources. This constitutionalizing trend is regional in character: Pakistan followed a comparable path through the Twenty-sixth Amendment, whose new Article 9A entrenches a fundamental right to a clean and healthy environment measured against international environmental norms (Haider, 2025). Where such rights are asserted, courts across the region are increasingly required to give them practical effect through environmental public-interest litigation, an exercise that demands careful judicial delineation of standing and of the genuine object of a suit, as comparative environmental civil public-interest litigation illustrates (Zhang & Haider, 2025). The ICJ's 2025 Advisory Opinion reinforced the connection between this right and climate obligations, holding that climate-induced marine environmental degradation directly affects the rights to life, health and an adequate standard of living (ICJ, 2025, paras. 388-412).

The contemporary climate litigation landscape has generated a body of precedent directly applicable to maritime environmental harm. *Urgenda Foundation v. State of the Netherlands* (Dutch Supreme Court, 2019) upheld a mandatory order requiring the state to reduce national GHG emissions by at least 25% by 2020 relative to 1990 levels, finding that European Convention on Human Rights (ECHR) Articles 2 and 8 rights to life and private life impose positive obligations requiring effective preventive measures against foreseeable climate risks. The Court accepted IPCC reports as a sufficient evidentiary foundation for harm foreseeability, removing the conventional tortious requirement of direct individual causation proof. *Milieudefensie et al. v. Royal Dutch Shell plc* (District Court of The Hague, 2021) ordered Shell to reduce global CO_2 emissions by 45% by 2030, applying the Dutch duty of care informed by international human rights standards and the Paris Agreement, extending enforceable climate obligations to a corporation's entire value chain, including product use emissions, with direct relevance to major shipping companies whose commercial activities generate significant GHG emissions. The European Court of Human Rights' Grand Chamber judgment in *Verein KlimaSeniorinnen Schweiz and Others v. Switzerland (2024)* found Switzerland in violation of Article 8 ECHR for systemic failures in its climate action legal framework, establishing for the first time that a regional human rights court could find a state in violation of the Convention for inadequate climate policies. Within South Asia, the Supreme Court of India's verdict in *M.K. Ranjitsinh v. Union of India* extended this rights-based reasoning to the tension between biodiversity conservation and renewable-energy expansion, articulating a right against the adverse effects of climate change with direct resonance for the Bay of Bengal's climate-vulnerable littoral (Zhang et al., 2025a).

First, Bangladesh's constitutional framework provides analogous grounds to Urgenda, Article 18A, read together with the right to life under Article 32 as elaborated through decades of public interest litigation (PIL) in the High Court Division, supports constitutional challenges to the state's failure to enforce MARPOL Annex VI standards or develop an adequate EEZ GHG regulatory framework. The Bangladesh Supreme Court has affirmed in environmental PIL proceedings that Article 32 encompasses the right to a healthy environment, following the jurisprudential path established by the Supreme Court of India under Articles 21 and 48A. The measurable contribution of such environmental judicial activism to environmental-protection outcomes across South Asia has itself been the subject of recent empirical analysis (Ahmad et al., 2025). This pathway is not merely doctrinally available in the abstract: it has direct precedent in the maritime domain. In *BELA v. Bangladesh (2009)*, brought by the Bangladesh Environmental Lawyers Association, the High Court Division ordered the closure of all Chittagong ship-breaking yards operating without the environmental clearance required under domestic law and in 2016 issued a contempt rule against both the government and yard owners for continued non-compliance with that order (NGO Shipbreaking Platform, 2024). A PIL petition challenging the Merchant Shipping Directorate's MARPOL enforcement adequacy would therefore extend an established line of domestic maritime environmental litigation rather than break new procedural ground. Second, the Milieudefensie principle of corporate duty of care extending to value chain GHG emissions applies to shipping companies domiciled in duty of care jurisdictions and operating carbon-intensive vessels through the Bay of Bengal. Climate attribution science (IPCC, 2022) and fuel consumption records from the MARPOL Annex VI Data Collection System provide the evidentiary foundation for proportional corporate liability claims of the type advanced in *Lliuya v. RWE AG*, in which the Higher Regional Court of Hamm, while ultimately dismissing the plaintiff's claim in May 2025 for want of sufficient flood risk to his specific property, nonetheless confirmed that major emitters can in principle be held liable under German civil law for climate-related harm occurring outside Germany, in proportion to their share of cumulative global emissions (Oberlandesgericht Hamm, 2025). The decision illustrates both the doctrinal opening such claims now enjoy and the substantial evidentiary burden, particularly as to localised causation, that claimants must still discharge. Bangladesh's own common-law tort jurisprudence supplies a parallel domestic avenue: the application of strict- and civil-liability principles to environmentally harmful conduct has recently been examined in the closely analogous South Asian setting (Ullah et al., 2026), furnishing a doctrinal basis for domestic claims against actors such as ship-breaking yard operators whose activities cause foreseeable coastal environmental harm. Third, the KlimaSeniorinnen framework for systemic regulatory inadequacy translates into a potential individual communications claim before the UN Human Rights Committee under the Optional Protocol to the International Covenant on Civil and Political Rights (ICCPR), to which Bangladesh is a party. General Comment No. 36 (2019) explicitly recognizes environmental degradation and climate change as threats to the right to life, triggering positive state obligations under Article 6 (HRC, 2019, para. 62).

REGIONAL MECHANISMS AND COMPARATIVE GOVERNANCE APPROACHES

The rights-based litigation strategies available to Bangladeshi coastal communities do not operate in institutional isolation. They intersect with a broader landscape of regional environmental governance whose limitations illuminate, by contrast, the relative accessibility and analytical promise of litigation centered approaches. A comparative review of three regional frameworks the Barcelona Convention system for the Mediterranean, the Nairobi Convention for the Western Indian Ocean, and the emerging Pacific regional climate architecture clarifies both the structural conditions under which regional environmental accountability has succeeded and the obstacles that Bay of Bengal states must overcome to replicate those successes.

The Barcelona Convention for the Protection of the Marine Environment and the Coastal Region of the Mediterranean (1995 consolidated text), together with its seven protocols, constitutes the most institutionally developed regional marine environmental treaty system outside the Arctic. Its Offshore Protocol, Hazardous Wastes Protocol and Land-Based Sources Protocol collectively address the range of pollution sources vessel-source, land-based and atmospheric that generate harm across the Bay of Bengal's semi-enclosed geography. The Barcelona system's relative effectiveness reflects two structural advantages absent in the Bay of Bengal context: the European Union's direct legislative competence over marine environmental standards applicable to EU Mediterranean member states effectively imports binding supranational law into the regional framework and a common institutional infrastructure, the United Nations Environment Programme's Mediterranean Action Plan, providing technical secretariat support, monitoring capacity, and compliance facilitation to a relatively homogeneous group of states sharing significant regulatory capacity. Neither advantage is replicated in the Bay of Bengal, where the institutional architecture of the South Asian Association for Regional Cooperation (SAARC) has not been mobilized for maritime environmental purposes and the extreme disparity in regulatory capacity between India and the smaller Bay of Bengal littoral states precludes the assumption of common technical baselines.

The Nairobi Convention for the Protection, Management and Development of the Marine and Coastal Environment of the Western Indian Ocean (1985) offers a closer and more instructive comparator. Functioning under the same UNEP Regional Seas Programme framework, the Nairobi Convention encompasses states at comparable levels of institutional development to Bangladesh and Myanmar and confronts analogous challenges of limited enforcement capacity, high dependence on blue-economy sectors, and acute coastal climate vulnerability. Yet the Convention's implementation record has been characterized by chronic underfunding, inconsistent national reporting, and the absence of binding compliance mechanisms that might transform its substantive environmental obligations into enforceable accountability. The Nairobi Convention experience confirms what the analysis of the Bay of Bengal framework has already established through the UNCLOS context: that a well-designed regional treaty framework without binding enforcement architecture and adequate financial resourcing produces cooperative statements rather than environmental accountability. For Bay of Bengal states contemplating regional framework negotiation, the Nairobi experience counsels against the assumption that treaty adoption is itself sufficient; the design of compliance mechanisms, financial arrangements and technical support structures is equally determinative of the framework's accountability value.

The Commission of Small Island States (COSIS), whose ITLOS request generated the transformative 2024 Advisory Opinion, models a third and analytically distinctive regional approach: one that substitutes strategic international legal advocacy for the construction of elaborate regional institutional machinery. By aggregating the legal standing and diplomatic resources of climate-vulnerable small island states into a focused litigation strategy, COSIS achieved a doctrinal outcome, the COSIS Advisory Opinion, that no individual member state could have produced acting alone. Its success invites replication. A Bay of Bengal Climate Accountability Coalition bringing together Bangladesh, the Maldives, Sri Lanka and potentially Myanmar under a coordinating secretariat supported by academic institutions and Northern NGO legal partners could, through joint submission to MEPC, coordinated Article 283 UNCLOS diplomatic exchanges, and shared PIL litigation frameworks, replicate the essential features of the COSIS model without requiring the negotiation of a comprehensive regional treaty. This coalition approach is politically more attainable in the near term than full regional treaty negotiation and legally more promising than unilateral state action in circumstances where individual states leverage over major flag states and shipping corporations is structurally limited.

CORPORATE ACCOUNTABILITY AND MARITIME SUPPLY CHAINS

The accountability deficit shared by inadequate IMO frameworks and state-directed litigation reflects a common underlying cause: the failure of existing legal mechanisms to reach the commercial actors whose business decisions most directly determine shipping's environmental performance. Over the past decade, that gap has begun to narrow through the emergence of mandatory corporate due diligence and ESG disclosure obligations, extending the reach of legal accountability from the state level to the corporate level.

The normative foundation is the UN Guiding Principles on Business and Human Rights (UNGPs), endorsed by the Human Rights Council in 2011 (Ruggie, 2011). The UNGPs articulate the corporate responsibility to respect human rights through human rights due diligence across operations, subsidiaries and value-chain relationships, including the environmental consequences of chartering decisions, fuel procurement choices, flag registry selection and end-of-life vessel disposal arrangements. Their limitation is the absence of domestic legal enforceability, a gap addressed by France and the European Union through binding mandatory due diligence legislation.

France's Law No. 2017-399 on the Duty of Vigilance (loi de vigilance), enacted in March 2017, became the first mandatory due diligence instrument, requiring large companies to establish, publish and implement a plan de vigilance identifying and mitigating human rights and environmental risks across their full global value chain. Failure to establish an adequate plan exposes companies to civil liability, injunctive relief and damages before French courts, with standing granted to any interested person, including foreign-affected communities represented by French civil society organizations. This standing provision is of direct relevance to Bangladesh's coastal communities: through French partner NGOs, they could participate in *loi de vigilance* proceedings challenging the adequacy of a French shipping company's environmental risk mitigation measures in the Bay of Bengal. The first significant environmental judicial test arose in *Notre Affaire à Tous et al. v. Total SE* (Paris Court of Appeal, 2023), establishing the template for analogous claims against shipping companies with inadequate GHG and marine pollution risk management.

The EU's Corporate Sustainability Due Diligence Directive (CS3D, Directive 2024/1760), adopted in May 2024, represented at the time of its enactment the most significant advance in mandatory corporate environmental due diligence law. As originally adopted, it required large EU companies and large non-EU companies with substantial EU revenues to integrate human rights and environmental due diligence across their full value chains, with an environmental scope covering GHG emissions "severely harmful to the environment or human health," biodiversity loss and hazardous substance use; shipping companies meeting the original thresholds of over 1,000 employees and €450 million in global net turnover were required to adopt due diligence policies, conduct risk assessments, take preventive and remedial measures, and maintain accessible complaints procedures. Article 29, as enacted in 2024, explicitly granted affected persons in third countries standing to bring civil liability claims before member state courts for damages caused by due diligence failures.

This position has since changed materially. The EU's *"Omnibus I"* simplification package, finalized through Directive (EU) 2026/470 published in the Official Journal on 26 February 2026 and in force from 18 March 2026, substantially narrowed the CS3D in three respects directly relevant to this chapter's argument. First, the applicable thresholds were raised to companies with more than 5,000 employees and €1.5 billion in worldwide net turnover (or, for non-EU companies, €1.5 billion in EU turnover with no employee threshold), removing a significant proportion of mid-sized shipping operators from the Directive's scope. Second, due diligence obligations were narrowed from the full value chain to direct, "Tier 1" business partners, extending to more remote links in the chain of activities only where there is plausible information of adverse impacts there. Third, and most significantly for the accountability pathway this chapter examines, the harmonized EU-wide civil liability regime was deleted in its entirety, including the third-country standing provision originally codified at Article 29. A Bangladeshi fishing community whose livelihoods are documented to have been damaged by marine pollution attributable to a shipping company's inadequate environmental risk management can therefore no longer rely on a dedicated CS3D standing provision to bring a claim in the member state where the company is domiciled; any remaining avenue lies in the general tort law of the relevant member state, where it survives, rather than in a harmonized EU-level cause of action. Section 6 further addresses this narrowing by recalibrating the corporate accountability tier of the proposed model accordingly. Companies remaining within scope must comply with the amended Directive following national transposition by 26 July 2028, with application from 26 July 2029.

The EU's Corporate Sustainability Reporting Directive (CSRD, Directive 2022/2464), as amended by the same Omnibus I package, complements the due diligence regime through mandatory sustainability disclosure under the European Sustainability Reporting Standards (ESRS), including ESRS E1 (GHG emissions and climate transition plans), ESRS E2 (marine pollution impacts) and ESRS E4 (biodiversity impacts in sensitive areas), directly applicable to Bay of Bengal transit operations near the Sundarbans, a UNESCO World Heritage Site. The Omnibus amendments raised the CSRD's own applicability threshold to companies with more than 1,000 employees and €450 million in net turnover, applying from financial years beginning on or after 1 January 2027, and thus had the practical effect of removing a substantial share of previously reporting companies from its scope as well.

Figure 2. Note. From Sundarbans, Bangladesh',by European Space Agency

(https://www.esa.int/ESA_Multimedia/Images/2016/07/Sundarbans_web). Contains modified Copernicus Sentinel data (2016), processed by ESA.

Market-based voluntary instruments provide complementary accountability pressure independent of flag-state enforcement. The Poseidon Principles (2019), adopted by ship finance banks representing approximately 50% of the global ship finance portfolio, require annual public disclosure of portfolio climate alignment against the IMO decarbonization pathway, creating financial incentives for fleet decarbonization as a condition of competitive loan pricing. The Sea Cargo Charter (2020), adopted by major commodity charterers, commits signatories to assess and disclose the climate alignment of chartered vessels, generating demand-side decarbonization pressure. These instruments create indirect but commercially meaningful decarbonization incentives for Bay of Bengal trading vessels, which are disproportionately older, carbon-intensive bulkers and tankers on regional commodity routes.

The Chittagong ship-breaking industry crystallises the corporate accountability gap in tangible form. South Korea and the United Arab Emirates topped the NGO

Shipbreaking Platform's 2025 list of flag and ownership jurisdictions selling vessels for South Asian beaching, with twenty-one, nineteen and seventeen vessels, respectively, dismantled on South Asian beaches that year (NGO Shipbreaking Platform, 2026). European shipping companies selling end-of-life vessels destined for beaching at Chittagong, where hazardous materials are routinely released into intertidal zones with documented ecological and human health consequences, may, on the reasoning of Milieudefensie, face residual civil liability claims under the general tort law of their home state for foreseeable environmental damage at recycling facilities, although the CS3D's own harmonized liability mechanism no longer applies following the 2026 Omnibus amendments discussed above, and the Directive's narrowed Tier 1 due diligence scope means that a selling shipowner's obligations toward its immediate counterparties, such as a cash buyer or recycling broker, are now more clearly engaged than its obligations toward the ultimate recycling facility itself, absent plausible information on adverse impacts there. NGO Shipbreaking Platform's (2024) detailed documentation of environmental conditions at Chittagong, and its annually updated dismantling data (NGO Shipbreaking Platform, 2026), provide the evidentiary foundation for demonstrating harm foreseeability and for assessing the adequacy of a selling shipowner's due diligence measures, notwithstanding the narrower legal basis on which any resulting claim must now proceed.

Three structural limitations of the corporate accountability turn require acknowledgement. The jurisdiction gap means mandatory due diligence obligations currently reach primarily companies domiciled in EU member states or with significant EU market exposure, leaving major actors domiciled in South Korea, Japan, and other states beyond effective reach, collectively operating a large proportion of vessels trading in the Bay of Bengal. The enforcement gap means that affected Bangladeshi communities must assemble resources, legal expertise, and partnerships with Northern civil society to initiate litigation in European courts, which structurally disadvantages the communities that are most urgently in need of accountability. The standard gap means due diligence obligations are procedural as much as substantive, requiring adequate risk management processes rather than specific environmental performance achievements. These limitations reinforce rather than undermine the case for the integrated multi-level approach proposed in section six.

TOWARD A MULTI-LEVEL MODEL OF MARITIME ENVIRONMENTAL ACCOUNTABILITY IN THE BAY OF BENGAL

No single accountability tier is adequate to the scale and urgency of the problem. The combined and coordinated deployment of international, domestic and corporate accountability mechanisms through a deliberately designed multi-level architecture produces accountability pressure qualitatively greater than the sum of its parts. States, corporations and international judicial bodies that are simultaneously confronted with international adjudication, domestic constitutional litigation, regional regulatory pressure and corporate due diligence claims in home-state courts face a risk environment that is materially more challenging to navigate through inaction than the soft-law framework has historically permitted.

The multi-level model is organized around three tiers: Tier 1 (international), directed at flag states through international adjudication, IMO treaty advocacy and regional framework agreements; Tier 2 (domestic and regional), directed at coastal states through constitutional litigation, legislative reform and enhanced port state control and Tier 3 (corporate), directed at shipping companies and commercial counterparties through mandatory EU due diligence obligations, market-based ESG instruments and civil society litigation. Three design principles govern the architecture. First, consider complementarity: each tier addresses accountability gaps that the other tiers cannot reach. International adjudication binds flag states but cannot easily reach corporate actors. Domestic litigation addresses coastal state regulatory failure but does not cover foreign flag states. Corporate due diligence engages shipping enterprises but is constrained by jurisdictional thresholds. Second, strategic sequencing: short-term domestic and corporate actions generate the evidentiary record, institutional precedents and political momentum necessary for more demanding long-term international strategies. Third, calibration to capacity: the model is operationally realistic for Bangladesh's institutional situation, identifying international alliances with COSIS member states, Northern environmental NGOs and academic institutions capable of providing technical and advocacy support.

Tier 1: International Accountability

For international adjudication, Bangladesh acting alone or in coalition with Bay of Bengal coastal states could initiate UNCLOS Annex VII arbitration against major flag states whose fleets GHG emissions demonstrably harm the Bay's marine environment. The COSIS Opinion has materially advanced the legal foundations, confirming the enforceable due diligence obligation and removing the IMO-compliance defense. A staged approach is proposed: in the short term, formal Article 283 UNCLOS diplomatic exchanges with identified flag states notifying due diligence adequacy concerns under the COSIS Opinion, preserving Part XV procedural rights and creating a documentary record for future contentious proceedings. At the IMO MEPC, coordinated Bay of Bengal state advocacy for adoption of a binding GHG market-based measure specifically a GHG fuel levy and formal designation of the Bay as a Particularly Sensitive Sea Area (PSSA) or Emission Control Area (ECA) under MARPOL Annex VI represents the most immediately actionable Tier 1 element. Existing ECAs in the Baltic, North Sea and North American coastal zones demonstrate that regional emission control designations are achievable when coastal states present a coherent scientific and legal case to MEPC. Long-term, negotiation of a Bay of Bengal Framework Agreement on Maritime Environmental Standards establishing coordinated port state control, mandatory EIA requirements, shared monitoring obligations and minimum pollution standards would close the cooperative deficit in UNCLOS Article 123.

Tier 2: Domestic and Regional Accountability

Three parallel domestic tracks are available. A plausible legislative pathway would amend the Merchant Shipping Ordinance 1983 to incorporate current MARPOL Annex VI requirements, namely the EEXI, CII and Data Collection System provisions. A complementary step would be the enactment of a dedicated Maritime Climate and Environmental Protection Act, setting enforceable EEZ GHG emission standards by reference to the COSIS due diligence standard and the constitutional obligation under Article 18A. Such an Act would supply the statutory benchmark against which port state control enforcement and judicial review could subsequently operate, a benchmark that, as Section 2 noted, does not currently exist in Bangladeshi law. Constitutional litigation offers a second track: a PIL petition before the High Court Division, invoking Articles 18A and 32, could challenge the Merchant Shipping Directorate's MARPOL enforcement adequacy and the absence of an EEZ GHG regulatory framework, requesting a mandatory, time-bound regulatory action plan. As discussed in Section 4, such a petition would extend rather than initiate Bangladesh's tradition of maritime environmental PIL, building directly on the precedent established in *BELA v. Bangladesh (2009)*; a petition along these lines could integrate the COSIS Advisory Opinion as an interpretive aid to the constitutional environmental protection standard and document the disproportionate impact of maritime climate harm on economically marginalized coastal communities. The third track, enhanced port state control, would increase MARPOL Annex VI inspection rates with a specific focus on CII documentation and SEEMP compliance. The Indian Ocean MoU's existing inspection infrastructure, discussed in Section 2, provides the institutional starting point for this expansion, which could generate direct deterrence effects, compliance data usable in future adjudication, and improved regional standing in IMO negotiations. Capacity building support for Annex VI technical inspections is available through the IMO's Integrated Technical Cooperation Programme.

Tier 3: Corporate Accountability

The corporate tier operates primarily through European home-state legal systems and commercial relationships, accessible to Bangladesh's civil society through strategic Northern NGO partnerships, though its legal foundations are narrower than they were before the 2026 CS3D Omnibus amendments discussed in Section 5. Civil society engagement with corporate due diligence processes, through the surviving CS3D consultation mechanisms and the French loi de vigilance, which retains its own civil liability regime independently of the EU-level mechanism that was deleted, remain accessible to affected community organizations representing Sundarbans-dependent fishing families and Chittagong coastal residents and continue to generate evidentiary records of corporate risk-assessment adequacy that could support subsequent litigation. Strategic litigation in European courts, in partnership with Northern environmental law organizations such as ClientEarth, Urgenda Foundation or Notre Affaire à Tous, may now proceed primarily under the general tort law of the relevant member state rather than a harmonized CS3D cause of action, with outcomes accordingly more dependent on the particular national legal regime invoked, as the partial, fact-specific outcome in Lliuya v. RWE AG illustrates. For Chittagong ship-breaking, advocacy directed at Poseidon Principles signatory banks and Sea Cargo Charter members to incorporate end-of-life vessel disposal standards into portfolio alignment criteria remains a comparatively more durable accountability strategy, since it does not depend on the EU due diligence civil liability mechanism at all.

Table 2. Multi-Level Maritime Environmental Accountability Model: Reform Pathways for the Bay of Bengal

Tier	Mechanism	Specific Action	Horizon	Expected Outcome
Tier 1: International	UNCLOS Annex VII Arbitration	Initiate formal Art. 283 diplomatic exchanges with major flag states citing COSIS Opinion paras. 222–234; build toward contentious proceedings, identifying specific due diligence failures.	Medium term (5-10 years)	Binding award confirming enforceable flag-state due diligence obligations for ship-source GHG harm in the Bay of Bengal EEZ.
Tier 1: International	IMO MEPC Advocacy	Coordinated Bay of Bengal state bloc submission for adoption of a binding GHG market-based measure (levy) and formal designation of the Bay of Bengal as a Particularly Sensitive Sea Area or Emission Control Area under MARPOL Annex VI.	Short to medium term (3-7 years)	Binding IMO emission standards calibrated to ecological sensitivity; conversion of IMO Strategy aspirational targets into enforceable MARPOL obligations.
Tier 1: International	Regional Framework Treaty	Negotiate a Bay of Bengal Framework Agreement on Maritime Environmental Standards under UNCLOS Art. 123, establishing coordinated port state control, mandatory EIA, shared monitoring and minimum vessel-source pollution standards.	Long-term (10-15 years)	Binding regional instrument to close the cooperative deficit in UNCLOS Art. 123; a common environmental floor for all shipping in a semi-enclosed sea.
Tier 2: Domestic	Legislative Reform	Amend the Merchant Shipping Ordinance 1983 to incorporate MARPOL Annex VI (EEXI, CII, DCS) requirements; enact a dedicated Maritime Climate and Environmental Protection Act referencing the COSIS due diligence standard and Art. 18A of the Constitution.	Short term (1-4 years)	Enforceable domestic regulatory standard for port state control and constitutional judicial review; integration of ITLOS jurisprudence into domestic law.

continued on following page

Table 2. Continued

Tier	Mechanism	Specific Action	Horizon	Expected Outcome
Tier 2: Domestic	Constitutional PIL Litigation	PIL petition under Art. 102, invoking Arts. 18A and 32, challenging Merchant Shipping Directorate MARPOL enforcement adequacy and absence of an EEZ GHG regulatory framework; request mandatory time-bound action plan.	Short term (1-3 years)	Judicial recognition of constitutional dimensions of maritime climate regulatory failure; mandatory remedial orders; precedent for future rights-based maritime claims.
Tier 3: Corporate	CS3D and Loi de Vigilance Engagement and Litigation	Engage Bangladesh's coastal community organizations in plan de vigilance and surviving CS3D due diligence consultations of European shipping companies; support civil liability proceedings under the loi de vigilance or general member-state tort law in European home-state courts, with Northern NGO partners, against companies with inadequate Bay of Bengal environmental risk management.	Short to medium term (2-5 years)	Continued community access to due diligence consultation and grievance mechanisms; potential civil liability under the loi de vigilance or general tort law, since the EU-level CS3D civil liability mechanism was removed by the 2026 Omnibus amendments; outcomes remain fact-dependent and as yet untested in the shipping sector specifically.

Note. CS3D = Corporate Sustainability Due Diligence Directive; ECA = Emission Control Area; MARPOL = International Convention for the Prevention of Pollution from Ships; PIL = public interest litigation; PSSA = Particularly Sensitive Sea Area; EEXI = Energy Efficiency Existing Ship Index; CII = Carbon Intensity Indicator; DCS = Data Collection System. Source: Compiled by the author.

Inter-Tier Interactions and the Logic of Cumulative Accountability Pressure

The strength of the multi-level model lies not only in the individual accountability mechanisms it deploys but also in the interactions between tiers that generate cumulative pressure exceeding what any single tier achieves alone. Three primary interaction channels operate within the architecture. First, evidentiary reinforcement: domestic PIL proceedings generate judicial findings providing the factual foundation for international adjudication claims; port state control inspection records document patterns of MARPOL non-compliance by specific flag-state fleets; CS3D and CSRD disclosures generate admissible corporate environmental performance data deployable in civil liability proceedings and IMO MEPC submissions. Second, interpretive cross-fertilization: the COSIS Opinions confirmation of the science-aligned due diligence standard strengthens the constitutional environmental protection standard under Article 18A as interpreted by domestic courts; domestic judicial recognition of maritime dimensions of the right to a healthy environment reinforces the factual context within which ITLOS would assess the materiality of GHG harm; the CS3D's definition of adverse environmental impacts by reference to international environmental conventions, incorporating UNCLOS and MARPOL standards directly into corporate due diligence obligations. Third, political and reputational amplification: a flag state simultaneously subject to an Annex VII arbitration notice and a CS3D civil liability claim in its home-state courts faces a risk environment materially more costly to navigate through inaction than the current soft-law framework permits.

The model's effectiveness depends on conditions actively cultivated. Legal expertise across international environmental law, the law of the sea, European corporate law and Bangladesh's constitutional law must be assembled through academic partnerships and civil society coalition-building. The asymmetry of resources between Bangladesh and target flag states and shipping enterprises is a structural constraint requiring mitigation through climate finance mechanisms, particularly the Loss and Damage Fund established at COP27 and operationalized at COP28, which appropriately finances international maritime climate accountability litigation as a legitimate climate response measure. The legitimacy and sustainability of the model require that communities most affected by maritime climate harm, such as Sundarbans fishing families, Chittagong coastal residents, delta agricultural communities are genuinely integrated into its design and implementation rather than treated as passive beneficiaries of elite-level legal advocacy.

Counterarguments and Limitations

The multi-level model proposed above invites three substantial counterarguments that merit direct engagement rather than dismissal.

The first is a sovereignty-based objection. Flag states may reasonably argue that UNCLOS Articles 91-94 and 211(5) deliberately preserve flag-state primacy over vessel regulation as a negotiated sovereign bargain and that an advisory opinion, however doctrinally significant, does not amend that bargain. The COSIS Advisory Opinion interprets existing obligations; it does not create new ones, and a flag state can plausibly maintain that its discretion over technical implementation remains intact so long as it can point to a colourable compliance program, however modest. This objection has genuine force: the Tribunal's rejection of an automatic IMO-compliance defence narrows the room for this argument but does not eliminate it, since flag states retain wide discretion over what counts as an adequate measure short of full compliance with best available science. The model's reliance on Article 283 diplomatic exchanges and, ultimately, Annex VII arbitration is precisely a response to this limitation: it accepts that interpretive pressure alone cannot override flag-state discretion and instead seeks to convert that discretion into a litigable question of fact in a forum where the due diligence standard, once invoked, must be defended on the merits.

The second is a state capacity objection. Bangladesh's ability to pursue Annex VII arbitration, sustain a multi-year PIL petition or coordinate a Bay of Bengal Climate Accountability Coalition depends on legal, technical and diplomatic capacity that is not currently assured. Annex VII arbitration in particular requires sustained access to law of the sea expertise, scientific attribution evidence, and the financial resources to litigate against well-resourced flag states and shipping interests over a period of years. This is not a hypothetical concern: it is the central reason the chapter identifies climate finance mechanisms, particularly the Loss and Damage Fund, as a necessary precondition for Tier 1 strategies, rather than treating litigation capacity as already in place.

The third is an enforcement objection that applies even where Bangladesh prevails. An ITLOS advisory finding, a favourable Annex VII award, or a successful PIL judgment does not self-execute against a flag state with no assets, registered vessels, or institutional presence within Bangladeshi jurisdiction. The history of international environmental adjudication offers limited precedent for direct enforcement against a non-compliant flag state, and the model's heavier reliance on reputational, political and market-based pressure, through Tier 3 mechanisms such as the Poseidon Principles and Sea Cargo Charter, reflects an implicit acknowledgment that formal legal victory and practical compliance are not the same thing. The model's claim is accordingly more modest than a guarantee of enforcement: it is that the cumulative reputational, financial and legal exposure generated across three tiers simultaneously raises the cost of continued non-compliance, even where no single tier can compel it outright.

These limitations do not undermine the case for pursuing the proposed pathways; they do, however, counsel against describing any individual pathway as more than presently accessible in principle, with outcomes that remain genuinely contingent on capacity, jurisdiction and political feasibility, a qualification maintained throughout the discussion above and revisited in the Future Research Directions below.

Future Research Directions

This chapter proposes a multi-level accountability model that is operationally actionable within current legal and institutional parameters. It identifies pathways that are presently accessible in principle; however, as the preceding discussion of counterarguments and limitations makes clear, their practical availability depends on jurisdiction, standing, and political feasibility. Nonetheless, the model's long-term effectiveness will depend on empirical, doctrinal and technical research that the existing literature has not yet adequately addressed. Four priority research directions emerge from the analysis.

The first is an empirical assessment of the COSIS Advisory Opinion's behavioral impact on flag state regulatory practice. Advisory opinions change how states view their legal obligations, but whether they lead to clear changes in domestic regulatory conduct like adopting more ambitious MARPOL Annex VI implementing legislation, increasing MEPC engagement for binding mid-term measures or speeding up voluntary fleet decarbonization is an empirical question that needs systematic comparative analysis across major flag registry jurisdictions from 2024 to 2030. Without this evidence base, the claim that advisory opinions harden soft-law commitments remains a doctrinal proposition rather than a demonstrated accountability mechanism.

The second is the development of climate attribution methodologies specific to Bay of Bengal shipping routes and their integration into litigation evidentiary frameworks. General IPCC attribution science establishes the causal link between sectoral GHG emissions and regional climate harms at a level of confidence adequate for legislative and regulatory purposes. It does not yet provide the vessel-class-specific, route-specific, and temporally disaggregated attribution data that would support proportional corporate liability claims of the type advanced in *Lliuya v. RWE* in the maritime context. Collaboration between climate attribution scientists, maritime environmental economists, and international law scholars to develop methodologically robust, court-admissible attribution frameworks for ship-source emissions would be a high-priority enabler of the Tier 3 litigation strategy.

The third is comparative analysis of domestic PIL jurisprudence in maritime environmental cases across South and Southeast Asian jurisdictions. While Bangladesh's PIL tradition is substantively promising, its application to maritime contexts is largely untested. Systematic analysis of how Indian, Sri Lankan and Thai courts have engaged with maritime environmental claims brought through PIL or analogous public interest mechanisms would identify transferable procedural innovations, common evidentiary challenges and judicial attitudes toward international law as an interpretive resource in domestic constitutional adjudication.

The fourth is economic modeling of the multi-level accountability model's implementation costs and environmental benefits for Bangladesh. Strategic decisions about resource allocation across tiers, the relative investment in domestic PIL litigation versus MEPC advocacy versus CS3D civil society engagement, require cost-effectiveness analysis that the legal scholarship has not provided. Interdisciplinary research integrating legal feasibility assessment with economic modeling of climate harm avoided, blue economy benefits preserved and accountability litigation costs would provide a rigorous empirical foundation for the strategic prioritization decisions that Bangladesh's government and civil society actors must make under significant resource constraints.

CONCLUSION

This chapter has traced the evolution of legal accountability for maritime climate harm and marine pollution from a soft-law architecture designed for cooperative flexibility toward a harder, more enforceable accountability landscape shaped by judicial innovation, rights-based litigation and mandatory corporate due diligence. The chapter's central claim can be precisely restated: the traditional insulation of flag states, coastal states and maritime corporations from meaningful legal responsibility for climate harm and pollution at sea is no longer defensible as a matter of international law.

Three developments anchor this trajectory, each of which is elaborated in full earlier in the chapter. The ITLOS and ICJ advisory opinions of 2024 and 2025 have hardened the due diligence standard governing flag-state conduct, foreclosing reliance on IMO-standard compliance alone and supplying, for the first time, an authoritative interpretive basis on which flag-state discretion can be contested. The human rights and climate litigation line running from Urgenda through Milieudefensie to KlimaSeniorinnen supplies a tested rights-based template that Bangladesh's Article 18A and Article 32 jurisprudence, already engaged in the maritime domain through *BELA v. Bangladesh*, can extend to climate-related maritime harm specifically. Mandatory corporate due diligence, principally the French loi de vigilance and, in its narrower post 2026 form, the EU CS3D, extends enforceable obligations to the corporate level in ways unavailable five years ago, though the 2026 Omnibus amendments have narrowed the EU-level civil liability route considerably, reinforcing rather than undermining the case for the integrated, multi-level approach this chapter proposes. The three-tier architecture set out in Section 6 synthesizes these three strands into a model generating cumulative accountability pressure through complementary mechanisms, strategic sequencing and inter-tier evidentiary and interpretive reinforcement, calibrated to Bangladesh's institutional reality and to the counterarguments and capacity constraints identified above.

What remains is the political will, institutional capacity and strategic coordination required to deploy these legal resources effectively, a challenge of a fundamentally different character from establishing the legal framework itself and one the counterarguments addressed in Section 6 suggest should not be underestimated. Two developments deserve attention as potential accelerants. The first is the emerging proposal to recognize ecocide as a fifth international crime under the Rome Statute of the International Criminal Court, building on the 2021 Independent Expert Panel's draft definition of the offence (Stop Ecocide Foundation, 2021); on a sufficiently expansive interpretation, such a crime could in principle encompass patterns of systematic regulatory negligence generating persistent marine environmental harm, though its adoption would require a Rome Statute amendment under Article 121 and remains, at present, a matter of advocacy rather than positive law. The second is the prospective extension of contentious ITLOS proceedings on maritime climate claims. The legal foundations for such proceedings have been substantially advanced by the COSIS Advisory Opinion, and the question is no longer whether maritime climate accountability can in principle be pursued through international adjudication but whether the political will, evidentiary capacity, and regional coordination identified in Section 6 as preconditions for such a claim can in fact be assembled, a matter this chapter has sought to clarify rather than to resolve in advance.

REFERENCES

Ahmad, I., Haider, A., & Ahmad, K. (2025). The impact of judicial activism on environmental protection and justice in Pakistan: A statistical analysis. *Journal of Asian Development Studies*. https://www.researchgate.net/publication/391771576

Bangladesh Environmental Lawyers Association (BELA) v. Bangladesh, Writ Petition No. 7260 of 2008 (High Court Division of the Supreme Court of Bangladesh 2009).

Birnie, P., Boyle, A., & Redgwell, C. (2009). *International law and the environment* (3rd ed.). Oxford University Press.

Boyle, A. (2012). Human rights and the environment: Where next? *European Journal of International Law*, *23*(3), 613–642. DOI: 10.1093/ejil/chs054

Corfu Channel Case (U.K. v. Alb.), Judgment, 1949 I.C.J. 4 (Apr. 9).

Council Directive (EU) 2022/2464 of 14 December 2022 as regards corporate sustainability reporting (CSRD), 2022 O.J. (L 322).

Council Directive (EU) 2024/1760 of 24 May 2024 on corporate sustainability due diligence (CS3D), 2024 O.J. (L 1760).

Council Directive (EU) 2026/470 amending Directives 2006/43/EC, 2013/34/EU, (EU) 2022/2464 and (EU) 2024/1760 ("Omnibus I"), 2026 O.J.

Dispute Concerning Delimitation of the Maritime Boundary Between Bangladesh and Myanmar in the Bay of Bengal (Bangl. v. Myan.), Judgment, 2012 ITLOS Rep. 4 (Mar. 14).

Haider, A. (2025). Review of the 26th Amendment to the Constitution of Pakistan and the fundamental right of a clean and healthy environment: A case study of new Article 9A. https://www.researchgate.net/publication/392602711

Harrison, J. (2017). *Saving the oceans through law*. Oxford University Press. DOI: 10.1093/law/9780198707325.001.0001

Human Rights Committee. (2019). *General comment No. 36 on Article 6 of the International Covenant on Civil and Political Rights, on the right to life* (CCPR/C/GC/36). United Nations.

Indian Ocean Memorandum of Understanding on Port State Control. (2025). *Annual report 2024*. Indian Ocean MoU Secretariat.

Intergovernmental Panel on Climate Change. (2022). *Climate change 2022: Impacts, adaptation and vulnerability. Contribution of Working Group II to the Sixth Assessment Report.* Cambridge University Press.

International Maritime Organization. (2020). *Fourth IMO GHG study 2020.*

International Maritime Organization. (2023). *2023 IMO strategy on reduction of GHG emissions from ships* (MEPC.377(80)).

International Maritime Organization. (2025). *Entry into force of the Hong Kong International Convention for the safe and environmentally sound recycling of ships.*

Loi n° 2017-399 du 27 mars 2017 relative au devoir de vigilance des sociétés mères et des entreprises donneuses d'ordre, Journal Officiel de la République Française [J.O.], Mar. 28, 2017.

Luciano Lliuya v. RWE AG, Case No. I-5 U 15/17 (Oberlandesgericht Hamm May 28, 2025).

Mayer, B. (2018). *The international law on climate change.* Cambridge University Press. DOI: 10.1017/9781108304368

Milieudefensie et al. v. Royal Dutch Shell plc, Hague District Court, C/09/571932/ HA ZA 19-379 (2021).

Nairobi Convention for the Protection, Management and Development of the Marine and Coastal Environment of the Western Indian Ocean, June 21, 1985.

Obligations of States in Respect of Climate Change, Advisory Opinion, 2025 I.C.J. Rep. 187 (July 23).

Peel, J., & Osofsky, H. M. (2015). *Climate change litigation: Regulatory pathways to cleaner energy.* Cambridge University Press. DOI: 10.1017/CBO9781139565851

Poseidon Principles. (2019). *Poseidon Principles: A global framework for responsible ship finance.*

Pulp Mills on the River Uruguay (Arg. v. Uru.), Judgment, 2010 I.C.J. 14 (Apr. 20).

Request for an Advisory Opinion Submitted by the Commission of Small Island States on Climate Change and International Law, Advisory Opinion, 2024 ITLOS Rep. Case No. 31 (May 21).

Responsibilities and Obligations of States Sponsoring Persons and Entities with Respect to Activities in the Area, Advisory Opinion, 2011 ITLOS Rep. 10 (Feb. 1).

Rothwell, D. R., & Stephens, T. (2016). *The international law of the sea* (2nd ed.). Hart Publishing.

Ruggie, J. (2011). Guiding principles on business and human rights: Implementing the United Nations "Protect, Respect and Remedy" Framework (Vol. A). UN Human Rights Council., HRC/17/31.

Sands, P., Peel, J., Fabra, A., & MacKenzie, R. (2018). *Principles of international environmental law* (4th ed.). Cambridge University Press. DOI: 10.1017/9781108355728

Sea Cargo Charter. (2020). *Sea Cargo Charter: Aligning shipping with climate goals.*

Shipbreaking Platform, N. G. O. (2024). *Bangladesh.* https://shipbreakingplatform.org/our-work/the-problem/bangladesh/

NGO Shipbreaking Platform. (2026). *Annual list of ships dismantled worldwide, 2025.*

State of the Netherlands v. Urgenda Foundation, Supreme Court of the Netherlands, ECLI:NL:HR:2019:2006 (2019).

Stop Ecocide Foundation. (2021). *Independent expert panel for the legal definition of ecocide: Commentary and core text.*

Tanaka, Y. (2023). *The international law of the sea* (4th ed.). Cambridge University Press. DOI: 10.1017/9781009025393

Tanhuanpää, T., Altarriba, E., & Rahiala, S. (2025). Transition towards Fossil-Free energy sources in the shipping sector: A Sustainability Index Analysis of methanol and liquefied Biogas. In *Lecture notes in intelligent transportation and infrastructure* (pp. 571–587). DOI: 10.1007/978-3-031-82818-8_43

Trail Smelter Arbitration (U.S. v. Can.), 3 R.I.A.A. 1905 (1941).

Ullah, Z., Saleem, M. M., & Haider, A. (2026). State's strict liability for managed forest fires under Pakistan's tort jurisprudence. https://www.researchgate.net/publication/400576266

United Nations Convention on the Law of the Sea, Dec. 10, 1982, 1833 U.N.T.S. 3.

United Nations Environment Programme. (1995). *Convention for the protection of the marine environment and the coastal region of the Mediterranean (consolidated text).* UNEP Mediterranean Action Plan.

United Nations General Assembly. (2022). Resolution 76/300: The human right to a clean, healthy and sustainable environment (Vol. A). United Nations., RES/76/300.

Verein KlimaSeniorinnen Schweiz and Others v. Switzerland [GC], App. No. 53600/20 (ECtHR Apr. 9, 2024).

Zhang, H., & Haider, A. (2025). China's first indoor smoking air pollution case: Remarkable but not justifiable. *Chinese Journal of Environmental Law*. https://www.researchgate.net/publication/397598122

Zhang, H., Haider, A., & Khan, A. (2025a). Biodiversity v. development: The Supreme Court's verdict in M.K. Ranjitsinh v. Union of India. https://www.researchgate.net/publication/395949557

Zhang, H., Haider, A., & Khan, A. (2025b). International trade and plastic waste in oceans: Legal and policy challenges. *Frontiers in Marine Science*. https://www.researchgate.net/publication/394575729

ADDITIONAL READING

Beyerlin, U., & Marauhn, T. (2011). *International environmental law*. Hart Publishing. DOI: 10.5771/9783845265582

Crawford, J. (2019). *Brownlie's principles of public international law* (9th ed.). Oxford University Press. DOI: 10.1093/he/9780198737445.001.0001

Faure, M., & Lixin, H. (Eds.). (2022). *Climate change liability and beyond*. Eleven International Publishing.

Nordquist, M. H. (Ed.). (2012). *UNCLOS 1982 commentary: A commentary* (Vol. 4). Martinus Nijhoff. DOI: 10.1163/9789004215627

Rajamani, L. (2016). Ambition and differentiation in the 2015 Paris Agreement: Interpretive possibilities and underlying politics. *The International and Comparative Law Quarterly*, *65*(2), 493–514. DOI: 10.1017/S0020589316000130

Saul, B., & Sherwood, A. (2019). Sea level rise and the international law of the sea. *International Journal of Marine and Coastal Law*, *34*(4), 586–637.

Scovazzi, T., & Treves, T. (Eds.). (1992). *World treaties for the protection of the environment*. Istituto per l'Ambiente.

Talmon, S. (2019). Determining customary international law: The ICJ's methodology between induction, deduction and assertion. *European Journal of International Law*, *26*(2), 417–443. DOI: 10.1093/ejil/chv020

Tanaka, Y. (2012). Reflections on the ITLOS judgement in the Bay of Bengal case. *International Journal of Marine and Coastal Law*, *27*(4), 681–712.

Viñuales, J. E. (2015). The Paris Agreement on climate change: Less is more. *Jahrbuch für Internationales Recht. German Yearbook of International Law*, *58*, 11–26.

KEY TERMS AND DEFINITIONS

Carbon Intensity Indicator (CII): A mandatory MARPOL Annex VI metric introduced in 2023 requiring ships to calculate and report their annual operational carbon intensity, rated from A (superior) to E (inferior), with underperforming ships required to adopt corrective action plans subject to flag state review.

Corporate Sustainability Due Diligence: A legally binding obligation, originally established under EU Directive 2024/1760 (CS3D) and subsequently narrowed by the 2026 "Omnibus I" amendments (Directive (EU) 2026/470), requiring large companies, currently those exceeding 5,000 employees and €1.5 billion in worldwide net turnover, to identify, prevent, mitigate and account for adverse human rights and environmental impacts within their own operations and direct business partnerships. The Directive's original harmonized EU-wide civil liability mechanism, including third-country standing under Article 29, was deleted by the 2026 amendments; enforcement now proceeds through national supervisory authority action and, where available, the general tort law of individual member states.

Due Diligence (Maritime Environmental): An obligation of conduct under international law confirmed as evolving with scientific understanding by the ITLOS Seabed Disputes Chamber Advisory Opinion (2011) and the COSIS Advisory Opinion (2024) requiring states to deploy adequate means and exercise best possible efforts to prevent marine environmental harm, calibrated against the best available science rather than fixed at a static threshold.

Flag State Discretion: The traditional UNCLOS principle by which the state under whose flag a vessel is registered exercises primary jurisdiction over that vessel, including responsibility for ensuring compliance with applicable international environmental standards. The practical consequence of flag-state primacy and the prevalence of open registries (flags of convenience) is that this jurisdiction is often exercised with minimal rigor, generating a structural enforcement deficit.

Marine Pollution (UNCLOS Definition): Defined under UNCLOS Article 1(1)(4) as "the introduction by man, directly or indirectly, of substances or energy into the marine environment... which results or is likely to result in such deleterious effects as harm to living resources and marine life, hazards to human health, hindrance to marine activities including fishing... impairment of quality for use of sea water and reduction of amenities." The COSIS Advisory Opinion (2024) confirmed that anthropogenic GHG emissions fall within this definition.

No-Harm Rule: A foundational norm of customary international law, articulated in the Corfu Channel Case (ICJ, 1949) and Trail Smelter Arbitration (1941) and confirmed as applicable to climate harm by the ICJ Advisory Opinion of 23 July 2025, prohibiting states from using or permitting the use of their territory or activities under their jurisdiction in a manner causing significant harm to other states or areas beyond national jurisdiction.

Particularly Sensitive Sea Area (PSSA): A designation by the International Maritime Organization under MARPOL recognizing that a maritime area requires special protection due to ecological, scientific, educational, cultural, historical, economic, or social significance and is vulnerable to damage by international shipping activities. Associated protective measures may include traffic separation schemes and emission control area requirements.

Port State Control: The right of port states under UNCLOS Articles 218-220 to inspect and, where necessary, detain foreign vessels in their ports that fail to comply with applicable international environmental standards, including MARPOL requirements. Coordinated through regional memoranda of understanding, including the Indian Ocean MoU port state control provides a supplementary enforcement mechanism to flag-state primacy.

Soft Law (International Environmental): Non-binding normative instruments including IMO GHG strategies, voluntary guidelines and aspirational targets adopted through MEPC resolutions that articulate environmental standards without creating legally enforceable obligations or providing direct accountability mechanisms for non-compliance.

Chapter 8
Environmental Accountability in Coastal Fisheries Management and Ecosocial Perspectives From the Black Sea

Serdar Yener
http://orcid.org/0000-0003-1413-7422
Sinop University, Turkey

Helena Belchior Rocha
ISCTE, University Institute of Lisbon, Portugal

ABSTRACT

This study examines environmental accountability in coastal fisheries management, focusing on the Black Sea region. Drawing on literature, policy analysis, and stakeholder insights, it reveals how coastal fishing communities—despite their dependence on marine ecosystems—are often excluded from decision-making processes that shape their livelihoods. The findings highlight persistent gaps between sustainability policies and their implementation, including ecological degradation, weak institutional coordination, and unequal power relations. Supported by the TÜBİTAK 2221 Sabbatical Research Program, the study proposes an ecosocial governance approach that integrates ecological sustainability, social justice, participatory decision-making,

DOI: 10.4018/979-8-2600-3678-5.ch008

and fair market conditions. By emphasizing community empowerment and shared responsibility, the research argues that more inclusive and transparent governance systems are essential for achieving sustainable fisheries management.

INTRODUCTION

Overfishing, climate change, pollution, industrial fishing, and the expansion of the aquaculture sector are increasing ecological pressures on marine ecosystems (FAO, 2024). Legal frameworks and practices for monitoring and protecting marine ecosystems are still insufficient. Furthermore, concerns about transparency and accountability in the management of marine ecosystems are growing (Gelcich et al., 2019). Coastal fishing, as a crucial component of the marine ecosystem, is one of the sectors most affected by ecological pressure (FAO, 2024; Jentoft et al., 2022). Despite being one of the sectors most affected by ecosystem degradation, coastal fishermen are not adequately involved in or represented in decision-making processes within the sector. While coastal fishermen are negatively impacted almost worldwide, the Black Sea has even more adverse conditions due to its unique characteristics. Being a closed sea, the Black Sea is more susceptible to ecological degradation, and fishermen along its coast are more affected by factors such as pollution, overfishing, and declining fish stocks. These factors significantly affect coastal fishermen operating in the Black Sea. Furthermore, the strategically fragile inter-country balance, resulting from political and diplomatic factors in recent years, makes the management of the Black Sea marine ecosystem problematic. Academic studies and stakeholder reports focusing on the Black Sea convey the real situation while emphasizing the importance of corrective measures. These studies highlight the uncontrolled increase in industrial fishing activities, the depletion of fish stocks, inadequate representation and participation, and pollution (Miret-Pastor et al., 2022). They also emphasize the inadequacy and lack of transparency in the governance mechanisms of the marine ecosystem. Field studies show that coastal fishermen, who share the same areas as industrial fishermen, complain about many negative factors such as fishing periods and quotas, incentives and subsidies, pollution, climate change and its consequences (Berkes, 2018). For example, some fishing methods used by industrial fishermen cause significant damage to the marine ecosystem. Pollution, overfishing, climate change, rising sea temperatures, and habitat degradation negatively affect sustainability in the Black Sea region (Pinsky et al., 2020). Similar ecological pressures have been identified in other coastal and delta ecosystems where resilience increasingly depends on integrated ecosystem governance and nature-based solutions (Dehalwar & Sharma, 2026; Pinsky et al., 2020). Coastal fishermen, who are most affected by all these negative factors, are the least represented party in decision-making processes that determine all practices affecting them. Similar to findings, environmental conditions effect human decisions and adaptive responses (Yadav et al., 2026).

Environmental accountability requires all stakeholders of the marine ecosystem to have sustainable, fair, and transparent management processes. Environmental governance can be defined as the inclusion of environmental stakeholders in decision-making processes and is concerned with whether stakeholders fulfill their environmental responsibilities. Environmental accountability, understood here as a multidimensional framework encompassing ecological, institutional, economic, and social dimensions (see Section 1.1), is a condition for sustainable marine governance. Dimensions like ecological accountability (ecosystem protection and monitoring), institutional accountability (regulation and implementation), economic accountability (fairness and market responsibility), and social accountability (participation, representation, and inclusion) are needed. Studies show that the most affected by ecological degradation are the least involved groups in decision-making processes (Gelcich et al., 2019). Environmental accountability encompasses the necessary legal, institutional, and social arrangements to ensure that stakeholders fulfill their responsibilities regarding sustainability (Kashyap, 2025). To strengthen environmental accountability in the marine ecosystem, monitoring systems, legal frameworks, ethical norms, and active participation in decision-making processes are ensured (Garcia et al., 2014). Although environmental accountability is promoted at the international level, it is observed that it is not sufficiently achieved at the local level. Fisheries management, as a fundamental component of the marine ecosystem, encompasses coastal fishermen, industrial operators, aquaculture producers, regulatory bodies, and scientists. Stakeholders in fisheries management may not be adequately represented due to power imbalances and unequal access to governance processes. Similar mechanisms of spatial exclusion and accessibility barriers have been documented in other resource-dependent communities (Yadav et al., 2025). Underrepresentation of stakeholders can hinder the achievement of sustainability goals in fisheries management. Strong governance, inclusivity, and collaboration among stakeholders are needed for sustainable fisheries management (Berkes, 2018). Similar governance challenges have been identified in infrastructure planning, where institutional coordination significantly affects long-term regional development (Sharma & Dehalwar, 2025). However, existing studies indicate that these requirements are not met and accountable environments are not created. One of the policies guiding fisheries management at the international level is the European Union's Common Fisheries Policy (CFP), which aims to achieve sustainability goals globally, primarily within the EU. The CFP integrates environmental and economic objectives, including scientific assessment of fish stocks, establishment of sustainable yield-based fishing quotas, monitoring and reporting of illegal fishing, and frameworks consistent with the European Green Deal and the EU Biodiversity Strategy (European Commission, 2023). Despite the establishment of the CFP framework, a lack of coordination among member states and stakeholders makes achieving the stated goals difficult. In addition, the lack of integration of local ecological knowledge from coastal fishermen and other stakeholders into decision-making processes, and the failure to utilize their feedback, exacerbates the representation problem.

The Black Sea basin stands out as one of the rare areas where the aforementioned problems can be observed from a geographical perspective. As a closed sea, the Black Sea, fed by the Danube River and other tributaries, contains pollution from almost all of Europe and neighboring countries. The Black Sea basin harbors many negative factors such as climate change, eutrophication, pollution, biodiversity loss, and invasive species. Furthermore, political crises and fragile balances between neighboring countries prevent the establishment of an international governance mechanism and the implementation of effective policies. This situation threatens the ecological balance and negatively impacts environmental accountability. Sinop, a city in Türkiye with the longest coastline in the Black Sea basin, also experiences these negative factors. Coastal fishermen in Sinop are also underrepresented in decision-making processes regarding issues such as declining fish stocks, the spread of industrial fishing activities, increasing pollution, and the setting of unfair fishing quotas. Furthermore, increasingly frequent political crises threaten and exacerbate the danger of collaborative decision-making processes. All these factors make the city of Sinop a unique area of study for environmental accountability. The limited participation of coastal fishermen operating in Sinop, as the most affected stakeholders, in decision-making processes makes this study valuable and forms a basis for future research. Moreover, coastal fishermen reported these negative aspects firsthand during the field studies of the TÜBİTAK project that formed the basis of this study. The studies show that the city of Sinop is an excellent research area where governance failures can be observed.

This study describes how environmental accountability mechanisms function for coastal fishermen and how the ecosocial model can contribute to achieving sustainability goals. The study will explain the governance challenges faced by coastal fishermen operating in Sinop, the ecological, institutional, economic, and social governance dimensions, and the integration of the ecosocial model as a solution tool. Institutional, ecological, economic, and social accountability dimensions will also be presented. The importance of factors such as participatory governance, social justice, and equitable economic balance for sustainable fisheries management will also be highlighted.

Several methods were used in this study. While the study is primarily based on a TÜBİTAK project titled "Application of the Ecosocial Model to Coastal Fishing Communities," a comprehensive literature review was also conducted.

Methodology

This study, which addresses environmental accountability from an ecosocial perspective, utilizes a qualitative case study method. The research was conducted using data obtained from semi-structured interviews with coastal fishermen living in Sinop, a city located on the Black Sea coast, and the results of a literature review. Existing practices and policies were also interpreted and integrated into the literature synthesis and semi-structured interviews. In this way, a multi-dimensional approach was used to achieve the research objectives. This perspective also recognizes that environmental conditions influence behavioural adaptation and livelihood decisions, an issue increasingly emphasized in environmental decision-making studies (Yadav et al., 2026).

This is seen as a comprehensive example of the problems and challenges faced by fishermen in the city of Sinop and along the Black Sea coast. The socio-economic dependence of coastal fishermen, marine pollution, climate change, inadequate governance and representation issues, and the decrease in biodiversity make the Black Sea a unique case.

The research methodology is based on data obtained from a literature review and grounded in different themes and disciplines such as environmental accountability, sustainability, and ecosocial theory. Numerous sources were utilized, including academic reports and other publications, practice documents, policy documents, marine management studies, and aquaculture and fisheries studies.

Secondly, policy-oriented document analysis was conducted using reports and frameworks developed by international and regional organizations; these included marine environmental assessments, sustainability frameworks, fisheries governance regulations, and ecosystem management guidelines. These materials were used to contextualize accountability mechanisms within broader environmental management systems and to identify recurring institutional and ecological concerns affecting coastal fisheries.

Third, the chapter incorporates contextual stakeholder insights generated through long-term interaction with fisheries-related activities in Black Sea coastal communities, particularly in the Sinop region of Turkey, through regional workshops, project collaborations, and field-based observations. Data collected from a field study funded by TÜBİTAK (Scientific and Technological Research Council of Turkey) under the 2221 Sabbatical Research Programme are used. These insights are treated not as formal interview data, but as contextual evidence enabling the interpretation of governance challenges, community concerns, and accountability gaps experienced by fisheries stakeholders. Where appropriate, these observations are used as demonstratives, not as claims of statistical generalizability.

Finally, the ecological, institutional, economic, and social accountability dimensions presented in this section have emerged through an analytical synthesis of the literature defined among policy debates and stakeholder concerns, ecosocial governance principles, fisheries governance discussions, and recurring themes.

In line with the logic of qualitative case studies, the aim of this chapter is not statistical generalization, but analytical and conceptual contribution. Therefore, the case study in the city of Sinop is used as a contextually based example to demonstrate how ecosocial environmental accountability can contribute to more participatory, equitable, and ecologically conscious fisheries management.

Although the shortcomings observed in this study are addressed, power asymmetries within the community have not been considered, and this can be seen as a limitation of the study. Coastal fishing communities are heterogeneous and experience internal competition and inequality (Asio et al., 2026). Furthermore, since fisheries governance challenges differ among Black Sea countries, governance solutions determined in a single context may not be fully translatable to the region. These limitations are considered not as weaknesses of ecosocial governance, but as conditions requiring critical attention during implementation.

After clarifying the analytical foundations of the chapter, the following section traces the conceptual development of environmental accountability in marine governance and explains why an ecosocial perspective provides a useful viewpoint for understanding fisheries governance challenges in the Black Sea context.

1. IMPROVING ENVIRONMENTAL ACCOUNTABILITY IN MARINE MANAGEMENT

Accountability has become a sought-after characteristic in all areas of life, guaranteeing stakeholder satisfaction. As a fundamental dimension of quality assurance, accountability is also a crucial factor in marine ecosystem management. The management and coordination of numerous components, such as coastal fishing, industrial fishing, maritime transport, offshore energy production, and aquaculture, through accountable processes has become a fundamental expectation for all stakeholders. Since the stakeholders of the marine ecosystem are dependent on the same resources, accountable processes must be established for the sustainable use of common and limited resources. The activities of stakeholders are scrutinized at local, national, and international levels. Sustainability pressure necessitates comprehensive monitoring of many processes, primarily accountability. The seemingly independent activities and impacts of stakeholders directly and indirectly affect other stakeholders or the marine ecosystem in general. For example, feed residues and other waste from aquaculture ponds affect the food chain of other organisms.

Similarly, unsustainable methods used by industrial fishermen lead to a decrease in fish stocks and consequently negatively impact the livelihoods of coastal fishermen.

1.1 Conceptualizing Environmental Accountability

Environmental accountability refers to the set of legal, institutional, economic, and social mechanisms through which stakeholders are held responsible for their actions affecting environmental systems. It encompasses transparency in decision-making, the ability to monitor and evaluate actions, and the obligation to justify and rectify environmental impacts. In the context of marine governance, environmental accountability integrates four interrelated dimensions: ecological accountability (protection and monitoring of ecosystems), institutional accountability (regulatory design and implementation), economic accountability (fairness in resource distribution and market practices), and social accountability (participation, representation, and inclusion of stakeholders). These dimensions operate interactively and form the analytical foundation of this study. This concerns all stakeholders sharing the common environment and is defined by many concepts such as participation, representation, governance, and marine management. While environmental governance emphasizes the participation of stakeholders using common environmental resources in decision-making processes, sustainability emphasizes the balanced and equitable use of environmental resources. Sustainability can be defined as the use of existing resources with ecological, social, and economic dimensions in mind, taking into account future generations. Decision-making processes in which all stakeholders participate or are represented ensure the fair distribution of resources. Environmental accountability focuses on transparency and fulfillment of responsibilities in the activities and processes carried out by environmental stakeholders. Accountability requires stakeholder participation in decision-making processes, environmental responsibility, inclusivity, and procedural and application justice. Because the areas of activity of the stakeholders in the marine ecosystem are independent of each other, the results they produce negatively affect each other in the long term, and their interaction is complex and leads to complex results. In addition, conflicts of interest and competition among stakeholder's fuel power inequality create problems of representation and participation, and exacerbate injustice. These results negatively affect accountability. The dimensions of environmental accountability, institutional, economic, ecological, and social accountability, are interconnected and interact. Conceptually, environmental accountability has gained more prominence in the context of ecological degradation. ecological degradation and environmental health risks increasingly threaten both ecosystem integrity and community well-being. During periods of increasing environmental problems, the lack of structural regulators was felt. Therefore, proactive environmental actions were shaped on a voluntary basis. The first legal and structural initiatives were explicitly declared in the 1972 Stockholm Declaration, emphasizing that states have legal responsibilities for environmental protection and highlighting the need for global cooperation (Bodansky, 2010). Increased environmental awareness has led to action at both state and international levels. The 1992 Rio Declaration on Environment and Development emphasizes the need for collective action by states on environmental issues. The Rio Declaration, which encourages states and international organizations to take preventive measures and promotes the polluter pays principle, also encourages the participation of relevant stakeholders in decision-making processes. The Rio Declaration aims to address the problem of representation at the local and national levels, fragmented governance structures, and the lack of coordinated activities. Although the Rio Declaration targets all activities and institutions related to the environment, fisheries management has been one of the most affected sectors. The fisheries sector consists of many components, including industrial fishermen, coastal fishermen, and aquaculture producers. Due to economic concerns, stakeholders in the fisheries sector prioritize short-term goals, leading to irreversible consequences. The fisheries sector is increasingly exposed to threats due to overfishing, inappropriate methods and materials, uncoordinated and illegal fishing. The capacities and behaviors of industrial fishermen and aquaculture producers in the sector often negatively impact the entire sector and the environment. At the same time, these economically powerful stakeholders dominate coastal fishermen (Demirci et al., 2026). Although initiatives aimed at the environment, and particularly at fisheries, have increased at national and international levels, many stakeholders in the sector still face numerous challenges. In particular, coastal fishermen are negatively impacted by asymmetric power pressures within the sector, quotas, and seasonal and regional fishing restrictions, and progress is insufficient (Rościszewski-Dodgson, & Cirella, 2026). The ecological, social, and economic challenges faced by fishermen demonstrate the necessity of an ecosocial model. The EU's Common Fisheries Policy (CFP) for the fisheries sector not only aims to improve fisheries management but also integrates broader initiatives such as the European Green Deal, the Blue Economy Strategy, and ecosystem restoration frameworks (European Commission, 2023). Furthermore, the EU's Maximum Sustainable Yield (MSY) policy in the fisheries sector encourages proactive fishing for the future in increasingly declining fish stocks and polluted marine ecosystems. Fishing activities are expected to be at a level that does not threaten fish stocks.

CFP encourages regulations for transparent monitoring of the sector's common resources while also requiring countries to take measures against poaching. Nevertheless, fish populations are seen as endangered in uncontrolled and unmonitored areas (European Parliament, 2024). As in all governance systems, participation and representation are determinants of fair, inclusive and accountable decision-making processes in fisheries management (Berkes, 2018).

1.2 Rocha's Ecosocial Model and Fisheries Management

The analytical perspective adopted in this section is influenced by Rocha's ecosocial model (2016), which emphasizes the inseparability of environmental systems, social justice, corporate responsibility, and human well-being. Originally developed within discussions of social policy and ecosocial well-being, the model argues that environmental degradation and social vulnerability are deeply linked and therefore require integrated governance responses rather than fragmented sectoral interventions.

One of the key contributions of the ecosocial perspective is its rejection of purely technocratic understandings of sustainability. Instead of viewing environmental problems as isolated ecological failures, ecosocial thinking emphasizes how governance structures, economic systems, social exclusion, and unequal access to decision-making shape environmental outcomes. From this perspective, sustainability relies not only on ecological protection but also on democratic participation, equality, institutional accountability, and social resilience.

This perspective is particularly useful in fisheries management because coastal communities experience environmental changes through daily livelihood insecurity, declining fish stocks, market instability, unequal regulatory burdens, and institutional exclusion. Small-scale fishermen frequently encounter conditions where environmental degradation is inextricably linked to social and economic vulnerability. Accordingly, an ecosocial perspective ensures that fisheries management is understood not merely as a technical management challenge, but as a governance issue encompassing justice, accountability, participation, and ecological citizenship.

Applying Rocha's ecosocial model to the Black Sea context reveals how failures in fisheries governance manifest through interconnected institutional, ecological, economic, and social mechanisms. Rather than interpreting fisheries decline solely as a biological or managerial issue, the ecosocial perspective makes it possible to analyze how accountability gaps reinforce inequalities and weaken long-term sustainability.

The ecosocial model emphasizes that communities are active actors in the decision-making process and that decisions should be evaluated from ecological, social, economic, and political perspectives (Rocha, 2016). The basic principles of the ecosocial model are listed below:

- Ecological Sustainability emphasizes that decisions and practices in society must be compatible with the capacities of the ecological environment. A lifestyle design compatible with the ecological environment is necessary for social and economic well-being (Rocha, 2016). In coastal fisheries, MSY fish stock calculations and applications, monitoring, and biodiversity conservation are essential.

- Social justice in the fisheries sector refers to ensuring that all actors have socially fair and transparent access to resources, with equitable resource distribution being a primary objective.
- Participatory and democratic governance emphasizes the representation and participation of all stakeholders in decision-making processes within the fisheries sector. The right to representation should be ensured in a manner consistent with the principles of social accountability.
- Intersectoral and Multilayered Integration emphasizes the need for coordination across all sectors (environment, economy, health, education) and at all levels (local, regional, national) related to fisheries management.
- The Community Empowerment and Capacity Building principle (Rocha, 2016) focuses on raising awareness among community members. The aim is to empower and increase the representation of fishing communities and groups.

The ecosocial approach is a suitable model for addressing the shortcomings in the fisheries sector and is compatible with other mechanisms.

2. THEORETICAL FRAMEWORK

2.1 Environmental Accountability Theory

Environmental accountability refers to the existence of legal, institutional, social, and ethical structural mechanisms in environmental activities and processes (Ranjan et al., 2026). Legal accountability defines the legal infrastructure in the fisheries sector. The existence of permit mechanisms, certifications, quotas, and planning required to operate in the fisheries sector strengthens legal accountability (Simataa et al., 2025). Institutional accountability refers to the definition and implementation of the roles of responsible public institutions in the fisheries sector. Comparable findings from regional infrastructure governance demonstrate that effective institutional coordination directly influences sustainable development outcomes (Sharma & Dehalwar, 2025). Mechanisms such as the definition of the regulatory roles of responsible public institutions, the effectiveness of coordination, and monitoring and reporting systems demonstrate institutional accountability. Social accountability shows the access, monitoring, and control of processes and practices in fisheries management by stakeholders in society.

2.2 Community-Based Resource Management

Community-based resource management defines the shared responsibilities of marine ecosystem stakeholders in common ecological resources. This leads to the development of mechanisms for sustainable consumption, use, and monitoring of common resources. It also emphasizes participatory decision-making processes regarding common resources. In fisheries management, the inclusion of all affected cooperatives, particularly coastal fishing cooperatives, in decision-making processes can be seen as community-based resource management (Basurto et al.,2017; Jentoft, 2007).

3. COASTAL FISHERIES MANAGEMENT AND ENVIRONMENTAL CHALLENGES

3.1 Global Challenges in Fisheries Management

The history of fishing, beginning with the history of humanity, has undergone many transformations along with human civilization (Safronova, 2025). Human dependence on the sea, wars, transportation, etc., has strengthened humanity's interaction with fishing. However, due to many reasons, primarily food concerns and logistical needs, unconscious activities endanger the marine ecosystem and reduce fish stocks. Overfishing and indiscriminate fishing reduce fish stocks and increase concerns for the future (FAO, 2024). As the population increases globally, the demand for seafood also increases and diversifies the pressure on marine ecosystems. The advanced technological equipment of industrial fishermen facilitates overfishing and threatens sustainability. The weakness of monitoring and control systems and legal infrastructure further exacerbates the danger. Overfishing and inappropriate practices by industrial fishermen threaten the marine ecosystem and biodiversity. Aquaculture, while creating alternative products to coastal and industrial fisheries, can increase pollution and threaten biodiversity. International competition, consumer preferences, political and geopolitical tensions, and regional insecurities negatively affect fishing activities and make market conditions unpredictable (Stiglitz et al., 2009). In addition, climate change is the biggest threat to fisheries, and rising sea temperatures are another factor threatening biodiversity and fish stocks (Pinsky et al., 2020).

3.2 Fisheries Management in Europe

The European Union's Common Fisheries Policy (CFP) offers more inclusive, fair, transparent, and sustainable frameworks compared to other regions (European Commission, 2023). Through tools such as MSY-based quota systems, scientific fish stock monitoring, and measures against overfishing, the CFP aims to achieve the goals of sustainable fisheries management. It also encourages monitoring of illegal fishing, tracking of vessel movements, and monitoring of harmful environmental practices. To strengthen the adaptation of fisheries management stakeholders to change and crises, the CFP provides support through programs such as the European Maritime, Fisheries and Aquaculture Fund (EMFAF). Furthermore, a roadmap has been established for achieving the goals of the EU Green Deal and the EU Biodiversity Strategy for 2030 (European Commission, 2020). Nevertheless, current crises, the increasing impact of climate change, pollution, regional tensions, and a lack of international coordination necessitate more sustainable fisheries management (European Parliament, 2024).

3.3 Fisheries Management in the Black Sea

The Black Sea is a shared area used by both EU member states and non-EU countries, making coordination very difficult. Furthermore, its geographical and physical characteristics as a closed sea with limited water exchange increase pollution and invasive species, threatening fish stocks. This situation threatens the Black Sea ecosystem (United Nations Environment Programme, 2021). Academic studies and observations indicate that fish stocks are declining, pollution is increasing, and the ecosystem is deteriorating in the Black Sea (FAO, 2024). However, coastal fishing has been the primary source of livelihood for communities living along the Black Sea coast for centuries. Nevertheless, industrial fishing and aquaculture threaten coastal fishing. Turkey, which has the longest coastline on the Black Sea, monitors fishing activities and environmental initiatives in the Black Sea through national legal frameworks. Despite being the group, most affected by these negative factors in the Black Sea, As established earlier, coastal fishermen remain structurally underrepresented and are among the least involved stakeholders in decision-making processes. The distrust of coastal fishing communities and cooperatives regarding fishing regulations and legal practices is increasing, and their representation is declining.

The insufficient participation and representation of coastal fishermen in decision-making processes in fisheries management, and their exposure to asymmetrical power pressures from industrial fishermen and aquaculture operators, are increasingly weakening coastal fishermen. Coastal fishermen need to increase their awareness, strengthen their capacities, and participate more in decision-making processes, both within their own communities and at higher levels. To empower coastal fishermen:

- Increasing participation in decision-making processes within fishing communities,
- Strengthening coastal fishing cooperatives and their governing bodies,
- To improve their capacity and competencies
- Creating opportunities for fair and transparent governance,
- It is necessary to develop resilience in the face of change and crises.

The ecosocial approach also emphasizes addressing these needs. Increasing stakeholder awareness and strengthening environmental accountability are necessary for the protection and sustainable use of the Black Sea's threatened resources. The aforementioned measures should be taken to protect fisheries and marine ecosystems against environmental threats, and to strengthen ecological resilience and governance.

4. ENVIRONMENTAL ACCOUNTABILITY IN BLACK SEA FISHERIES

Coastal fishing communities, despite being the stakeholders most affected by ecological degradation and bearing the least responsibility for it, are the least represented groups in decision-making processes (Basurto et al. 2017). Meanwhile, industrial fishermen, aquaculture producers, marine logistics companies, and other major actors and stakeholders, who bear the most responsibility for the degradation of the marine ecosystem, dominate decision-making processes and alienate coastal fishermen. Although all these stakeholders use the same resources, the relationships within the stakeholder structure are asymmetrical, and the governance system has shortcomings. However, accountability in fisheries management requires the active participation of many factors, including coastal fishermen, industrial fishermen, aquaculture operators, policymakers, and NGOs. Furthermore, climate change, pollution, technological advancements, power asymmetries, political pressures, and regional crises increase the challenges and obstacles. These challenges were also observed in field studies and semi-structured interviews conducted with coastal fishing communities in Sinop and reflect broader weaknesses in environmental accountability as defined in this chapter, particularly in terms of participation, transparency, and institutional effectiveness.

Coastal fishermen complained about quota allocations, planning of fishing and activity areas, market regulations, and stated that they could not participate in decision-making processes. These negative factors undermine the trust of coastal fishermen in policymakers and practices, fostering a sense of cynicism. Coastal fishermen complain of the lack of established governance mechanisms and state that they are not adequately represented. Cooperative managers declare that industrial fishermen dominate fishing practices. All these factors support the emphasis of the ecosocial approach used in this study. In an ecosystem where coastal fishermen are victimized by opaque and unfair processes, an ecosocial approach can offer a solution. Strengthening the position of coastal fishermen and improving their ecological, economic, and social standing is possible. The issues most affecting coastal fishermen can be clarified through new studies, allowing them to look to the future with hope. Otherwise, sustainable fisheries management would be difficult to achieve. In this respect, field studies like this one are considered very valuable.

4.1 Industrial Fishing, the Spread of Aquaculture and Ecological Pressures

Industrial fishing methods and tools are putting pressure on all stakeholders in the marine ecosystem, especially coastal fishermen. The development of technological tools allows for the detection of fish schools from very far distances, the deployment of nets over vast areas using trawling, and the use of large fishing vessels in mid-water fishing operations, leading to overfishing and depletion of fish stocks. Furthermore, the strong economic capacity of industrial fishermen creates asymmetrical power structures in fishing practices. Industrial fishermen primarily use mid-water trawling and similar methods to sweep fishing grounds, damaging the ecology. These effects are clearly visible in the Black Sea. In Türkiye, the influence of industrial fishermen on policymakers is also observed in the determination of fish quotas and fishing grounds.

Aquaculture is another area of activity that affects ecology. While aquaculture is used to support the dwindling fish supply due to declining fish stocks, unplanned aquaculture industries can also threaten natural fishing grounds. In particular, the types, quality, and raw materials of feed used by aquaculture producers can affect the ecology. In addition, the feeding of native fish species with aquaculture feed in areas where aquaculture ponds are located can affect native fish species and populations. Furthermore, since native fish species are used as feed in the aquaculture sector, it can lead to an increase in natural fishing and even overfishing. In all interviews conducted with Black Sea fishermen, the fishermen brought up these problems, stating that some fish species, such as whiting and turbot, have started to be fed with artificial feeds under the aquaculture ponds. It was also emphasized that artificial feeds and some drugs, such as antibiotics, used in aquaculture can cause diseases in native species. In addition, it was stated that seasonal migratory fish species such as bonito are changing their migration routes by moving away from the aquaculture ponds due to the noise and visual pollution created there.

4.2 Market Regulation and Socioeconomic Vulnerability

Coastal fishermen generally have limited means and methods of bringing their catch to market. It is known that coastal fishermen usually deliver their catches to cooperatives they are members of the cooperatives sell them to buyers through negotiation or auction and then pay the coastal fishermen. In this scenario, coastal fishermen may receive payment for their catch several days later. In another scenario, coastal fishermen who are not members of any cooperative sell their fish to wholesalers in the existing fish market and receive payment upfront. In this scenario, the coastal fisherman's bargaining power is weaker, but they can still receive payment upfront. It is observed that cooperatives generally sell the fish they buy from coastal fishermen to markets in neighboring provinces or other regions. Businesses in fish markets are generally considered to be financially strong because they also conduct industrial fishing activities. As a result, it is emphasized that the price level of fish caught by coastal fishermen often remains low. Coastal fishermen operating in Sinop clearly state this for the year 2026. In 2026, not only coastal fishermen but also industrial fishermen using the purse seine method will experience economic difficulties. Because the biggest economic contribution of industrial fisheries in Sinop comes from anchovy fishing. For 2026, anchovy prices remained very low and did not receive the value they deserved. Many industrial fishermen were forced to close 2026 with loans. From the perspective of coastal fishermen, although industrial fishermen appear strong, they too are experiencing problems due to market conditions in 2026. Despite all these negative aspects, cooperatives ensure the collection of resources for fishermen, increase their bargaining power, and create stable markets. Another factor increasing the vulnerability of coastal fishermen is the increasing average age of coastal fishermen, and the younger generations moving away from coastal fishing due to economic fragility and uncertainty. Furthermore, the decrease in fish stocks and the climate change that supports this are other determining factors.

4.3 How Policy Can Support Empowerment

The policies of relevant public authorities such as the Ministry of Agriculture and Forestry, the Coast Guard Command, the Port Authority, the Ministry of Environment, Urbanization and Climate Change, and local municipalities play important roles in coastal fishing communities. Based on Rocha's ecosocial model (2016), empowerment in fisheries management can be supported through several policy pathways:

- When the Ministry of Agriculture and Forestry is developing implementation circulars that determine the annual activities of coastal fishermen, including coastal fishermen and other relevant stakeholders in the decision-making process can lead to more effective processes. Coastal fishermen should not be treated as external stakeholders in the formation of fisheries policies, but rather integrated into decision-making systems to address structural empowerment.
- Formalizing and implementing governance mechanisms that strengthen the participation of coastal fishermen in decision-making processes is a crucial factor. Mechanisms that strengthen the participation of coastal fishermen in decision-making processes can also foster cooperation and solidarity among all stakeholders.
- Capacity-building programs should be implemented for coastal fishermen to institutionalize and enhance their professional standing. This deficiency was also evident in interviews conducted with coastal fishermen.
- Considering the current situation of coastal fishermen, diversification and making more functional the economic mechanisms that will strengthen the existing cooperative system are necessary to correct their fragile structures. Such measures meet the economic and ecological empowerment requirements stated by Rocha (2016), reinforcing sustainability and social justice.
- Ensuring that coastal fishermen have access to official and up-to-date information from all platforms they are affiliated with, especially their cooperatives, can facilitate the establishment of accountable governance systems.

All these suggestions can meet the requirements of an ecosocial model. Including coastal fishermen in active decision-making processes contributes to the conservation of common resources through cooperation and solidarity and ensures the sustainability of the profession in the future.

5. ENVIRONMENTAL ACCOUNTABILITY CHALLENGES

In Turkey, all maritime-based economic activities, including coastal fishing and industrial fishing, fall under the responsibility of the Ministry of Agriculture and Forestry. The Ministry of Agriculture and Forestry monitors fishing activities through routinely updated circulars. Furthermore, it is known that licenses for fishing boats and equipment are registered by the port authority. Port layout and usage are also under the control of the port authority. The Coast Guard Command monitors whether fishing activities are carried out in accordance with the law and their safety aspects while navigating the sea. Although the decision-making processes of all official authorities in fisheries management are designed to include fishermen, it is observed that the planning of meeting locations and times, announcements, transportation, accommodation, and many related factors can lead to governance gaps. This situation was clearly stated in interviews with fishermen. All these shortcomings reveal weaknesses in accountability mechanisms that negatively affect the participation and empowerment of coastal fishing communities. Environmental accountability is affected by gaps in institutional, ecological, economic, and social accountability dimensions. Ecological accountability gaps, understood as failures in ecosystem monitoring and protection (see Section 1.1), are particularly visible in the Black Sea context. Institutional accountability defines the dimensions of transparency, implementation, and governance. Economic accountability addresses issues such as fish markets and equitable access to resources, while social accountability focuses on participation, representation, and inclusion in decision-making processes. All the dimensions of environmental accountability listed above interact with each other.

5.1 Corporate Accountability

Institutional accountability can be compromised by coordination, conflicts of authority, and weak implementation capacity among authorities and institutions responsible for fisheries management (Van Santen, 2024). In the case of the Black Sea, the fact that the surrounding countries consist of both EU and non-EU member states, along with other determinants, affects this situation. Institutional challenges are considered the biggest challenge in marine governance. Inconsistent implementation policies, inadequate and inconsistent fishing monitoring systems, and insufficient cross-border cooperation exacerbate institutional shortcomings. Another problem is that, despite the existence of fisheries governance mechanisms in different seas and fishing grounds around the world, there is no integrated governance system established by coastal states in the Black Sea, and geopolitical tensions and conflicts make its realization impossible. From the perspective of Rocha's ecosocial model (2016), this kind of fragmentation negatively impacts structural strengthening.

Corporate accountability challenges often stem from gaps between formal regulations and daily practices. These gaps are highlighted by all the fishermen interviewed, and discrepancies are noted in the regulatory and monitoring activities of the institutions responsible for fishing operations. These conflicts create uncertainty, increasing stress and anxiety about the future.

5.2 Ecological Accountability Gaps

Ecological accountability varies depending on reliable scientific data on the ecological system, monitoring systems, and transparent reporting procedures. In the case of the Black Sea, its geopolitical fragility, tensions, and wars make monitoring ecological systems and taking necessary measures difficult. Nature-based approaches are increasingly recognized as complementary mechanisms for improving ecological resilience and long-term sustainability in coastal systems (Dehalwar & Sharma, 2026). Currently, the Black Sea is in a fragile ecological state, its ecological balance is being disrupted, and pollution is increasing. In addition, uncontrolled and intense maritime traffic is causing invasive species from different seas and oceans to be transported into the Black Sea. Cooperation and solidarity-based measures against the encountered ecological problems are also not being implemented. These difficulties prevent the creation of effective transboundary policies and local practices, and the ecology is progressively deteriorating. Furthermore, global challenges such as climate change are exacerbating local negative impacts. Rocha's ecosocial model (2016) suggests that ecological accountability can be achieved through the cognitive empowerment of stakeholders. Monitoring the activities of coastal fishermen, and especially fishermen themselves, and tracking changes in the sea can facilitate ecological monitoring. Interviews with fishermen reveal declines in fish stocks, marine pollution, decreased biodiversity, and chemical threats. Coastal fishermen can be considered the strongest stakeholders in ecological monitoring. In this respect, simple, implementable monitoring applications involving coastal fishermen can provide real-time ecological data for the Black Sea.

5.3 Economic Accountability Gaps

Economic accountability requires regulatory mechanisms in fish and seafood markets to be fair, transparent, and accessible. This is clearly stated in interviews with coastal fishermen. For example, it was stated that this year's anchovy prices did not meet expectations and did not even cover the cost of the materials used. It was stated that a crate of anchovies would not even cost 400 TL in 2026, but the cost of the crate itself would be at that price. On the other hand, it was stated that the price of red mullet, considered a very valuable fish species, did not meet expectations and did not even reach 250 TL per kilogram. However, in 2025, red mullet was sold for around 400-600 TL. Such examples show that uncertainty and lack of transparency are high in pricing in fish markets. On the other hand, it is stated that fish markets are dominated by industrial fishermen and that the power of cash is used negatively. In addition, limited access to credit, unstable demand, and price fluctuations further weaken bargaining power. Rocha's ecosocial model (2016) considers economic empowerment as a requirement for effective environmental management. The volatile livelihoods of coastal fishermen and difficulties in accessing the market make it difficult for them to participate in decision-making processes. All coastal fishermen take out loans and start the next fishing season in debt. Starting the season with debt for an uncertain future creates anxiety and despair.

5.4 Social Responsibility Gaps

Social accountability refers to the participation and questioning of decision-making processes within the governance systems of communities, such as cooperatives, to which coastal fishermen belong. Coastal fishermen who are not included in decision-making processes concerning issues that affect them feel excluded and isolated. Interviews with coastal fishermen indicate that coastal fishing communities and cooperatives are ineffective and weak, and that they need to be strengthened. Coastal fishermen do not trust either their own communities or other official authorities, and their expectations for the future are weakened. It is stated that industrial fishermen with strong economic resources have a say in decisions such as quota applications and fishing seasons that currently affect coastal fishermen. Rocha (2016) suggests that social accountability can be used to eliminate relational power imbalances, inequalities in voice, influence, and recognition between communities and powerful institutional or economic actors.

6. ECOSOCIAL GOVERNANCE MODEL FOR SUSTAINABLE FISHERIES

The challenges and governance gaps encountered in coastal fisheries stem from weaknesses in ecological sustainability and institutional regulatory approaches. Effective and sustainable fisheries management requires governance models that integrate social and economic justice, ecological sustainability, and participatory, inclusive decision-making processes. The ecosocial governance model offers a comprehensive conceptual framework that can address these multifaceted needs. The ecosocial model also emphasizes integration between governance mechanisms and sectors. According to the ecosocial approach, social justice, ecological sustainability, and economic resilience interact within socio-ecological systems (Radzi, 2023). As the ecosocial model emphasizes, coastal fishermen, being dependent on marine resources, should be included in marine-related decision-making processes. The interests of all stakeholders in fisheries management should be balanced in the management model. Indeed, coastal fishing communities clearly demonstrate that social and ecological systems are deeply interconnected and that governance arrangements should reflect this complexity. In fisheries management, the ecosocial approach revolves around four interconnected pillars: ecosystem sustainability, social justice, participatory decision-making, and fair, transparent markets.

Figure 1. Eco social governance model for sustainable fisheries

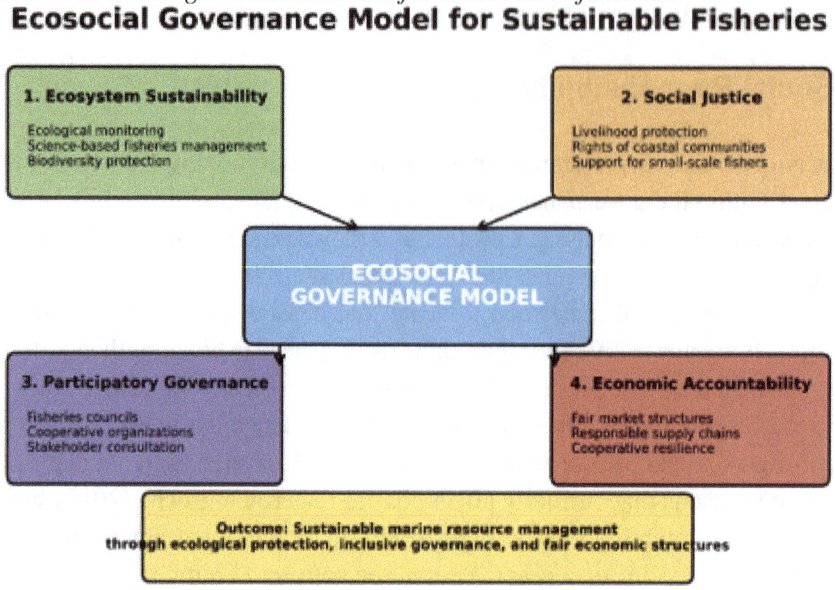

Source: Author's statement

As can be seen from the figure, the first fundamental element of the ecosocial governance model is ecosystem sustainability. Ecosystem sustainability refers to the conservation and management of marine ecosystems and fish stocks. Ecosystem sustainability includes biodiversity conservation, monitoring systems, and preventive measures. Scientific monitoring is necessary to understand the holistic monitoring of the marine ecosystem, the migration movements of fish species, climate change, and its effects. Research results show that ecological sustainability depends on long-term monitoring, adaptive management, and integration. The unsustainability of ecosystems weakens both food security and the livelihoods of coastal communities. Social justice, the second dimension of the ecosocial governance model, emphasizes the need to protect the livelihoods and rights of coastal fishing communities in fisheries management. Fisheries management encompasses coastal fishing communities, industrial fishermen, aquaculture operators, market actors, and logistics actors. Protecting the rights of weaker stakeholders within a group of stakeholders whose interests often conflict ensures social justice. Social justice problems include:

- Unequal access to marine resources
- Exclusion from the decision-making process
- Loss of cultural heritage
- Power asymmetries in markets and governance.

The third dimension of the ecosocial model, participatory decision-making, refers to the involvement of coastal fishing communities in decision-making processes that concern them and their representation in governance systems. Co-management, one of the most studied participatory models, has shown effectiveness in increasing cohesion, improving ecological outcomes, and strengthening community capacity. In fisheries management, shared governance mechanisms encourage the effective participation of all stakeholders in decision-making processes, and this requirement is also emphasized in the ecosocial model. Rocha's ecosocial model adds depth to this perspective by confirming that participation without empowerment is insufficient. It is emphasized that fisheries management needs to be strengthened and its capacities increased, both scientifically and in other aspects. Coastal fishermen should be seen as constructive actors in fisheries management. The fourth dimension in the ecosocial governance model focuses on establishing fair and transparent markets that will provide economic empowerment;

- Transparent pricing
- Elimination of intermediaries
- Fair access to markets and distribution channels
- Strong cooperatives
- Supportive financial mechanisms

Interviews with coastal fishermen show that markets are generally dominated by industrial fishermen who possess strong economic resources and political power. It is stated that economically and politically powerful industrial fishing communities influence official institutional practices, quotas, and fishing practices. It is even stated that monitoring such industrial fishermen is difficult, even in local monitoring processes. This demonstrates that political powers can intervene at every stage of the process. Economic empowerment is considered the foundation of the ecosocial model. It is emphasized that economic empowerment is essential for coastal fishermen to be recognized and respected as a respectable profession. Another factor affecting economic empowerment is the lack of recognition of coastal fishing as a profession. Coastal fishing is still not officially recognized as a profession.

Governance systems are presented below:

- Structural power imbalances refer to the dominance of industrial fishermen and aquaculture farmers in decision-making processes, with limited participation for smaller stakeholders such as coastal fishermen. The way to prevent this is to establish participatory governance mechanisms.

- Relational power imbalances are used to describe unfair and unequal dominance in decision-making processes. Asymmetrical relationships and opaque processes increase relational power imbalances.
- Cognitive power imbalances refer to the weakening of traditional knowledge due to factors such as climate change and pollution, causing coastal fishermen's predictions about the future to become invalid. However, the integration of scientific knowledge can reduce these cognitive power imbalances and increase the fishermen's adaptability.

Applying the ecosocial model to fisheries management within an ecosocial governance model requires the integration of ecosystem sustainability, social justice, participatory governance, and fair markets. The components of the ecosocial model reinforce both ecological and social outcomes. This integrated approach is compatible with multidimensional, inclusive, and justice-oriented governance models in fisheries research. When these components are brought together, they offer a pathway to develop accountable, resilient, and equitable fisheries management systems that can address the complex socio-ecological challenges faced by Black Sea coastal communities.

CONCLUSION

This research is based on interviews with coastal fishermen in Sinop, a city synonymous with coastal fishing in Türkiye, which has the longest coastline on the Black Sea. The interviews revealed that while the problems faced by coastal fishermen are generally related to depletion of fish stocks, climate change, and pollution, ecological, economic, and institutional pressures are also significant. The research is based on field interviews designed based on a literature review. In-depth interviews with coastal fishermen revealed that they are exposed to challenges arising from various interconnected problems. These problems generally appear to stem from sociological and ecological determinants. The Black Sea's unique characteristics, pollution, climate change, regional wars and tensions, fragile international relations, and weak cooperation are prominently featured. It has also been observed that studies of this kind targeting coastal fishermen are insufficient. A significant finding is that fishermen themselves have reported almost no contact with academics for such studies. From a fisheries management perspective, it is evident that practices and policies from other parts of the world are not applicable in the Black Sea, and local and international cooperation is weak. Evidence from the Black Sea shows significant gaps between global policy objectives and local implementation. However, similar implementation failures have been reported in many small-scale fisheries worldwide, where top-down policies often fail to consider local realities and community knowledge.

Ecological monitoring remains inadequate, and institutional limitations weaken environmental accountability. These observations align with global findings indicating that inconsistent monitoring, fragmented regulatory structures, and insufficient ecological data undermine sustainable fisheries governance (Alsaleh & Wang, 2025; Levin et al., 2014).

The findings of this study demonstrate that environmental accountability in fisheries management cannot be considered solely as a legal, regulatory, or environmental/ecological issue. As defined earlier, environmental accountability operates through interconnected ecological, institutional, economic, and social dimensions, all of which are challenged in the Black Sea context, therefor this requires interactions between ecological conditions, legal/institutional regulations, economic conditions, and social participation. Since all these factors are considered from an ecosocial perspective, environmental accountability can be strongly ensured through feedback processes. For example, weak legal/institutional accountability jeopardizes ecological monitoring activities and leads to dangerous losses of ecological resources. This reduces institutional trust among coastal fishermen and makes their lives fragile as their economic resources are at risk. At the same time, it reduces trust in management among coastal fishermen, which increases feelings of cynicism. However, non-inclusive and non-participatory management introduces governance problems in the sector. These negative consequences create situations lacking legitimacy and accountability.

One of the main findings of this study is that sustainability cannot be achieved solely through legal or technical regulations, quota systems, monitoring technologies, or ecological restrictions. Furthermore, governance mechanisms that include all stakeholders, particularly coastal fishermen, ensure inclusivity, transparency, legitimacy, accountability, and consequently, environmental accountability. The ecosocial approach emphasizes the need for comprehensive participatory governance practices because fishing communities suffer from underrepresentation issues stemming from power imbalances. Conflicts of interest and inequalities arising from factors such as gender, race, and religion can lead to underrepresentation and prevent coastal fishermen from fully expressing themselves within governance mechanisms.

The comprehensive perspective of the ecosocial model emphasizes that fisheries problems cannot be solved with simple solutions but can be improved through governance mechanisms that ensure inclusive, participatory decision-making processes. The ecosocial approach envisions the integrated development of justice, corporate responsibility, participation, and ecological sustainability through an interactive process. The findings further indicate that ecological sustainability should be considered together with environmental health and community resilience, consistent with broader environmental management literature.

Ecological, social, institutional, and economic accountability can be achieved through mechanisms that enable stakeholder participation via policy-oriented analytical studies. Identifying the challenges and obstacles faced by coastal fishing communities is important. It is necessary to define issues such as economic problems, market inequality, procedural injustice, non-transparent governance processes, participation/representation issues, and environmental degradation faced by coastal fishermen.

Furthermore, it appears that the participatory governance approach is not entirely ideal for coastal fishing communities. Internal inequalities are observed within communities, particularly in coastal fishing cooperatives. Power imbalances, lack of representation, and access problems in community representation can occur, and these drawbacks may restrict the homogeneous representation of coastal fishing communities in other platforms. Therefore, representation and participation mechanisms should be well-structured from the grassroots level, and feedback channels should always be open. This issue was also highlighted in interviews conducted in the Sinop region, the focus of this study. Overall, all participants expressed concerns about representation and governance.

Fishermen in the Black Sea highlighted several additional concerns, including ecological degradation, market pressures on seafood, and the unregulated expansion of industrial fishing and aquaculture. The issues raised by coastal fishermen also exacerbate ecological problems and social inequalities. Interviews highlighted the weakness of coastal fishing cooperatives, their limited participation in decision-making processes, and their economic vulnerability. Coastal fishermen are losing trust in both their communities and official authorities. This situation is becoming more dramatic with the depletion of fish stocks. The interviews also revealed weaknesses in governance. Participation and inclusion are fundamental principles of governance not only for coastal fishing communities but in all areas of society. This study, which establishes the conceptual foundations of environmental accountability and the ecosocial analytical framework guiding this section, examines how these governance dynamics become visible in the context of Black Sea fisheries, paying particular attention to accountability failures in ecological, institutional, economic, and social areas.

Although based on the Black Sea context, the governance challenges addressed in this study reflect broader tensions affecting global fisheries, including unequal participation, fragmented governance, ecological uncertainty, and market vulnerability. Consequently, the analytical framework proposed here can be transferred to other marine regions facing similar tensions between ecological sustainability, economic pressure, and governance legitimacy. Future fisheries governance research may benefit from examining how ecosocial accountability operates across different coastal systems and institutional settings. In many coastal regions, including the Black Sea, spatial competition, the expansion of aquaculture, and the lack of recognition of local ecological knowledge contribute to social and spatial inequalities. This study proposes an ecosocial governance framework as an analytical tool to address these interconnected ecological and social challenges, for example:

- Ecological sustainability,
- social justice,
- community empowerment,
- participatory governance and
- Fair market relations.

Importantly, Rocha's ecosocial model (2016) offers a complementary perspective by explicitly addressing structural, relational, and cognitive power imbalances that hinder community engagement and weaken accountability. Focusing on treating communities as co-producers of environmental well-being, Rocha's (2016) framework provides a conceptual foundation for empowering fishermen and strengthening local governance structures.

The findings clearly demonstrate that environmentally accountable fisheries management requires the following:

- Very strong ecological monitoring,
- corporate transparent governance,
- fair market conditions and
- Socially inclusive and participatory decision-making systems.

These principles converge across international scientific literature, global guidelines, and local experiences in the Black Sea. Consequently, sustainable fisheries management should be based on empowerment, shared responsibility, and ecosystem management, aligning ecological and social objectives through collaborative, inclusive, and transparent management mechanisms.

REFERENCES

Alsaleh, M., & Wang, X. (2025). The integration of ecosystem-based management in EU fisheries governance: Challenges and opportunities. *Economics of Governance, 26*(4), 527–558. DOI: 10.1007/s10101-025-00338-y

Asio, M. V. S., Cvitanovic, C., Lau, J., & Diedrich, A. (2026). Collaborative governance in marine and coastal management: A systematic review. *Ocean and Coastal Management, 272*, 108020. DOI: 10.1016/j.ocecoaman.2025.108020

Basurto, X., Virdin, J., Smith, H., & Juskus, R. (2017). *Strengthening governance of small-scale fisheries: An initial assessment of theory and practice*. Nicholas Institute for Environmental Policy Solutions.

Berkes, F. (2018). Evolution of co-management: Role of knowledge generation, bridging organizations and social learning. *Journal of Environmental Management, 90*(5), 1692–1702. DOI: 10.1016/j.jenvman.2008.12.001 PMID: 19110363

Bodansky, D. (2010). *The art and craft of international environmental law*. Harvard University Press.

Das, J. (2023). Blue economy, blue growth, social equity and small-scale fisheries: A global and national review. *Studies in Social Science Research, 4*(1), 38–58. DOI: 10.22158/sssr.v4n1p38

Dehalwar, K., & Sharma, S. N. (Eds.). (2026). *Deltas resilience: Nature-based solutions for sustainable development in India*. Springer Nature. DOI: 10.1007/978-3-032-07240-5

Demirci, A., Şimşek, E., & Demirci, S. (2026). Weak representation, fragmented governance: Evidence from four fisheries conflicts in Türkiye. *Marine Policy, 192*, 107176. DOI: 10.1016/j.marpol.2026.107176

Dietz, T., Ostrom, E., & Stern, P. C. (2003). The struggle to govern the commons. *Science, 302*(5652), 1907–1912. DOI: 10.1126/science.1091015 PMID: 14671286

Dixon, M., Grilli, G., Stewart, B. D., Bark, R. H., & Ferrini, S. (2024). The importance of rebuilding trust in fisheries governance in post-Brexit England. *Marine Policy, 161*, 106034. DOI: 10.1016/j.marpol.2024.106034

European Commission. (2020). *Fisheries innovations under the European Maritime and Fisheries Fund 2014–2020*. https://oceans-and-fisheries.ec.europa.eu/document/download/0eb81df7-8d30-4eb0-8e00-2fb187f29859_en

European Commission. (2023). *The common fisheries policy.* https://wodnesprawy.pl/en/common-fisheries-policy-evaluation-consultations/

European Parliament. (2024). *EU oceans and fisheries policy: Latest developments and future challenges.* https://oceans-and-fisheries.ec.europa.eu/policy_en

Food and Agriculture Organization. (2022). *Guidebook for evaluating fisheries co-management effectiveness.* DOI: 10.4060/cc2228en

Food and Agriculture Organization. (2024). *The state of world fisheries and aquaculture 2024.* https://openknowledge.fao.org/items/ceda292f-ac05-4254-b8b5-695156007fab

Gelcich, S., Martínez-Harms, J., Tapia-Lewin, S., Vásquez-Lavín, F., & Ruano-Chamorro, C. (2019). Co-management of small-scale fisheries and ecosystem services. *Conservation Letters, 12*(4), e12637. DOI: 10.1111/conl.12637

Jentoft, S. (2007). Limits of governability: Institutional implications for fisheries and coastal governance. *Marine Policy, 31*(4), 360–370. DOI: 10.1016/j.marpol.2006.11.003

Jentoft, S., Chuenpagdee, R., Bugeja, A., & Isaacs, M. (2022). *Blue justice: Small-scale fisheries in a sustainable ocean economy.* Springer., DOI: 10.1007/978-3-030-89624-9

Kashyap, A. (2025). *Sustainable global trade: Balancing economic growth with environmental accountability. SSRN.* DOI: 10.2139/ssrn.5550500

Miret-Pastor, L., Molina-García, A., García-Aranda, C., & Herrera-Racionero, P. (2022). Social vulnerability and adaptive capacity in small-scale fisheries under climate change. *Aquaculture and Fisheries, 7*(6), 593–604. DOI: 10.1016/j.aaf.2022.03.004

Pinsky, M. L., Fenichel, E., Fogarty, M., Levin, S., McCay, B., St. Martin, K., Selden, R. L., & Young, T. (2020). Fish and fisheries in hot water: What is happening and how do we adapt? *Population Ecology, 63*(1), 17–26. DOI: 10.1002/1438-390X.12050

Radzi, H. (2023). An overview of environmental, social and governance (ESG) and company performance. In *European Proceedings of Social and Behavioural Sciences* (pp. 1111–1122). DOI: 10.15405/epsbs.2023.11.90

Ranjan, D., Singh, S., Singh, A., Misra, V., Singh, C., & Kumar, D. (2026). Recent advancements in sustainable aquaculture: Innovative techniques and future prospects. *Frontiers in Aquaculture, 5*, 1770106. DOI: 10.3389/faquc.2026.1770106

Rocha, H. M. B. C. C. L. (2016). *Social work and environment: Ecological sustainability of vulnerable communities (Serviço social e ambiente: A sustentabilidade ecológica das comunidades vulneráveis)* (Doctoral dissertation, ISCTE – Instituto Universitário de Lisboa). https://hdl.handle.net/10071/11964

Rościszewski-Dodgson, M. J., & Cirella, G. T. (2026). Quota constraints, market pressures, and structural change in Polish Baltic fisheries. *Marine Policy, 189*, 107120. DOI: 10.1016/j.marpol.2026.107120

Safronova, A. (2025). Experts faced with a drying wetland: History of fisheries in the Volga River delta, 1930–1962. *Water History*, *17*(2), 201–228. DOI: 10.1007/s12685-024-00357-2

Sharma, S. N., & Dehalwar, K. (2025). Transit-oriented development and economic development. *Transportation in Developing Economies*, *11*(2), 23. DOI: 10.1007/s40890-025-00245-1

Simataa, C. B., Persendt, F. C., & Gomez, C. (2025). Illegal, unreported and unregulated (IUU) fishing in Africa: A systematic review. *Aquaculture, Fish and Fisheries*, *5*(4), e70107. DOI: 10.1002/aff2.70107

Stiglitz, J. E., Sen, A., & Fitoussi, J.-P. (2009). *Report by the Commission on the Measurement of Economic Performance and Social Progress*. European Commission.

United Nations. (1992). *Rio declaration on environment and development.*

United Nations Environment Programme. (2021). *Making peace with nature: A scientific blueprint to tackle climate, biodiversity and pollution crises.* DOI: 10.18356/9789280738377

Van Santen, G. (2024). Fisheries governance: The search for effective management and the illusion of control. *Global Perspectives*, *5*(1), 94988. Advance online publication. DOI: 10.1525/gp.2024.94988

Yadav, K., Dehalwar, K., & Sharma, S. N. (2025). Factors affecting first and last mile accessibility. *GeoJournal*, *90*(6), 298. DOI: 10.1007/s10708-025-11546-8

Yadav, K., Dehalwar, K., & Sharma, S. N. (2026). Exploring the environmental determinants of mode choice in first and last mile connectivity. *Innovative Infrastructure Solutions*, *11*(4), 204. DOI: 10.1007/s41062-026-02614-0

ENDNOTE

1 This study was funded by TÜBİTAK (Turkish Scientific and Technological Research Council) under the 2221 Sabbatical Research Program.

Chapter 9
Article 9-A and the Constitutionalization of Administrative Climate Governance in Pakistan

Ibrar Ahmad
http://orcid.org/0009-0001-0846-0218
Southwest University of Political Science and Law, China

Aftab Haider
http://orcid.org/0009-0009-3339-5057
Southwest University of Political Science and Law, China

ABSTRACT

Pakistan ranks 179th out of 180 countries in the 2024 Yale Environmental Performance Index, yet it possesses some of the most advanced climate jurisprudence in the Global South. This paradox reflects not a lack of environmental laws, but the persistence of a colonial administrative structure designed for extraction rather than conservation. This chapter argues that the Supreme Court's decision in Mehar Badshah v. Government of Khyber Pakhtunkhwa, delivered shortly after the insertion of Article 9-A into the Constitution in October 2024, establishes a doctrine of Administrative Environmental Reasonableness. The doctrine creates a constitutional standard for evaluating climate governance through administrative capacity, transparency, responsiveness, and accountability rather than legislative enactment alone. The chapter further examines the constitutional tension between Article 9-A and the 18th Amendment's devolution of environmental powers, proposing Pakistan as a potential Global South model for climate governance through administrative reform.

DOI: 10.4018/979-8-2600-3678-5.ch009

1.0 INTRODUCTION

On 21 October 2024, Pakistan incorporated Article 9A into its constitution. The new provision ensures the right to a clean, healthy, and sustainable environment for all the citizens of the country (Constitution of the Islamic Republic of Pakistan, 1973, art. 9-A). Three days later, on 24 October 2024, the Supreme Court of Pakistan decided the case of *Mehar Badshah v Government of Khyber Pakhtunkhwa* (PLD 2025 SC 36)[1]. This was the first case in which the new right was used by the court. The Court did not mandate payment of damages by a polluter and did not instruct a development project. It did not provide compensation for victims. Rather, it called upon the provincial Forest Department to give up its colonial attitude. The Department was directed to publish its yearly budget including all 7,422 employees. It was directed to mention the exact method of measuring forest cover and ordered to differentiate between legal timber permits and illegal seizures. The Court relied on its previous judgment in Province of Sindh v. Shahzad Hussain Talpur (2022 SCMR 439) to emphasize that the name and designation of the officer signing the document must be mentioned in every official document. The Court found that anonymity cannot provide freedom from constitutional accountability (Iqbal, 2022). These demands are not the traditional environmental solutions. In conventional litigation, the goal is to prevent injury or to compensate the injured parties. *Mehar Badshah vs Government of Khyber Pakhtunkhwa* requires changing the agency that allows harm. The Court focused on the constitutional issue of the quality, capacity, and transparency of the bureaucracy itself. This is the first time in Global South climate jurisprudence.

Mehar Badshah has been welcomed as a landmark environmental rights case. They describe Article 9A as a transformative constitutional platform for ecological governance. These evaluations are right but superficial. They are unaware of the underlying structural change. The Court does not simply compel compliance with environmental standards by a violator. It is exercising a constitutional right to operate in a colonial administrative state. The forest, irrigation and environmental bureaucracies were created by the British Raj to extract resources and maximize revenue in Pakistan. Their working plans are based on timber output rather than on protecting the ecosystem. They take advantage of opaqueness and patronage in their field operations. The nigahban system is a good example of this legacy: poorly paid guards with no accountability upwards, involved with local timber interests. These agencies were devolved to the provinces in 2010 under the 18th Amendment, but were not redesigned. It gave constitutional ownership to institutions that maintain the same old colonial extraction design. The outcome of this approach is decentralized incompetence. There is now a Forest Department and Environmental Protection Agency in each province. However, they haven't been reorganized for ecological stewardship. The disconnect between the promise of the constitution and the reality of administration is not an implementation failure. It's a design fault.

This chapter proposes that Article 9-A develops a doctrine of Administrative Environmental Reasonableness. It is a constitutional standard that can be tested in court and that evaluates the state's climate governance. That is not based on state legislation, but on its administrative capacity, transparency, and responsiveness. It is the first attempt in the Global South to constitutionalise the governance of climate change by means of bureaucratic reform and not by statutory expansion. The doctrine turns the focus of the judiciary from the environment to the machinery that creates the environment. It makes the quality of the administrative state itself a justiciable constitutional question. The doctrine renders administrative capacity justiciable and thus gives a new definition to the content of Article 9-A. The right to a clean environment is not a right to be free from pollution but a right to a pollution-free state. The chapter highlights the Court's jurisprudence in four prongs: capacity adequacy, methodological transparency, administrative responsiveness, and designated accountability. These points give courts constitutional sources to ask questions of the administrative state.

The chapter proceeds in four analytical steps. Section II is an analysis of the colonial pathology of Pakistan's environmental bureaucracy. It shows that the Forest Department and provincial Environmental Protection Agencies are not only under-resourced, but constitutionally incompatible with environmental governance, as their institutional design is made for extraction. It follows the devolution of incapacity in the 18th Amendment and reveals the ritualism of hard law. The doctrinal trajectory between Leghari's procedural remedies and Mehar Badshah's structural reform mandate is explored in Section III. It breaks down Administrative Environmental Reasonableness into four parts as a constitutional test. It sets this doctrine in contrast with the ad hoc public interest litigation in India, the explicit rights in Kenya, and the intergenerational jurisprudence in Colombia. Section IV addresses the ongoing federalism issue between Article 9-A and the 18th Amendment. It states that the Court has been an accidental harmonizer of federalism, and suggests the Council of Common Interests as a constitutional solution. The Global South implication is discussed in Section V. It claims that administrative constitutionalism is a less costly and more politically viable climate approach than statutory expansion in post-colonial states. The chapter concludes that the true test of Article 9-A is not whether the air in Pakistan becomes cleaner, but whether its colonial bureaucracies have been restructured into climate-responsive institutions.

This chapter examines the emerging literature on environmental constitutionalism and climate governance. Bookman (2024) differentiates between substantive and procedural environmental rights and introduces systemic environmental constitutionalism, noting that most constitutional systems merely recognize rights without establishing effective implementation structures (Bookman, 2024). Climate constitutionalism has become a global trend, but legal systems are more likely to adapt than transform, as they incorporate climate into the current legal system without changing the administrative structure. Niehaus (2023) examines climate litigation 'from below,' with courts playing an increasingly important role in international climate law, but does not focus on administrative capacity as a constitutional principle, but rather on litigation as a corrective (Niehaus, 2023). Baber (2024) focuses on implementation and suggests that the success of the European Green Deal depends more on institutional capacity than on normative objectives (Quirico & Baber, 2024). Petersmann (2024) takes this one step further and classifies the enforcement gaps as a structural element of multi-level governance (Petersmann, 2024). This chapter addresses a particular research gap: no existing scholarship has constitutionally considered administrative capacity – including budget, staffing, procedures, and accountability of environmental agencies – as a justiciable standard, distinct from substantive rights and procedural remedies. The proposed doctrine of Administrative Environmental Reasonableness responds to this by elevating the quality of the administrative state to a constitutional question, rather than a managerial or statutory one.

2.0 THE COLONIAL PATHOLOGY

2.1 The Extractive Architecture

The Supreme Court in the Mehar Badshah case characterized the failure of the Forest Department as a "colonial mindset". This was a factual assessment, not just a rhetorical flourish, highlighting an institutional design. The Court noted that the Department's forest policy is based on the concept of forests as a resource to be exploited, which is a legacy of the former colonizers (Supreme Court of Pakistan, 2025). This is an insight that has pinpointed the fundamental issue of Pakistan's environmental challenges: the bureaucracy was built for extraction, not for conservation and governance.

There is ample empirical evidence to support this opinion. Pakistan has approximately 5% of forest cover in its total land area, which is one of the lowest in Asia (Khan, 2025). From 2001 to 2025, 9.6 thousand hectares of tree cover were lost, 95% of which was in Khyber Pakhtunkhwa and the former Federally Administered Tribal Areas (Global Forest Watch). Three-quarters of the loss (7.2 thousand hectares) is due to logging, which directly correlates with the jurisdiction of the Forest Department examined in Mehar Badshah (Vizzuality, 2026). The data of the Department show that the institutional approach is more of an authorization rather than a conservation approach. It employs 7,422 staff and has issued timber cutting permits for 6,066,185 cubic feet of timber in the past five years while 130,255 cubic feet of illegal timber has been seized (Supreme Court of Pakistan, 2025). The ratio of authorization to enforcement is about 47:1. This ratio shows that it is not an environmental protection policy, but a revenue collection policy. This ratio is a direct reflection of an organizational principle that was directly transferred from British Indian forest management. Colonial forestry was more interested in revenue generation and strategic provision of timber than in protection of the ecosystem (Sivaramakrishnan, 1999). The forest working plans, which still form the backbone of the Department's planning, are still largely geared towards timber production and not the entire spectrum of ecosystem services. They are still outdated, sectoral and not ecologically oriented (Khan, 2025). This focus is not due to the lack of bureaucratic dynamism, but to institutional memory in a system oriented towards resource extraction.

Administrative structure and environmental outcomes are not necessarily related. There are other factors that need to be taken into account when it comes to Pakistan's environmental problems. Limited resources are a major constraint: Provincial environmental budgets are inadequate to meet legal requirements (World Bank, 2022), and international climate finance for Pakistan is far lower than the estimated USD 7–14 billion required annually for adaptation (Ministry of Climate Change, 2025). Political economy factors also play a role: timber mafias have local political backing, and elected officials may be more interested in short-term economic benefits than in ecological sustainability (Khan, 2025). Post-colonial economic dependence also limits policy options, as debt repayment and structural adjustment reduces fiscal space for reform. The central thesis is that administrative structure alone is insufficient to guarantee positive environmental results: an agency with adequate resources and political will to be an ecological steward cannot be if its structure does not change to enable it to do so. Mehar Badshah's 47:1 ratio of authorisation to enforcement is not just about budget — it's about the institutional mindset of permits and not prevention. The framework of decentralised despotism, as outlined by Mamdani (2018), explains why this approach continues to be in place even after devolution, since the institutional design was simply transferred without reorienting purpose away from colonial legacy. The central argument is that the administrative structure affects the impact of resources and political will on environmental outcomes. Thus, the constitutional inclusion of administrative capacity addresses a basic structural problem that cannot be solved by more money or political pressure.

This is due to the decentralised despotism of Mamdani. Colonial governments ruled over rural areas with unaccountable local authorities, who regarded natural resources as a patrimony, rather than a public trust (Mahmood Mamdani, 2018). The backbone of the Department is the nigahban system, which is an example of this model. These officers are poorly paid, have no independent oversight, have limited accountability upwards, and are closely linked to local timber interests (Khan, 2025). The Court implicitly recognized this lack of transparency by stating that an undated, unsigned report was submitted to a provincial constitutional inquiry by a district-level officer (Supreme Court of Pakistan, 2025). This anonymity is not a regulatory failure; it's a structural characteristic of a system designed to be unaccountable.

The environmental effects are significant and quantifiable. *Natural montane* forests in the Himalayan, Karakoram, Hindu Kush, and Sulaiman ranges have degraded, worsening the floods in 2010, 2022, and 2025 (Khan, 2025). In 2022, 33 million people were affected by floods, with more than USD 30 billion in damages and losses (Adil et al., 2025). During the monsoon floods of 2025, the Department's Khyber Pakhtunkhwa watersheds claimed over seven hundred lives (Khan, 2025). Natural forests are useful for slope stabilization, runoff control, and sediment control. Plantations of monocultures fail to provide these hydrological benefits (Khan, 2025). But the Department's colonial history is more interested in timber volume than watershed resilience.

This significant failure is worsened by concealment of the methodology. According to official figures, natural forest areas have shrunk from 3.59 million hectares to 3.32 million hectares, a drop of 270,000 hectares or 27,000 hectares per year (Ministry of Climate Change, 2025). Nevertheless, Global Forest Watch shows that there was a net gain of 95,000 hectares of tree cover between 2000 and 2020 (Vizzuality, 2026). This is because official figures combine plantation trees with natural forests, which overestimate coverage and conceal hydrological collapse. The Court's demand for clarification on "how is forest cover measured/determined" was thus not just informational but a constitutional demand for transparency in methodology (Supreme Court of Pakistan, 2025). Table 1 shows the structural mismatch in Pakistan's key environmental agencies, which is consistent: a colonial legacy of revenue extraction, a post-18th amendment legal mandate for environmental regulation, a constitutional mandate for rights through Article 9-A, and a lack of administrative redesign to connect these.

Table 1. The Colonial Design vs. Constitutional Mandate Matrix

Environmental Agency	Colonial Design Function	Post-18th Amendment Legal Mandate	Structural Incompatibility
Provincial Forest Departments	Timber extraction, revenue maximisation, imperial yield (Sivaramakrishnan, 1999)	Forest conservation, anti-logging enforcement, watershed protection	7,422 KPK personnel organised for extraction, working plans focus on timber yield, not ecosystem services (Khan, 2025)
Provincial Environmental Protection Agencies	No colonial precursor, created post-PEPA 1997	Environmental regulation, EIA/IEE approval, NEQS enforcement	Attached institutions of provincial departments; structurally unable to scrutinise parent government projects (Malik, 2025)
Provincial Irrigation Departments	Revenue collection through canal water taxation, colonial land settlement	Water resource management, irrigation governance	Colonial canal infrastructure; no climate-resilience adaptation in planning cycles (Ministry of Climate Change, 2025)
Ministry of Climate Change	No colonial precursor, created 2012	National climate policy coordination, international reporting	"Limited influence on outcomes... because of capacity and resource constraints", lacks authority to direct provincial agencies (World Bank, 2022)

Note. Data compiled from (Supreme Court of Pakistan, 2025), (Khan, 2025), (Malik, 2025), (World Bank, 2022), (Ministry of Climate Change, 2025), (Sivaramakrishnan, 1999).

2.2 The 18th Amendment as Devolution Without Decolonization

The 18th Amendment (2010) removed the Concurrent Legislative List and transferred the responsibility of *environmental pollution and ecology* to the provinces. This change was considered a move towards democratic decentralization. Nevertheless, it was devolution without decolonization. Constitutional powers were transferred to provincial institutions, but these institutions maintained their colonial structure. Provinces were granted legal authority, but no institutional changes were made. Their Forest Departments, Environmental Protection Agencies, and Irrigation Departments were still in the same form as they were during the colonial period. They were extractive, and provincial jurisdiction was a cover.

The design flaws were compounded by the fiscal framework. The provincial budgets are largely dependent on federal allocations, and have limited revenue generation capacity (World Bank, 2022). The World Bank notes that virtually no investment in building capacity below the provincial level occurs and that no effective devolution of responsibilities or budgets to local or municipal governments occurs (World Bank, 2022). Provinces have constitutional authority over environmental governance but lack the financial means to exercise it. Consequently, environmental requirements have turned into unfunded mandates, which are constitutional obligations that are not backed up by specific funding. The independence of the institutions was essentially unattainable. According to the Human Rights Commission of Pakistan, Environmental Protection Agencies (EPAs) in all provinces are attached institutions of provincial departments (Malik, 2025). An attached institution cannot investigate or take sanctions against its parent department. However, the most environmentally important projects like CPEC corridors, irrigation projects, and urban development authorities are government-owned or restricted. EPA verifies compliance for the same entity it reports to. This arrangement constitutionally makes it difficult to enforce EIAs on major state projects.

Data and policy are part of the coordination gap. The World Bank highlights weak federal-provincial coordination and disjointed spending plans (World Bank, 2022). According to Pakistan's Third National Communication, Pakistan lacks a single national Measurement, Reporting and Verification (MRV) system. Data on greenhouse gases is "not available in a single database or repository" (Ministry of Climate Change, 2025). The UNFCCC depends on Tier-1 secondary data because provincial institutions are not able to generate primary data. The 18th amendment created four distinct provincial environmental data regimes, none of which are encouraging to a working national system.

This lack of capacity is reflected in the Balochistan Climate Change Policy 2024. It acknowledges that 40% of trees in the Koh-e-Sulaiman Range were burnt in 2022. It also mentions that 1.3 million people were displaced due to floods in the province that year (Government of Balochistan & United Nations Development Program, 2024). A province with such ecological destruction did not have the institutional capacity to create its own climate policy. As reported in the earlier Collaborative Adaptation Research Initiative in Africa and Asia Working Paper, climate change "remained foreign terms in the provinces," and as of 2016 there were no adaptation policies in place (Jo-Ellen Parry, 2016). The Third National Communication, 10 years later, indicates that capacity limitations in the provinces continue. Four provincial environmental agencies, all colonial relics, were established by the 18th Amendment, none of which can effectively manage the environment.

2.3 Ritualism and the Failure of Procedural Remedies

As the Pakistan has a well-developed legal system to protect the environment. The Pakistan Environmental Protection Act, 1997, provides for the creation of Environmental Tribunals, National Environmental Quality Standards and criminal penalties. The 2017 Climate Change Act establishes a Federal Climate Change Authority, Council and Fund. Each of the four provinces has its own environmental legislation, and EIAs are mandatory above a certain threshold. This is formal hard law. But hard law requires a Weberian administrative state[2], with hierarchical authority, trained personnel, clear mandates, and binding accountability. Pakistan does not have all of these requirements. As a result, enforcement becomes ritualistic: legal procedures are followed, but do not achieve good environmental outcomes. This chapter outlines the continued inflation of normative standards without administrative effectiveness.

The Climate Change Act 2017 is a good illustration of this mechanism. It was intended to create three federal institutional bodies, but according to the World Bank, these institutions have "never come into existence" due to a lack of technical capacity, funding, and institutional support (World Bank, 2022). While the law is in place, the institutions are not. This is a clear example of statutory inflation. Similarly, the provincial system of Environmental Protection Agencies has similar issues. The Lahore High Court in 2015 pointed out the deficiencies in the regulation and enforcement of EPAs. The Human Rights Commission of Pakistan adds that these deficiencies "persist even today" (Malik, 2025). Many big development schemes are granted without independent scrutiny. The law is in place, but the political will to implement it against the government is not there.

The Leghari Climate Change Commission (2015-2018) exemplifies procedural change as a mere bureaucratic performance. The Commission stated that it had achieved 66% of its priority actions (Mahaseth & Goyal, 2021). However, it also acknowledged that no completion report on any of the Priority Actions was requested or submitted. These numbers were self-reported by departments and were not properly documented, reviewed, or verified. Percentages without evidence are not good governance, but only a facsimile of it. Mehar Badshah showed lip service at the highest level. The Khyber Pakhtunkhwa government responded to a Supreme Court constitutional order with an undated, unsigned report from a Division Forest Officer instead of the Provincial Secretary (Supreme Court of Pakistan, 2025). The Court referred to Province of Sindh v. Shahzad Hussain Talpur (2022 SCMR 439) to highlight that "the designation and name of the person must always be disclosed as it also affixes responsibility" (Supreme Court of Pakistan, 2025). The necessary form was completed, but there was a lack of accountability.

The Third National Communication of Pakistan has identified major gaps in the measurement infrastructure. There is no national MRV system in place. The data on GHGs is not centralized in a database or repository, and most of the categories of the inventory do not have emission factors for the country (Ministry of Climate Change, 2025). Although Pakistan has set ambitious Nationally Determined Contributions under the Paris Agreement, the country is not methodologically capable of tracking its progress. The statutory framework has grown, but the administrative structure to implement it has not changed significantly. Table 2 shows this trend in five institutional mechanisms.

Table 2. The Enforcement Ritualism Index

Ritualistic Indicator	Legal Existence	Administrative Reality	Source
Environmental Tribunals	Established under PEPA 1997 and all four provincial equivalents	Convict only low-level industrial polluters, government-linked project violations routinely unsanctioned	(Malik, 2025) (World Bank, 2022)
EIA/IEE Procedures	Mandatory under PEPA and provincial environmental protection laws	Approved for politically sensitive projects without independent scrutiny, process chronically politicised	(Kathmandu School of Law, 2022), (World Bank, 2022),
Climate Change Commission (*Leghari*)	Judicially created 2015 under WP 25501/2015 (Lahore High Court)	Dissolved 2018, reported 66% action completion with zero verification documents ever filed or requested, replaced by toothless Standing Committee	(Mahaseth & Goyal, 2021)
National MRV System	Committed under NDC 2021 and Climate Change Act 2017	Does not exist, GHG monitoring data siloed across provincial agencies, no single national data repository, Tier-1 secondary data used for UNFCCC submissions	(Ministry of Climate Change, 2025)
Federal Climate Change Council	Established under Climate Change Act 2017	Potentially *ultra vires* the Federal Legislative List post-18th Amendment; lacks constitutional authority to coordinate or direct provincial environmental agencies	(Malik, 2025), (World Bank, 2022)

Note. Data compiled from *(World Bank, 2022)*, (Malik, 2025), (Mahaseth & Goyal, 2021),(Ministry of Climate Change, 2025), (Kathmandu School of Law, 2022).

This pattern represents the change of doctrine in Mehar Badshah. The Court did not tell the legislature to make more stringent environmental laws, nor did it create a new commission or tribunal. Rather, it directed the Forest Department to make public its budget, staffing, and measurement methodology (Supreme Court of Pakistan, 2025). These are not legislative mandates, but requests aimed at enhancing administrative capacity. The Court noted that the environmental problems in Pakistan are not due to the absence of laws, but due to the failure of the administrative state to implement the laws. Article 9-A is not compatible with the administrative system of Pakistan. Article 9-A, therefore, is not a mandate to establish new rules for institutions that are not able to enforce existing rules, but rather a constitutional requirement for administrative reform.

3.0 DOCTRINAL EVOLUTION AND ADMINISTRATIVE RATIONALITY

3.1 The Leghari Legacy and Procedural Compliance

The significance of the key doctrinal insight of *Asghar Leghari v. Federation of Pakistan* (WP 25501/2015, Lahore High Court) is not in its remarkable procedural innovations. Rather, it points to the unasked constitutional question. The Court created a Climate Change Commission, appointed focal persons, and made continuous mandamus orders (Asghar Leghari vs. Federation of Pakistan, 2025). It did not ask if ministries had the right institutional set-up to manage the environment, but only whether they fulfilled their duties. The Commission reported 66% progress on priority actions, but noted that it never asked for verification documents, as specified in Section II. The Commission was dissolved in 2018, and the institutions created by the Climate Change Act 2017 have "never come into existence" (World Bank, 2022). Leghari showed that procedural actions alone, without structural reform, lead to bureaucratic performativity. The Supreme Court's 2024-2025 jurisprudence is a major difference from past procedural rulings. The first sign of this change is in the language that is used. Leghari used supervisory verbs like constituted" a Commission, "summoned" ministries for "assistance," and monitored the "implementation" of the Framework (Mahaseth & Goyal, 2021). These verbs suggest an administrative state that can be made to comply by appropriate supervision. In contrast, Mehar Badshah employed interrogative verbs, commanding the Forest Department to "abandon" its colonial mindset, "disclose" its budget, "account" for each employee, and "explain" its measurement methods (Supreme Court of Pakistan, 2025). These verbs do not imply capacity, but rather determine if capacity exists.

This linguistic change is a fundamental change in doctrinal perspective. Leghari was appointed as the administrator and was responsible for the implementation of the same. On the other hand, Mehar Badshah asks for transparency and doubts the design of the administration. The Court is no longer satisfied with activity reports, but wants to know how the structure is, how much money is spent, who is working, and how. The emphasis has moved from the environmental results to the quality of the administrative system, which is now considered a constitutional problem. The timing of these events is crucial: Article 9-A was introduced on October 21, 2024, and the Court's ruling on Mehar Badshah came three days later on October 24, 2024. It was the first time that a judge applied an unqualified constitutional right to a clean environment, and it was not about the environment; it was about the ability of the administration. The Court noted that, from the beginning of this new constitutional era, the practical test of Article 9-A would be related to administrative capability, and that the right is useless unless the government demonstrates that it has the institutional capacity to enforce it. The Court has observed that Leghari's failure led to the dissolution of the Commission, since it was not empowered to restructure the Forest Department. It now recognizes that procedural oversight is not enough without a change in the internal structure of the administrative state. The 2024 change is therefore not merely incremental but a fundamental doctrinal change from policy compliance to strengthening administrative constitutional capacity.

3.2 Administrative Environmental Reasonableness: A Four-Pronged Constitutional Test

Before outlining the four components, it's essential to clarify the doctrine and differentiate it from current legal frameworks. Administrative Environmental Reasonableness is a constitutional standard that calls for environmental agencies to have adequate institutional capacity, transparency, prompt responses, and clear accountability. Lack of any of these elements is a justiciable violation of Article 9A. This doctrine is different from the traditional administrative reasonableness review in three important ways. First, it looks at the institutional structure of the agency, including its budget, staffing, and measurement systems, instead of the individual decisions, such as permits or penalties, that the agency makes (Bookman, 2024). Second, it goes beyond the traditional environmental constitutionalism that focuses on substantive rights to a clean environment or procedural rights such as access to information and participation (Viola, 2022); (Wang, 2024). It includes institutional capacity as an integral part of these rights, rather than as an external precondition. Third, it is different from case-specific and retrospective climate litigation remedies, such as orders to reduce emissions, to pay compensation, or to cancel projects, which are triggered by harm (Niehaus, 2023); (Jenart, 2024). Rather, it is a framework that is structural and forward-looking, requiring continuous evidence of institutional effectiveness. These three differences, in comparison with administrative law, environmental constitutionalism and climate litigation, point to the novelty of this doctrinal approach and to its connection with existing law.

This chapter is based on the 2024–2025 jurisprudence and legal developments, outlines Administrative Environmental Reasonableness as a four-part constitutional test. The doctrine does not have a single express judicial pronouncement, but is implicit in the Court's orders. The four components are based on the specific needs of Mehar Badshah and the constitutional principles that they represent.

First the Capacity Adequacy refers to the environmental agency's ability to monitor and enforce environmental standards, through sufficient budget, trained staff and technical infrastructure. The Court particularly asked for information regarding the annual budget of the Forest Department, the number of employees, and the amount spent on reforestation (Supreme Court of Pakistan, 2025). This is a request of constitutional obligations and not managerial concerns. It takes into account the possibility of violations of Article 9-A due to lack of funding or personnel. The agency may not assert a lack of resources as an excuse for not performing a constitutional obligation. Thus, the sufficiency of state resources is a question that can be subject to legal review.

Second, Transparency of methodology requires the agency to make environmental data available that are verifiable, regularly updated, and follow good scientific practice. The Court inquired, "How is forest cover measured/determined?" (Supreme Court of Pakistan, 2025). This is not a technical inquiry but a constitutional one, which means that accounting practices that mask the loss of Pakistan's natural forests with plantation gains are not acceptable (Ministry of Climate Change, 2025). The government says it is constitutionally compliant if it counts monoculture plantations as forest cover and these plantations are not used to protect watersheds. Transparency in methodology helps to clarify what is accepted as evidence and sets a justiciable constitutional standard.

Third, Administrative responsiveness requires that agencies respond to violations in a constitutionally reasonable time. The KPK Forest Department's enforcement-to-authorization ratio of approximately 47:1 (as discussed in Section II) indicates a lack of institutional attention to documented deforestation (Supreme Court of Pakistan, 2025). The Court's message is that a failure of the administration to act on a constitutional right, which is unqualified, is a constitutional violation. Delay goes beyond mere inefficiency. Incorporating the reasonableness standard from administrative law into the realm of constitutional rights, administrative responsiveness makes the timeliness of enforcement a constitutional issue.

Fourth, Administrative accountability means that environmental governance functions are attributed to a specific state actor at an appropriate hierarchical level. The Court highlighted that documents should "affix responsibility" by disclosing the identity of the signatory, citing Province of Sindh v. Shahzad Hussain Talpur (2022 SCMR 439) (Iqbal, 2022). Anonymously produced bureaucratic outputs are constitutionally inadequate for compliance with Article 9-A. This aspect deals with the opacity of the nigahban system in the colonial era, where decisions are protected from accountability upwards. Assigning responsibility makes traceability of administrative actions a constitutional standard. In addition, Table 3 provides an overview of the evolution of doctrine in five analytical dimensions.

Table 3. The Doctrinal Evolution from Leghari (2015) to Mehar Badshah (2024)

Dimension	Leghari (2015)	Mehar Badshah (2024)	Doctrinal Significance
Judicial Question	Is the government implementing climate policy?	Does the government have institutional machinery to govern?	Shift from policy compliance to administrative capacity
Remedy Type	Procedural: Commission, focal persons, continuing mandamus	Structural: budget disclosure, personnel audit, methodology scrutiny	From oversight of existing institutions to interrogation of their architecture
Accountability Mechanism	Coordination and self-reported percentage completion	Named responsibility; provincial-level hierarchy; verified disclosure	From activity monitoring to institutional architecture review
Constitutional Basis	Article 9 (implied right to life)	Article 9-A (explicit, unqualified environmental right)	From negative/implied to positive/explicit constitutional obligation
Empirical Outcome	66% claimed completion; zero verification documents; Commission dissolved	Undated, anonymous DFO report; Court demands Secretary-level accountability	Both reveal administrative incapacity; only *Mehar Badshah* treats incapacity itself as the constitutional violation

Note. Data compiled from (Mahaseth & Goyal, 2021); (Supreme Court of Pakistan, 2025); (Gomez & Çalı, 2025); (World Bank, 2022).

3.3 Comparative Positioning: Pakistan's Administrative-Structural Innovation

Administrative Environmental Reasonableness in the Global South is incredibly different. In India, public interest litigation under Article 21 is typically ad hoc and aimed at specific remedies. The Supreme Court frequently mandates project shutdowns or payments, but it does not routinely examine agencies' budgets, staffing, or approaches. The National Green Tribunal is a mechanism to deal with specific environmental disputes, but does not establish a general doctrine of administrative reasonableness (Rajamani L, 2007). The Constitution of Kenya 2010 provides for an environmental right in Article 42 and the National Environment Management Authority as a semi-autonomous body, but enforcement is weak, and the right and body were established at the same time, with capacity gaps. In 2018, Colombia's case Future Generations v. Ministry of the Environment acknowledged the Amazon as a rights-bearing entity and mandated intergenerational action plans, but failed to consider agencies' budgets, staffing, or measurement methods. Lozada Gómez and Çalı (2025) refer to this as "limited implementation—procedural compliance without substantive action" (Gomez & Çalı, 2025). Pakistan's approach is unique, as it makes the quality of the state itself justiciable, not just its outputs or rights.

This claim of uniqueness should respond to potential counterarguments. The National Green Tribunal (NGT), established by the 2010 Act, has some supervisory powers through its expert committees and can review environmental compliance at the regional level (Rajamani L, 2007). This may be considered to be a type of administrative reasonableness review. But the NGT's jurisdiction is activated by certain disputes, not by a constitutional obligation to show institutional capacity. It is not a proactive demonstration of competence, but rather a response to failure. As mentioned by Lozada Gómez and Çalı (2025), the case of Future Generations v. Ministry of the Environment (2018) in Colombia required intergenerational action plans with institutional impacts, such as structural reforms. However, the ruling did not consider agencies' budgets, staffing or measurement methods—the heart of Administrative Environmental Reasonableness. The Belgian Climate Case, as explored by Jenart (2024), demonstrates how multi-level constitutionalism can establish climate commitments at different levels of government, albeit in the context of emission limits and not agency capacity (Jenart, 2024). A relevant comparison is provided by Müller Gómez (2024), who shows that local units are often lacking in incentives and capacity to implement international environmental treaties, and that federalism and decentralized governance can be a hindrance (Müller Gómez, 2024). While administrative capacity is a problem in federal systems, Pakistan is unique in that administrative capacity is a constitutional standard itself that is subject to judicial review. The claim here is not that Pakistan is the first country in the Global South to have an environmental agency whose institutional design is the subject of a constitutional challenge, but that it is the first country where a court has directly considered the institutional design of an environmental agency (its budget, staffing, methods and accountability) as a constitutional question under an explicit environmental right, not as a managerial or statutory issue.

Administrative Environmental Reasonableness is a constitutional principle, not a statutory rule, clarifying how Article 9-A should be interpreted. It emphasizes that the right to a clean environment is not only the right to be protected from pollution, but also the right to an effectively managed state, which is legally empowered to prevent pollution. This doctrine focuses the courts' attention from the environmental results to the administrative processes that produce them. It gives the judiciary the power to review the competence of the administration, thus redefining the meaning of Article 9-A and providing a new constitutional structure for climate policy in the Global South.

4.0 THE FEDERALISM CRISIS

4.1 The Constitutional Collision

The problem with Article 9-A and the 18th Amendment is not a drafting issue, but one of sequencing. As outlined in Section II, the 2010 18th Amendment abolished the Concurrent Legislative List, making environmental issues solely provincial. Subsequently, the 2024 26th Amendment introduced Article 9-A, which provides an unconditional federal constitutional right that is binding on all state actors. But this resulted in a federal right without the legislative authority to establish the administrative machinery. The Climate Change Act of 2017 highlights this constitutional ambiguity: it set up the Pakistan Climate Change Council and Authority as federal entities, despite the removal of "environment" from the Federal Legislative List following the 18th Amendment. The Human Rights Commission of Pakistan cautions that these institutions might have constitutional problems as environment is no longer included in the federal legislative list (Malik, 2025). According to the World Bank, federal-provincial coordination mechanisms are "weak," and spending plans are not coherent (World Bank, 2022). The Third National Communication of Pakistan highlights that stakeholders are working in "silos" (Ministry of Climate Change, 2025).

This raises a doctrinal issue which no scholar has yet addressed: what constitutional theory would allow a federal court to enforce a federal right against a provincial administrative body, where the federal legislature is prevented from legislating on the subject of the federal right? Article 9-A provides for the right, but the federal mechanism to enforce it is removed by the 18th Amendment, and the judiciary will address this gap.

4.2 Mehar Badshah as Judicial Federalism by Default

In Mehar Badshah, the Supreme Court, as a federal institution, ordered the Khyber Pakhtunkhwa provincial Forest Department to make its budget, staffing and reporting system transparent. As outlined in Section II, the Court specified six types of administrative disclosures and dismissed the Department's anonymous, undated district-level response (Supreme Court of Pakistan, 2025). The Court drew upon the case of Province of Sindh v. Shahzad Hussain Talpur (2022 SCMR 439) to highlight the importance of having named persons and proper hierarchical accountability in constitutional environmental questions (Iqbal, 2022).

This is a constitutionally unusual situation. The Court is not deciding whether or not the provincial laws meet federal standards, but rather the administrative capacity of the provinces. It's making use of a federal constitutional right to look into the workings of a provincial department. This is problematic, according to the traditional federalism principles: generally, federal oversight of provincial laws is acceptable, but federal intervention in provincial administration is not (Watts, 2008). But the Court is constrained to exceed its usual limits due to the limited constitutional options. If Article 9-A is subject to judicial review, but the federal government has no legislative authority on environmental matters, then only the judiciary can strike a balance between the administrative capacity of the provinces and the rights of the federal government. The Court has unwittingly become a mediator in federalism, a constitutional role that the legislature cannot legally play. This is not cooperative federalism, it is judicial federalism by default. This conflict is shown in Table 4.

Table 4. The Federalism Collision Matrix

Constitutional Element	Federal Level	Provincial Level	The Collision
Legislative Power	No power over environment (18th Amendment, 2010)	Exclusive jurisdiction	Federal government cannot legislate administrative capacity standards for provincial agencies
Constitutional Right	Article 9-A (unqualified, 26th Amendment, 2024)	No equivalent provincial right	Federal right binds all state actors. Federal government cannot create implementing machinery
Administrative Machinery	Climate Change Council/ Authority (potentially *ultra vires*)	EPAs and Forest Departments (colonial, under-resourced)	Federal machinery is constitutionally doubtful provincial machinery is functionally dysfunctional
Judicial Enforcement	Supreme Court issues structural Article 9-A orders	High Courts hold constitutional jurisdiction	Supreme Court orders provincial restructuring without legislative codification
Fiscal Coordination	NFC Award transfers funds unconditionally	Provinces control agency budgets	

Note. Data compiled from (Malik, 2025); (World Bank, 2022); (Supreme Court of Pakistan, 2025); Constitution of Pakistan, 1973 (as amended to 2024).

4.3 The Unacknowledged Federalism Revolution

Article 9-A is a federally inspired protection that gives federal courts the power to evaluate the quality of provincial governance. This is a very important change in federalism that has been neglected by many scholars. It has great doctrinal significance. In general, federalism can be divided into two categories: federal review of provincial legislation (constitutional) and federal interference in provincial administration (controversial). But Article 9-A merges these concepts. The Court is not considering the question of whether the forest laws of Khyber Pakhtunkhwa are in line with the federal standards, but whether the provincial Forest Department has the institutional capacity to discharge its constitutional duties. This is a good example of administrative constitutionalism operating on the federal frontier.

The comparison shows the anomaly of Pakistan. In India, environmental issues are included in the Concurrent Legislative List, which means that both the central and state governments can enact laws on the subject, but this is not the case in Pakistan. The national standards are set by the Central Pollution Control Board in India, and the National Green Tribunal Act of 2010 gives a federal body the power to resolve environmental conflicts (Rajamani L, 2007). This is because India's federal environmental infrastructure is constitutionally empowered. By contrast, Kenya's 2010 Constitution created Article 42 (environmental rights) and the National Environment Management Authority as semi-autonomous federal machinery at the same time (*Constitution of Kenya*, 2010). Nevertheless Pakistan didn't do either at first. It has only established the federal environmental right in 2024, 14 years after the abolition of its federal machinery for the enforcement of this right.

Müller Gómez (2024) provides a comparative theoretical analysis which shows the uniqueness of the challenges of federalism in Pakistan (Müller Gómez, 2024). Based on joint decision-making theory, the study demonstrates that federal systems can have implementation problems in environmental governance, as subnational units have administrative control and federal units have normative authority. This leaves a gap in the structure between the standard setters and the implementers. This pattern is found in other federal systems, such as the EU environmental directives implemented by the states in Germany, the federal climate policy in Canada, where the provinces control natural resources, and the federal environmental policy in Australia, where the states manage federal environmental standards, all of which have similar problems. The Pakistan situation is remarkable not in kind but in order. Federal environmental rights and administrative structures were established at the same time or under federal control in Germany, Canada and Australia. The 18th Amendment (2010) in Pakistan transferred the authority of the environment to provinces, which was way before the 26th Amendment (2024) provided a federal constitutional right that is binding on provinces. This means a federal right without federal implementing machinery, which Müller Gómez foresees will lead to maximum implementation deficits (Müller Gómez, 2024). There is no constitutional motivation for the provincial units to build capacity for a right that they did not create, and the federation has no power to enforce capacity building. This analysis highlights that the Council of Common Interests is not only a functional forum but the only constitutional forum that can reconcile federal rights and provincial capacity.

The order was bad. The federal Parliament could have created implementing institutions if Article 9-A had been passed prior to the 18th Amendment. Likewise, had there been capacity building at the provincial level before devolution, the provinces would have been able to exercise their rights independently. Rather, incapacity was delegated in 2010, and a federal right was constitutionally guaranteed in 2024. The institutional debt that Article 9-A now needs to address stems from this fourteen-year structural gap. Thus, Article 9-A is not only an environmental right, but it also contains an implicit federalism principle. It holds that provincial shortcomings in the protection of a federal fundamental right are a federal constitutional concern. This changes the Supreme Court's role from being just an environmental court to being a protector of federalism, a role it has not explicitly recognized but that the logic of Article 9-A renders unavoidable.

4.4 The Resolution: CCI as Climate Governance Harmonizer

The Council of Common Interests (CCI) is the only constitutional body with the power to solve this conflict. It was established under Article 153 and comprises the Prime Minister, provincial Chief Ministers, and federal ministers. The 18th Amendment expanded its powers: it now oversees the implementation of policies that impact multiple provinces and develops binding inter-governmental positions on national issues. The CCI should not centralize environmental governance, because it is not constitutionally feasible and it would violate the democratic principles of the 18th Amendment. Rather, it should set a federal standard of Administrative Environmental Reasonableness and leave it to the provinces to determine how to achieve it. The CCI would establish a national protocol with minimum standards for capacity, transparency, and accountability of provincial environmental agencies. Provinces would maintain their legislative and administrative independence. Compliance would be verified by the CCI. This is an inter-governmental agreement, not a federal mandate, and is compatible with the Constitution and is based on mutual consent.

The fiscal structure is conducive to the feasibility of this resolution. By 2030, Pakistan requires USD 350 billion for climate investments (World Bank, 2022) and USD 7–14 billion annually for adaptation financing (Ministry of Climate Change, 2025). The federal government has control over international climate fund disbursements (e.g., Green Climate Fund allocations, UNFCCC contributions, NFC Award supplements) and can tie these to provinces' certification of administrative capacity. Provinces not meeting transparency and staffing requirements would not be eligible for conditional climate finance. This is conditional cooperative federalism, not coercion, but an economic system that is constitutionally compliant. Without this resolution, Article 9-A is a weak federal right that is enforced by judicial improvisation in the absence of provincial capacities. Through it, Pakistan has a unique constitutional solution for the Global South: climate rights guaranteed by building administrative capacity, not by centralizing power or proliferating laws.

This proposal is not descriptive, but rather normative, and the implementation of this proposal has identifiable challenges that must be addressed. First, there will probably be political opposition from the provincial governments. Administrative Environmental Reasonableness requires disclosure of budgets, staffing and methodologies, which provincial governments may see as an encroachment on their autonomy. The intergovernmental nature of the Climate Change Initiative (CCI) is designed to reduce this risk by basing compliance on mutual consent instead of federal mandates, but the history of negotiations around the National Finance Commission (NFC) Award demonstrates that conditional fiscal arrangements are often resisted by provinces even when they are constitutionally possible (World Bank, 2022). Second, the CCI has its own capacity limitations. It meets rarely, has no permanent technical secretariat and has concentrated mainly on questions of economic distribution and not administrative standards. The CCI would require institutional support, perhaps in the form of a permanent environmental technical committee, similar to the technical working groups of the NFC, to effectively harmonize climate governance. Third, there is a risk that the CCI protocol could fall into the same ritualism discussed in Section 2.3. This may result in a repeat of the failures of the Leghari Commission if compliance is based on self-reported progress percentages only and not independently verified. Baber (2024) cautions in particular against this in the implementation of environmental constitutionalism: "Normative frameworks without enforcement mechanisms lead to procedural compliance without substantive change (Quirico & Baber, 2024). The proposed protocol should include third-party verification of administrative capacity, for example, by an independent environmental commissioner or the Office of the Auditor General, to address this risk.

This enforcement deficit is a major structural weakness of environmental constitutionalism worldwide, as Petersmann (2024) points out, without institutional means to check compliance, constitutional rights are largely symbolic (Petersmann, 2024). The CCI proposal seeks to address this by tying capacity certification to eligibility for climate finance, which will provide real-world consequences for non-compliance that are not just dependent on judicial enforcement. But the proposal is still a constitutional possibility, not a foregone conclusion. It requires political will, institutional learning and ongoing civil society engagement, which are not guaranteed and must be fostered.

5.0 THE GLOBAL SOUTH IMPLICATION

As demonstrated in Section 2.0, the gap between legislative ambition and administrative capacity is a structural feature of environmental governance across the Global South. This pattern extends beyond Pakistan. The National Green Tribunal and Public Interest Litigation (under Article 21) are in support of India's climate legislation. However, despite these legal successes, problems such as ecosystem degradation, pollution, deforestation, and water scarcity persist throughout these areas. The issue is not a lack of legal ambition, but rather a lack of administrative resources to effectively implement and enforce these laws.

Project Bajho reviews the climate law documents of South Asia, where each SAARC country has its own climate legislation (Kathmandu School of Law, 2022). Nevertheless, the region continues to be heavily impacted by global warming and climate change (Kathmandu School of Law, 2022). The institutional framework of the SAARC has not been very effective. This mismatch between legal framework and environmental results is not unique to Pakistan, but rather a symptom of a more general problem in post-colonial legal systems: they tend to import international environmental standards without having the administrative capacity to implement them.

Pakistan is a case in point, with hard numbers. According to the 2024 Yale Environmental Performance Index (Yale Center, 2024), it is ranked 179th out of 180 countries. The institutions mandated by its Climate Change Act 2017 have "never come into existence" (World Bank, 2022). According to World Bank, the climate and environmental risks could reduce Pakistan's GDP by 18-20 percent by 2050. This is not due to uncontrollable weather events, but to administrative failures. It highlights the problem of capacity in the government. Pakistan's Article 9-A jurisprudence offers a new way to address challenges faced by the Global South. It focuses on administrative constitutionalism rather than climate constitutionalism—where courts use constitutional rights to push for restructuring the administrative state itself. This is not a new law, new ministry, or new treaty. Rather, it invites courts to consider administrative issues such as budgets, staffing, procedures, and accountability as constitutional issues. This judicial power to reform can be more effective than legislative action in post-colonial countries where bureaucracies are more likely to be concerned with extraction than with governance. Bureaucracies that are opposed to legislative change are more susceptible to Supreme Court directives for transparency and accountability in real time.

The comparative analysis emphasizes the innovative aspects of the doctrinal approach. The Public Interest Litigation (PIL) in India under Article 21 of the Constitution evokes strong judicial responses, which are often ad hoc and case-specific; the Supreme Court often suspends projects or orders compensation without setting up permanent administrative standards (Rajamani L, 2007). The Constitution of Kenya 2010 provides for an environmental right in Article 42, and the National Environment Management Authority is a semi-autonomous body, although its implementation is weak (Republic of Kenya, 2010). The Future Generations decision in Colombia (2018) is based on a rights-based, intergenerational approach, which includes an intergenerational pact and the Amazon as a rights-bearing entity. It did not, however, review the budgets, staffing or procedures of existing environmental agencies (Lozada Gómez & Çalı, 2025). While constitutional rights have expanded in each jurisdiction, there is no constitutional mandate for the administrative capacity needed to enforce them. Pakistan is unique in the sense that it has systematised the constitutionalisation of administrative capacity in the context of environmental right.

6.0 CONCLUSION

The true litmus test of Article 9-A is whether Pakistan's institutions, such as forest departments, EPAs, irrigation bureaus, etc., are fundamentally transformed into climate responsive agencies or not, and not just whether air quality improves in the next 10 years. It also depends on whether the Supreme Court's idea of Administrative Environmental Reasonableness has become a sustainable and systematized mechanism to hold the post-colonial state accountable. The Court in the Mehar Badshah case set the administrative capacity as a constitutional standard that is justiciable (Supreme Court of Pakistan, 2025). It pointed out four areas of operation based on its jurisprudence: capacity adequacy, methodological transparency, administrative responsiveness, and accountability. The Court directed the KPK Forest Department to provide information on its budget, staffing (7,422 employees), and methodology for measuring forest cover (Supreme Court of Pakistan, 2025). It was the first time in Global South jurisprudence that a court considered the quality of the administrative state as a constitutional matter in and of itself. The doctrine it established is not only legislative goals, but institutional mechanisms are justiciable. It places the burden of proof on the state to prove administrative adequacy, rather than merely statutory compliance. Three days after the incorporation of Article 9-A, the Court acknowledged that the test of the right would be on the administrative, not environmental, side.

The issues that remain are more important than what has been accomplished, particularly the federalism issue between Article 9-A and the 18th Amendment, which, at present, has no legislative solution under the current constitution. The Court is putting in place structural requirements without legislative backing. The HRCP warns that the federal Climate Change Council might be ultra vires (Malik, 2025). Moreover, according to the World Bank, federal-provincial coordination mechanisms are "functionally inoperative" (World Bank, 2022). The cost of climate adaptation in Pakistan is estimated at USD 7-14 billion per year (Ministry of Climate Change, 2025). Nevertheless, the administrative system is not reformed annually, and the disconnect between constitutional promises and ecological realities continues to grow. Article 9-A is in danger of becoming a symbolic right, one that the judiciary declares but the administration cannot implement, unless the Council of Common Interests resolves the issue. If the Court only addresses procedural issues—such as the creation of additional commissions, focal points, or reporting targets—Pakistan could end up repeating Leghari's mistakes. The Leghari Commission reported that 66% progress has been made on priority actions, but no completion reports have ever been requested (Mahaseth & Goyal, 2021). The federalism issue arising from the 18th amendment and the limited scope of Article 9-A administration may lead Pakistan to lose the chance to create a pioneering constitutional model for climate governance through administrative reforms. The true choice isn't between centralization and provincial autonomy, but between administrative reasonableness and superficial rituals. Climate justice in the Global South is not about more laws, but about courts that are prepared to evaluate the effectiveness of the bureaucracy.

REFERENCES

Adil., ... (2025, February 12). *Climate Risk Index 2025 | Germanwatch e.V.* https://www.germanwatch.org/en/93013

Asghar Leghari vs. Federation of Pakistan. (2025). *Asghar Leghari vs. Federation of Pakistan | UNEP Law and Environment Assistance Platform.* https://leap.unep.org/en/countries/pk/national-case-law/asghar-leghari-vs-federation-pakistan

Bookman, S. (2024). Demystifying Environmental Constitutionalism. *Environmental Law (Northwestern School of Law), 54*(1), 1–77. https://www.jstor.org/stable/48778824

Constitution of Kenya. (2010). https://new.kenyalaw.org/akn/ke/act/2010/constitution/eng@2010-09-03

Gomez, M. L., & Çalı, B. (2025). *FROM LITIGATION TO IMPLEMENTATION: FRAMING SMART REMEDIES IN RIGHTS-BASED CLIMATE LITIGATION.*

Government of Balochistan & United Nations Development Program. (2024). *Balochistan Climate Change Policy.* UNDP. https://www.undp.org/pakistan/publications/balochistan-climate-change-policy

Iqbal, N. (2022). *Supreme Court orders avoidance of term 'competent authority'.* Dawn. https://www.dawn.com/news/1671469

Jenart, C. (2024). The Belgian Climate Case and rising (sea) levels: Human rights, separation of powers and multi-level constitutionalism. *Maastricht Journal of European and Comparative Law, 31*(3), 445–461. DOI: 10.1177/1023263X241287408

Jo-Ellen Parry. (2016). *Review of Current and Planned Adaptation Action in Pakistan | International Institute for Sustainable Development.* https://www.iisd.org/publications/report/review-current-and-planned-adaptation-action-pakistan

Kathmandu School of Law. (2022). *Project Bajho: A Study of the Climate Change Laws and Policies in South Asian Countries.* https://ksl.edu.np/resource/home/resource_detail/50

Khan, M. I. (2025). *Deforestation.* Forest Degradation, and Flood Risk in Pakistan.

Mahaseth, H., & Goyal, S. (2021). Asghar Leghari and Environmental Justice. *Transformative Climate Change Litigation Judgements One Step at a Time., 1*, 124–129.

Mahmood Mamdani. (2018, January 5). *Citizen and Subject | Princeton University Press*. https://press.princeton.edu/books/paperback/9780691180427/citizen-and-subject

Malik. (2025, November 16). The Promise of Article 9A - Bridging constitutional rights and environmental governance. *Human Rights Commission of Pakistan*. https://hrcp.wittywebsolutions.com/the-promise-of-article-9a-bridging-constitutional-rights-and-environmental-governance/

Ministry of Climate Change. (2025). *Pakistan's Third National Communication on Climate Change (TNC)—Climate Change Laws of the World*. https://climate-laws.org/documents/pakistans-third-national-communication-on-climate-change-tnc_e262

Müller Gómez, J. (2024). Federalism, Decentral Governance, and Joint Decision-Making: Bad News for the Implementation of International Environmental Agreements? *Publius, 54*(2), 407–433. DOI: 10.1093/publius/pjae003

Niehaus, M. (2023). Constitution, Constitutionalisation and Constitutionalism. In Niehaus, M. (Ed.), *Global Climate Constitutionalism "from below": The Role of Climate Change Litigation for International Climate Lawmaking* (pp. 71–104). Springer Fachmedien., DOI: 10.1007/978-3-658-43191-4_3

Petersmann, E.-U. (2024). European Economic and Environmental Constitutionalism as Driver for UN and WTO Sustainable Development Reforms. In G. Z. Capaldo (Ed.), *The Global Community Yearbook of International Law and Jurisprudence 2023: Global Law, Politics, Ethics, Justice* (p. 0). Oxford University Press. DOI: 10.1093/oso/9780197795392.003.0004

Quirico, O., & Baber, W. (Eds.). (2024). *Implementing Climate Change Policy: Designing and Deploying Net Zero Carbon Governance*. Cambridge University Press., DOI: 10.1017/9781009341493

Rajamani, L. (2007). *The Right to Environmental Protection in India: Many a Slip between the Cup and the Lip?* https://www.climate.ox.ac.uk/publication/1041472/dimensions

Sivaramakrishnan. (1999). *Modern Forests | Stanford University Press*. https://www.sup.org/books/asian-studies/modern-forests

Supreme Court of Pakistan. (2025, January 25). PLR 2025 SC 5 = PLD 2025 SC 36 = 2024 SCP 372. *PakistanKanoon.Com*. https://pakistankanoon.com/pakistan-law-reporter/plr-2025-sc-5/

Viola, P. (2022). The Ongoing Foundation of the Constitutionalisation of Climate. In Viola, P. (Ed.), *Climate Constitutionalism Momentum: Adaptive Legal Systems* (pp. 41–82). Springer International Publishing., DOI: 10.1007/978-3-030-97336-0_3

Vizzuality. (2026). *Pakistan Deforestation Rates & Statistics | GFW*. https://www.globalforestwatch.org/dashboards/country/PAK?category=undefined

Wang, E. L. (2024). Environmental Constitutionalism in China: A Constitution without Constitutionalism? *Journal of Environmental Law*, *36*(2), 187–202. DOI: 10.1093/jel/eqae012

Watts, R. L. (with Queen's University & Queens's University). (2008). *Comparing federal systems* (Third edition). McGill-Queen's University Press.

World Bank. (2022). *Open Knowledge Repository*. https://openknowledge.worldbank.org/entities/publication/614ddc2b-ca31-53c9-b59c-6bf12a56d336

Yale Center. (2024). *2024 Environmental Performance Index—Pakistan*. Environmental Performance Index. https://epi.yale.edu/country/2024/PAK

ENDNOTES

1 Mehar Badshah v. Government of Khyber Pakhtunkhwa (2025 PLD 36 SC) was a watershed Supreme Court case that explicitly applied Article 9-A (the right to a clean, healthy, and sustainable environment) to penalize institutional failure and environmental degradation in Pakistan.

2 The Weberian administrative state refers to the classic model of public administration theorized by Max Weber, characterized by strict hierarchy, fixed jurisdictional competencies, exhaustive written rules, and a reliance on generalist, career-track civil servants. In the context of South Asian environmental governance, this path-dependent legacy manifests as a rigid, procedural formalism that prioritizes bureaucratic compliance and file-routing over substantive, data-driven ecological outcomes.

Chapter 10
Victim or Violator?
Bangladesh's Dual Legal Identity After the ICJ's Climate Change Advisory Opinion

Mehenaj Binte Amin
http://orcid.org/0009-0008-1192-9567
Green University of Bangladesh, Bangladesh

ABSTRACT

This chapter examines the obligations arising for Bangladesh under existing climate law regimes in the post-ICJ Climate Advisory Opinion period. By identifying the applicable treaty and customary international law obligations, the chapter argues that Bangladesh has, in a strict doctrinal sense, acquired a dual legal identity to comply with its climate commitments; simultaneously a climate-vulnerable rights-holder and a climate-violating duty-bearer. The chapter identifies substantive and procedural gaps in Bangladesh's implementation of these obligations, and delineates the legal consequences, including attribution, causation, and the tripartite responsibility framework, that Bangladesh may face for such breaches under the Opinion's post-2025 jurisprudential architecture. The chapter establishes that Bangladesh can no longer evade responsibility under the veil of vulnerability, and must undertake the legislative and institutional action the Opinion requires to discharge its obligations under customary and treaty-based climate law.

1. INTRODUCTION

On 23 July 2025, the International Court of Justice delivered its Advisory Opinion on the *Obligations of States in Respect of Climate Change* (hereinafter referred to as

DOI: 10.4018/979-8-2600-3678-5.ch010

the "Opinion"), widely regarded as the most consequential judicial pronouncement on international climate law to date. Its significance lies not merely in the answers it provides, but in its potential to create *legal tipping points* that will shape the functioning and decision-making of domestic authorities, including the executive, judiciary, and legislature, within the bounds of their own constitutional mandates (Wewerinke-Singh & Viñuales, 2026)

The Opinion was structured around two broad, interconnected questions referred to the Court: first, 'what obligations do states have under international law in relation to climate change'; and second, 'what legal consequences arise where those obligations are breached'. Recognising the interdependence of these questions, the Court proceeded sequentially, first identifying the substantive obligations owed by states, and only then addressing the consequences flowing from any "act or omission" in breach of them (para 94).

Environmental law, apart from treaty obligations, was for a long time regarded as soft law. However, emerging trends show that environmental protection has gradually become a matter of customary international law among nations worldwide (Amin, 2022). By answering the two aforementioned questions, the Court confirms in the Opinion that, apart from treaty obligations, States are also bound by certain obligations arising under customary international law. This chapter demonstrates how other principles and duties are interconnected within these customary obligations, thereby limiting the scope for States to evade responsibility.

By presenting the Advisory Opinion in a coherent and systematic manner, this chapter chronologically analyses the Court's findings, providing a comprehensive account of its interpretation of States' climate obligations. Alongside examining these findings, the chapter assesses whether Bangladesh's existing legal and policy framework is aligned with the obligations articulated by the Court. The chapter, thus, by connecting the context and progress of Bangladesh, argues that the findings on these two questions, taken together, impose a dual legal identity upon Bangladesh. On the one hand, Bangladesh remains a paradigmatic climate-vulnerable State, entitled to invoke a right to environmental protection under the obligations articulated in the Opinion. On the other hand, the Opinion's treatment of the legal consequences of State responsibility for failures to prevent environmental harm, protect the climate system, and fulfil climate-related obligations, places Bangladesh in the position of an obligatory saviour of the nature. Just as climate degradation makes Bangladesh a victim of climate change, it may likewise become a violator if it fails to fulfil its corresponding obligations.

This chapter interrogates both dimensions of this identity in turn, assessing Bangladesh's position as both a rights-holder and a duty-bearer under the framework established by the Opinion.

This chapter employs a doctrinal legal methodology, with the ICJ Advisory Opinion serving as its primary source of analysis. It also draws upon relevant secondary literature, including scholarly commentaries and academic analyses of the Opinion, to support and contextualise the arguments advanced. In addition, the chapter scrutinises Bangladesh's legislative and policy frameworks to assess the country's compliance with the legal principles articulated in the Opinion. Relevant secondary sources are further examined to strengthen and substantiate the analysis.

This chapter proceeds as follows. Section 2 discusses the obligations identified by the ICJ under both treaty law and customary international law and examines the extent to which Bangladesh has complied with those obligations. This section following the Opinion dissects the obligations under treaty law by mentioning the sections that recognises the obligations directly. It also argues which laws are customary international law and how they are recognised as such. Section 3 builds upon Section 2 by discussing how the ICJ Opinion links and extends the scope of these obligations by connecting various principles and duties that operate within them. This section demonstrates that, in the post-Opinion era, a State remains bound by these obligations even where gaps exist in policy-making or in the implementation of those policies under treaty and customary international law. Section 4 deals with the legal consequences arising from the breach of these obligations. It clarifies the wording mentioned in the Opinion 'obligations' and 'responsibility' to better understand the 'consequences' mentioned in the Opinion. It connects with Section 5 on Bangladesh's dual legal identity by demonstrating how certain activities of private actors may give rise to State responsibility for Bangladesh. Section 5 then examines Bangladesh's dual position as both a climate-vulnerable State entitled to the rights and protections recognised in the Opinion and a State responsible for complying with the obligations articulated by the Court. Finally, the chapter argues that the Opinion is not merely a clarification of existing international law. Rather, by recognising the *erga omnes* character of certain climate obligations and reinforcing the legal consequences of their breach, it significantly strengthens the international legal framework for climate accountability and advances the pursuit of climate justice.

2. OBLIGATIONS UNDER THE ICJ

Obligations in the Opinion are defined in paragraph 102 as "the obligations of States under international law to ensure the protection of the climate system and other parts of the environment". However, by referring to 'international law', the court here included treaty laws, customary international law and relevant principles. This part of the chapter, therefore, firstly discusses which treaty laws are recognised by courts and applicable to Bangladesh; secondly, which obligations arise out of these treaty laws for Bangladesh; lastly, what customary international environmental law is recognised by the courts and their implications for Bangladesh;

2.1 Environmental Treaties Under International Law

To identify the first issue named obligations under international law regarding climate change, the court needed to identify the "law" itself. In this regard, the court set the 'UN Charter as the pillar of all the treaties' and emphasised on the treaties which impose obligations that are 'directly connected' to broader protection of environmental harm (para 115). In the Opinion, the Court mentioned that the United Nations Framework Convention for Climate Change (UNFCCC), the Kyoto Protocol and the Paris Agreement are 'the principal legal instruments regulating the international response to the global problem of climate change' (para 116). Bangladesh is a party to all three.

However, as there was a concern before the Court whether principles of *Lex posterior* apply to the three directly connected International Environmental Law instruments or not, before identifying the obligations, it was necessary for the Court to show how none of the treaties excludes each other under the principle of *Lex posterior* and therefore, obligations under these three treaties are still applicable to the states.

a) Application of Lex Posterior Principle

The Court prescribed that, according to Article 30, paragraphs 2 and 3 of the Vienna Convention on the Law of Treaties, two conditions need to be fulfilled to apply the principle of *Lex posterior*; one is 'incompatibility with the prior or later Act', and the other is 'where the treaty parties are different'. The court singled out that neither of the two conditions is applicable for UNFCCC, Kyoto Protocol and Paris Agreement.

Firstly, the UNFCCC is a foundational Treaty and provides a framework for all other treaties (para 191). Article 2 of UNFCCC legitimises the Conference of the Parties (COP) 'to adopt related legal instruments' in accordance with the relevant provisions of the UNFCCC 'including adopting protocols to the Convention' (Article 17). The Kyoto Protocol thus becomes a 'related' instrument not invalidating or being incompatible with UNFCCC, as the preamble explicitly states, "the Protocol is '[i]n pursuit of the ultimate objective of the [UNFCCC]' as stated in its Article 2". Additionally, the parties to the two instruments are the same. Therefore, no question of applying *lex Priori* to exclude UNFCCC or the Kyoto Protocol, either of them.

On the Other hand, the Paris Agreement, although not a protocol to UNFCCC, but outcome of a process initiated by the decisions taken in the COP decision 1/

CP.17 (2011) (para 193). Besides, the preamble to the Paris Agreement states that the Agreement is adopted "[i]n pursuit of the objective of the [UNFCCC], and being guided by its principles". Additionally, Article 2 of the Agreement exclusively states-

"[t]his Agreement, in enhancing the implementation of the Convention, including its objective, aims to strengthen the global response to the threat of climate change".

All of these prove that although the Paris Agreement is independent, it is enacted in alignment with the UNFCCC to fulfil its objectives and the treaty parties are same. Here also, the two conditions are not met to apply the principle of *lex Priori*. Therefore, neither the Paris Agreement nor the Kyoto Protocol modifies or abrogates UNFCCC and vice versa.

While deciding the treaty obligations, another principle came forward that is *lex specelis*. the court tried to understand whether UNFCCC, Paris Agreement and Kyoto Protocol form *lex specelis* or not.

b) Principle of Lex Specelis

According to the ILC Commentary on Article 55 of the Articles on State Responsibility, "to be *lex specialis*, there should be some level of *inconsistency* between the laws with 'a discernible intention' that one provision is to exclude the other" (para 4).[1] The court neither found such inconsistency nor did it identify a discernible intention with other environmental treaty laws(para 168,169). The Court identified that

"the preambles of the UNFCCC and the Paris Agreement themselves contain references to other rules and principles. The States parties to these treaties have thereby recognized that climate change constitutes a problem for whose solution other rules and principles also play a role..... However, the object and purpose of the climate change treaties do not contradict other rules or principles of international law or suggest that these treaties are intended generally to replace such other rules or principles"

Therefore, these climate laws do not exclude other laws, rules, principles, and customary rules applicable to environmental protection as *lex specialis*.

Therefore, The Court mentioned several other Conventions which are relevant to climate change and the obligations of those treaties are binding on all the parties. Some mentionable treaties here for Bangladesh can be Montreal Protocol, Ozone Layer Convention, Biodiversity Convention, the Desertification Convention, United Nations Convention on the Law of the Sea (UNCLOS), etc. (paras 316, 325, 336). The court emphasing on the obligations under these treaties conclude that 'they are obligations that contribute to ensuring the protection of the climate system as a whole and are therefore directly relevant and binding to the parties'

Regarding the Human Rights Obligation, the Court also identified ICCPR and ICESCR as the directly relevant treaties of which Bangladesh is a party. Compliance with those Treaties is a must for Bangladesh to fulfil its obligations toward climate Change.

However, due to world limit constraints, this chapter will focus on the three major treaties for the time being to check if Bangladesh is complying with the three major climate treaties.

2.2 Obligations Under the Treaty Laws

In broader terms, the Court identified two obligations under the three major treaties: one is mitigation, and another is adaptation. However, obligations under other treaty laws are also relevant when the context involves those laws. This chapter neither discusses all the separate obligations arising under specific treaty regimes nor intends to do so, given the extensive existing literature on the subject. Instead, it focuses on the broader obligations identified by the Court and their relevance to Bangladesh.

2.2.1 Obligation of Mitigation

a) Under the UNFCCC

Firstly, the Court has identified that 'mitigation lies at the heart of the UNFCCC's objective' (para 200). Article 4, paragraph 2 (a) of UNFCCC 'seeks to achieve the mitigation objectives in two ways: first, by limiting anthropogenic GHG emissions by sources, and second, by preserving and enhancing sinks and reservoirs of GHGs'. Article 1 of UNFCCC further clarifies 'the source' as "any process or activity which releases a greenhouse gas, an aerosol or a precursor of a greenhouse gas into the atmosphere" and 'a sink' as "a process, activity or mechanism that removes GHGs from the atmosphere; and a reservoir is part of the climate system that enables GHGs to be stored". According to the Sixth Assessment Report of Working Group III of the Intergovernmental Panel on Climate Change (IPCC), forests and oceans are the largest natural sinks in the climate system.

In upholding the obligation of limiting GHG, Bangladesh submitted its NDC 3.0 in September 2025. It committed to an unconditional 6.39% reduction and a conditional 13.92% reduction in economy-wide GHG emissions by 2035. It also adopts a conditional economy-wide greenhouse gas reduction target of 13.92 percent by 2035, alongside targets to achieve 25 percent of total power mix from renewable energy sources and enhance energy efficiency by 19.2 percent' (NDC 3.0, 2025)

On the other hand, to preserve the forest, recently Bangladesh adopted the Forest and Tree Conservation Act 2026, imposing heavy penalties for cutting endangered tree species. It also adopted a National REDD+ Strategy, Bangladesh Climate Change Strategy and Action Plan (BCCSAP), and National Action Plan (NAP), all of which, although works as 'aspirational commitments', doesn't include any enforceable duties (Mahmud, 2025).

On top of that, the Marine Pollution Ordinance, 1989; the Environment Protection Act 1995; the Water Act 2013; Water Rules 2018- none serve the purpose of carbon sink protection, but form the general marine-environmental statutory baseline.

Hui et al. (2025) proposed frameworks to prohibit the deliberate disposal of plastic waste at sea, mandating that "Parties control marine dumping, thus complementing Basel and UNCLOS". Although Bangladesh is a party to UNCLOS, it do not have any framework for disposal of plastic waste in the Bay of Bengal. These gaps shows absolute unalignment to protect Ocean, one of the largest carbon sink. The Opinion requiring the state parties 'a must for environment protection' now acts as a guardian for Bangladesh to fill these gaps as much as necessary.

Secondly, Article 4 (1)(a) of the Framework Convention requires Bangladesh as a 'party' to 'develop, periodically update, publish and make available to COP the national inventories of anthropogenic emissions by sources and removals by sinks of all greenhouse gases'. Bangladesh has submitted three National Communications to date. The Third National Communication (NC3) submitted in 2018 contains its most comprehensive national GHG emission inventory, using 2012 as the base year. Nonetheless, none of the Bangladesh's policy regarding forest and sea, establish a blue carbon protection mechanism, mangrove or seagrass carbon accounting framework, or coastal ecosystem conservation regime tied to GHG removal

Also, according to Article 4(1)(b), the Convention requires parties to 'Formulate, implement, publish and regularly update national and regional programmes containing change by addressing anthropogenic emissions by sources"

Complying with this obligation, Bangladesh took its principal national mitigation programme, the Mujib Climate Prosperity Plan (MCPP), launched at COP26 in 2021. It sets a quantified target of 30% renewable energy by 2030, 40% by 2041, along with 11 pillars to combat climate change. Still, those policies remain still as aspirations, not producing any binding effect.

The UNFCCC also requires a non-Annex I party to submit bi-annual reports. Bangladesh, for the first time, submitted its Biennial Update Reports on 25 November 2023. This update report has been submitted after 9 years of the original Report. Although Bangladesh, as a Least Developed Country (LDC), currently benefits from greater flexibility in the timing of these submissions than other non-Annex I Parties, its graduation from LDC status in 2026 will require it to comply with more stringent reporting timelines, making timely submission a greater legal and institutional challenge.

As all these aforementioned obligations are legally binding and Bangladesh cannot escape from these liabilities anymore. If done, Bangladesh too, like other developed countries, can be tagged as a 'violator' and shall be invoked with consequential responsibilities as discussed in Part 04 of the Chapter.

b) Under Kyoto Protocol

The Kyoto Protocol has concretised certain obligations under the UNFCCC by requiring quantified emission reduction commitments for Annex I parties listed in Annex B of the Protocol. The most interesting part is Court affirmed that the Kyoto Protocol is still relevant for two reasons,

1. substantive provision in Annex B of the Protocol
2. interpretative aid for identifying obligations.

The Court notes that the "absence of a new commitment period does not deprive the Kyoto Protocol of its legal effect" (para 221)

As Bangladesh is not an Annex I Party, this finding bears on Annex I mitigation commitments rather than Bangladesh's own obligations directly. However, surely it does benefit Bangladesh as a climate 'victim' to claim its rights through financial aid from the Annex I parties.

c) Under Paris Agreement

Although exempted by the Kyoto Protocol obligation to mitigate, Bangladesh is not exempted under the Paris Agreement from fulfilling some obligations. The Court mentioned significance of Paris agreement by stating that "Indeed, the Agreement is the most recent legally binding universal instrument addressing the issue of climate change and its provisions set out obligations in wide range of areas" (para 223) The Court considers that the aim of "[h]olding the increase in the global average temperature to well below 2°C above pre-industrial levels" and pursuing efforts "to limit the temperature increase to 1.5°C above pre-industrial levels" obligates state parties under this Agreement to limit emission. Bangladesh is not an exception to this. Bangladesh is thereby bound to comply with this binding objective of the Paris Agreement.

At the fifth session of the Conference of the Parties serving as the meeting of the Parties to the Paris Agreement (CMA 5, 2023), Parties were required to submit

> *"Ambitious, economy-wide emission reduction targets, covering all greenhouse gases, sectors and categories and aligned with limiting global warming to 1.5°C, as informed by the latest science, in the light of different national circumstances"*

Bangladesh complied with the same, as mentioned earlier, by communicating 3rd NDC in September 2025.

Also, Article 4. Paragraph 1 of the Paris Agreement talks about the obligation of mitigation, stating "Parties aim to reach a global peaking of greenhouse gas emissions as soon as possible . . . and to undertake rapid reductions thereafter", to achieve net zero emissions by balancing anthropogenic GHG emissions and their removal by sinks by the second half of the century. According to the Climate National Pathway Explorer's Climate analysis in September 2025, the 2035 target of Bangladesh "will require greater ambition to align the country with 1.5°C aligned pathways, "which reflects that Bangladesh has *formally complied* with the procedural CMA.3 request but has not met the substantive 1.5°C-alignment standard the Agreement calls for.

Additionally, The Paris Agreement imposes an obligation on a state party under Article 4 para 3 that NDCs must 'represent a *Progression*..... and reflect its *highest possible ambition'*. In the Opinion, the requesting parties sought direction regarding 'how to measure *progression* that "reflect [a party's] *highest possible ambition*" making it a prominent question for the court to answer (para 240).

The court clarified that 'progression' refers to 'a party's NDCs' to become more demanding over time' (para 241). On the other hand, the court comments that the lack of definition of 'highest possible ambition' in the Agreement doesn't leave the determination of the NDCs entirely to the discretion of the parties. Therefore, the ambition must be 'adequate' to be capable of achieving the temperature goal. Now the question is the 3rd NDC prepared by BD is adequate or not.

The Centre for Policy Dialogue (Bangladesh's leading independent think tank) published a targeted critique in 2025 arguing NDC 3.0 to be *not* adequately ambitious by stating that, "Ironically, Bangladesh's NDC 2.0 targets were much 'ambitious' compared to that of NDC 3.0." This reporting questions the *progression* of that is required under the commitment of the said obligation and non-compliance makes Bangladesh a violator.

Additionally, similar to UNFCCC, Article 5 of the Paris Agreement too invokes obligation on Bangladesh "to conserve and enhance, as appropriate, sinks and reservoirs of greenhouse gases . . ., including forests," where Bangladesh is still lagging behind (as discussed earlier).

The Court notes that these "are obligations addressed to all parties as a whole" (para 231). Therefore, this ICJ opinion, creating *opinio juris*, leaves no gate open for Bangladesh to evade these responsibilities under the veil of a 'vulnerable country' or 'climate victim'. Therefore, it is high time Bangladesh adopted legislation on this issue from the good practices around the world.

An example here can be Fiji's Climate Change Act 2021. Unlike Bangladesh's Conservation Act that codifies only a general environmental protection framework empowering the Department of Environment, setting pollution standards, and enabling Environmental Conservation Rules, Fiji's Climate Change Act 2021 is purpose-built around GHG accounting. It defines and legally operationalises "removals by sinks," creates transferable "carbon sequestration property rights," sets carbon budgets, and directly cites Article 4 of the Paris Agreement and Article 13 in its own text (Section 47, Section 25(h), Sections 52-58).

Taken together, Bangladesh's record under the UNFCCC and the Paris Agreement's mitigation architecture reveals a consistent pattern: procedural compliance without substantive enforceability. Having examined this pattern in the mitigation context, the following section turns to Bangladesh's corresponding duty to adapt under the ICJ's Advisory Opinion.

2.2.2 Obligation of Adaptation

The Court observes that adapting to the adverse effects of climate change is, along with mitigation, a major area of action for parties under the Framework Convention (para 209). Besides, under the Paris Agreement Article 7 paragraph 2, 'adaptation is recognised as a global challenge'. Adaptation is defined by the IPCC as "the process of adjustment to actual or expected climate and its effects, to moderate harm or exploit beneficial opportunities" (IPCC 2023 Glossary, p. 120). As well, aobligations related to adaptation is codified under several paragraphs under UNFCCC Article 4.

Paragraph 1(b) requires parties to "[f]ormulate, implement, publish and regularly update national and, where appropriate, regional programmes containing measures to . . . facilitate adequate adaptation to climate change".

Paragraph 1(e) establishes an obligation for parties to "[c]ooperate in preparing plans...develop and elaborate appropriate and integrated plans for coastal zone management, water resources and agriculture, and for the protection and rehabilitation of areas."

Paragraph 1(f), requires parties to "[t]ake climate change considerations into account, to the extent feasible, in their relevant social, economic and environmental policies and actions" .....and to "employ appropriate methods, for example impact assessments, formulated and determined nationally", with a view to minimizing adverse effects that adaptation projects or measures could have on the economy, on public health or on the quality of the environment.

All these provisions under UNFCCC are legally binding for using the wording as 'shall'. Similarly, the Paris Agreement also makes it binding under para 9, stating, "[e]ach Party shall, as appropriate, engage in adaptation planning processes and the implementation of actions, including the development or enhancement of relevant plans, policies and/or contributions".

The court also identifies five concrete duties associated with paragraph 9 which countries 'must take'. With the word must, the obligations became non-derogable. The five obligations are-

a) “Implementation of adaptation actions, undertakings and/or efforts
b) the formulation and implementation of national adaptation plans
c) the assessment of climate change impacts and vulnerability
d) monitoring, evaluating and learning from adaptation plans, policies, programmes and actions
e) building the resilience of socio-economic and ecological systems, including through economic diversification and the sustainable management of natural resources” (para 257)

Complying with these provisions, Bangladesh in the National Action Policy (NAP) 2023–2050 has undertaken 113 adaptation interventions across 8 sectors. Similarly, in Bangladesh Climate Change Strategy and Action Plan (BCCSAP) 2009, adaptation is set as one of its six pillars. Another significant step is the Coastal Zone Policy (CZPo), which was approved by the government on 17 January 2005, and provides the directives and framework for integrated coastal zone management. Nine strategic priorities, subsequently set as the Coastal Development Strategy (CDS) 2006, have been adopted that focus on the implementation of the coastal zone policy.

Despite these policy frameworks, significant implementation gaps persist. First, most of these instruments, NAP, BCCSAP, CZPo, and CDS, function as policy documents rather than binding legal instruments, meaning none of the 113 adaptation interventions or the six BCCSAP pillars carry statutory enforcement mechanisms. Second, weak inter-agency coordination undermines implementation on the ground. Prior studies on coastal adaptation governance in Bangladesh reveal overlapping mandates, inadequate coordination between ministries, and limited information sharing across agencies, which often result in fragmented implementation(Amin & Salekin, 2026). Third, no single law or policy effectively regulates marine and coastal resource management as a whole; instead, fragmented laws and policies create jurisdictional overlap, followed by interest conflicts in most cases (Islam & Chowdhury, 2022). Therefore, problems arise as over a dozen government agencies operate in the coastal zone with overlapping and sometimes conflicting mandates. For example, port operations, ship-breaking, mangrove conservation, and MPA establishment-these all are under separate authorities. (Hossain, 2018).

Institutional fragmentation is described as common in the coastal region, since although a dozen or more departments operate there, there is hardly any linkage between them (Islam et al., 2009). One study frames this dynamic sharply: because many organisations share responsibility for coastal resources without any single body holding overall accountability, the practical result is that everybody's asset becomes nobody's asset (Khan, 2020). Even the Ministry of Water Resources, designated as the lead coordinating ministry with WARPO as the lead agency, faces weaknesses in its coordination and harmonisation mechanisms among concerned authorities, in investment prioritisation, and in resolving coastal conflicts, which means coastal communities are often unable to realise the full benefit of adaptation measures meant to protect them (Khan, 2020).

Therefore, Bangladesh's inadequate adaptation governance reinforces its dual identity under the ICJ Advisory Opinion. While the country remains a victim of climate change and is entitled to international support, it simultaneously emerges as a violator by failing to establish and implement an effective institutional framework capable of fulfilling its international adaptation obligations.

2.3 Customary International Law

After discussing the treaty obligations, this part of the chapter discusses which obligations are given *a customary* nature, making them applicable and binding to all states regardless of being a treaty party or not. The Court, for the first time, recognised two customary obligations in the context of climate change, namely the "duty to prevent significant harm to the environment" and the "duty to co-operate for the protection of the environment" (para 131). Both of these Customary International Law (CIL) obligations were first established as treaty obligations. The following section presents the Court's reasoning, along with Bangladesh's compliance and the future implications for Bangladesh.

2.3.1 Duty to Prevent

The Court stated that Participants generally agree that States have a duty under customary international law to prevent significant harm to the environment. Therefore, there was no debate to accept the *duty to prevent* as CIL (para 132)

However, the court clarifies that in the climate context, the duty to prevent arises from the *Legality of the Threat or Use of Nuclear Weapons* case, where the Court explicitly stated:

"[t]he existence of the general obligation of States to ensure that activities within their jurisdiction and control respect the environment of other States or of areas beyond national control is now part of the corpus of international law relating to the environment" (para 134).

However, the court argues.[2] that this obligation needs to be done with *due diligence,* as established in the *Pulp Mills case*[3] where it is required that "the obligation of [States] is rather to employ all means reasonably available to them, to prevent [harm] so far as possible"[4]

a) Due Diligence

The court validated the ITLOS Advisory Opinion, where it was identified that "obligation of due diligence is particularly relevant in a situation in which the activities in question are mostly carried out by private persons or entities" (Climate Change, Advisory Opinion, ITLOS Reports 2024, p. 90, para. 236).

Duty to prevent, thus, connected with due diligence as customary law, brings a strict legal obligations to be carried out for Bangladesh. Any incident where the government could act in due diligence to prevent environmental crimes shall make Bangladesh a violator if not undertaken with proper care and vigilance.

Previously, private industrial actors' environmentally harmful conduct in Bangladesh was routinely left unattributed to state responsibility because of institutional oversight failure and ineffective regulatory enforcement (Belal & Roberts, 2010). Manufacturers widely avoided installing effluent treatment plants despite regulatory requirements, and industrial polluters largely ignored government enforcement notices. Of 903 industrial units identified by the Department of Environment as "mostly polluting," only 196 responded to a formal government notice (Khan & Belal, 1999). Report of Belal et al (2015) shows that

> *"[R]egulations... in reality are routinely flouted due to lack of enforcement by the relevant agencies which appear to be corrupt, weak and ineffective. They also lack strong political will and necessary resources such as manpower, skills and knowledge to be able to implement the relevant laws."*

This negligence can now, after the Opinion, backfire to define Bangladesh as a violator, as Bangladesh can no longer deny private actors' liability by passively deflecting responsibility. If left unaddressed, these gaps do more than weaken domestic policy. They expose Bangladesh to genuine legal risk and the world to genuine climate harm, eventually being accountable for the 'omission' from Bangladesh's part. Certainly, a state that fails to convert its treaty obligations into enforceable domestic law cannot indefinitely shield itself from international accountability; third states or affected parties may eventually invoke these same failures before international courts, reversing Bangladesh's position from vulnerable claimant to respondent state, a climate violator.

b) Precautionary Measures

The court again associated the duty to prevent with the precautionary measures. This is further affirmed by the court, where the court, by mentioning the pulp mills case.[5], stated that due diligence "entails not only the adoption of appropriate rules and measures, but also a certain level of vigilance in their enforcement and the exercise of administrative control" and emphasised that a 'heightened degree of vigilance and prevention is required'(para 138).

Bangladesh evidently falls short in this area as well. For example, section 12 of the Environment Conservation Act 1995 prohibits the establishment of any industrial unit or project without obtaining an Environmental Clearance Certificate (ECC) from the Director General of the Department of Environment (DoE), while section 20 provides for the formulation of rules governing the review and approval of Environmental Impact Assessments (EIAs). In practice, however, implementing agencies often treat the EIA process as "merely a clearance formality" and prioritise the physical progress of projects over environmental safeguards (Naher, 2025). A notable example is the Rampal Power Plant, which is located within 14 kilometres of the Sundarbans, an ecologically sensitive area and a significant carbon sink recognised under the UNFCCC framework. As Hassan observes, "They started development of the site before taking any [required environmental clearance]" (Hassan, 2023). Following the ICJ Climate Advisory Opinion (2025), Bangladesh can no longer evade its obligation to exercise due diligence by ensuring that environmental impact assessments are conducted effectively and that projects likely to cause significant environmental harm are not authorised or implemented without adequate environmental scrutiny. Its continued failure to do so reinforces its position as a *violator* of its international environmental obligations.

c) Proving Significant Harm

According to the Pulp Mills case, for the duty to prevent to arise, there must be a risk of significant harm to the environment[6].

According to the *Climate Change* Advisory Opinion, ITLOS, a risk of significant harm depends on "both the probability or foreseeability of the occurrence of harm and its severity or magnitude," and should therefore be "determined by, among other factors, an assessment of the risk and level of harm combined."[7] The Court in the Opinion further held that "it is necessary to take into account the risks which current activities might pose in the future, including in the long term" (para 275).

On the first part, foreseeability, Bangladesh does fairly well. Its National Adaptation Plan (2023–2050) plans 113 costed interventions all the way through mid-century, and its NDC 3.0 openly forecasts that emissions will rise from 252 MtCO2eq in 2022 to 418.40 MtCO2eq by 2035 if nothing changes. So Bangladesh clearly sees the risk coming and has put a number on it.

The problem is the second part: severity, and what Bangladesh actually does about the risk it has identified. The ITLOS standard doesn't just ask states to forecast harm; it asks them to respond in proportion to how serious that harm is. But Bangladesh's own forecast shows emissions growing by nearly two-thirds by 2035, while its actual mitigation commitment, only a 6.39% unconditional cut and 13.92%

conditional cut, is quite modest next to a risk that large. So, Bangladesh has done the diagnosis (spotting and measuring the risk) without matching it with a response that fits the size of that risk. That gap, strong on foreseeing the danger, weak on acting proportionately to it, demonstrates the "unevenness" that leaves Bangladesh unable to prevent significant harm making it a climate violator.

The Court also took the view that a risk of significant harm may be present where significant environmental harm results from the cumulative effect of acts undertaken by various States and private actors subject to their respective jurisdiction or control. Bangladesh has developed a comprehensive Climate Change Measurement, Reporting, and Verification (MRV) platform, hosted by the Department of Environment under the Ministry of Environment, Forest and Climate Change, to meet Paris Agreement transparency requirements. It integrates GHG inventory data across sectors with stated QA/QC (quality assurance/control) processes. However, this measurement capacity has not translated into source-specific regulatory action. There is no distinct regulatory instrument targeting, for example, CNG vehicle emissions or household cooking-fuel emissions as such; these remain folded into broad sectoral aggregates used for UNFCCC reporting rather than domestic regulatory design. Moreover, Bangladesh's own government has acknowledged, in a conference presentation in 2024 on Bangladesh's progress in NDC 3.0 and BTR1 preparation, that its MRV platform, though established, "is yet to [be] functional." Here, the whole commitment becomes hollow without translating into concrete environmental protection, representing documentation without action, and measurement without steps taken.

d) Transboundary Harm as a Global Harm

The Court further recognised that the duty to prevent environmental harm is closely connected with the principle of preventing transboundary harm (para. 160). Read together with the Court's observation that climate change is "global in nature" (para. 216), the Opinion suggests that the prevention of significant environmental harm can no longer be viewed solely through the traditional lens of neighbouring States. Rather, activities undertaken within one State that contribute to cumulative global environmental degradation may engage the responsibility of all States, thereby requiring greater coherence in domestic regulation and international cooperation.

The global shipbreaking industry illustrates this challenge. Under the Basel Convention's Ban Amendment, OECD and EU member States are prohibited from exporting hazardous waste, including end-of-life ships containing toxic substances such as asbestos, lead, and mercury, to non-OECD countries, including Bangladesh. Nevertheless, the practice continues. According to *Maritime Magazine*, "roughly 30% of the tonnage broken down on the shipbreaking beaches of Bangladesh, India, and Pakistan is owned by EU-domiciled shipping companies." As reported by DNV (2025), many EU-owned vessels have circumvented the practical effect of the Ban

Amendment by changing their flag and relying on the regulatory framework of the Hong Kong International Convention for the Safe and Environmentally Sound Recycling of Ships, which does not impose an equivalent export prohibition. A similar pattern can be observed in the export of ageing and high-emission vehicles, many of which no longer comply with environmental standards in their countries of origin but continue to be exported for use in Bangladesh.

These practices demonstrate that preventing significant environmental harm in the era of climate change cannot be achieved through fragmented regulatory approaches. In light of the ICJ Advisory Opinion, they reinforce the need for greater harmonisation and effective international implementation of rules governing the transboundary movement of environmentally hazardous materials and technologies, so that environmental risks are not merely transferred from one jurisdiction to another.

2.3.2 Duty to Co-Operate

The court described the duty to cooperate as a heart of the UN Convention, stating "[t]o achieve international co-operation in solving international problems of an economic, social, cultural, or humanitarian character[8]". The Court in the Opinion held that "international co-operation is indispensable in the field of climate change, and that the customary duty to co-operate for the protection of the environment is reflected in several provisions of the climate change treaties, including the UNFCCC" (para 215).

The rationale for co-operation in the UNFCCC may be found in the treaty's preamble, which provides that "the global nature of climate change calls for the widest possible cooperation by all countries and their participation in an effective and appropriate international response" (sixth preambular paragraph). The Court considers that paragraph 1 (c), (d), (e), (g), (h) and (i), and paragraphs 3 to 5 of Article 4 of the UNFCCC give precise content and scope to the duty to co-operate.

The Court notes that the Paris Agreement too establishes obligations of co-operation with respect to specific issue areas, such as adaptation, and loss and damage (Article 7, paragraphs 6 and 7; Article 8, paragraph 4). Additionally, Article 12 establishes an obligation to "cooperate in taking measures, as appropriate, to enhance climate change education, training, public awareness, public participation and public access to information".

The duty to cooperate is not confined to the climate treaties but is also reflected in other multilateral environmental agreements, including Article 5 of the Convention on Biological Diversity and Article 197 of UNCLOS. The ICJ recognised that this obligation has also developed as a rule of customary international law. Consequently, the duty to cooperate may bind States independently of treaty membership where it reflects customary international law.

In interpreting UNCLOS, the Court emphasised that "States parties must therefore take their obligations under UNCLOS into account when implementing their obligations under the climate change treaties and other relevant environmental treaties and under customary international law" (para. 335). In support of this interpretation, the Court endorsed the reasoning of the ITLOS Advisory Opinion on Climate Change (2024), which held that although Article 193 of UNCLOS recognises the sovereign right of States to exploit their natural resources pursuant to their environmental policies, that right is expressly qualified by their duty to protect and preserve the marine environment (para. 344; see also ITLOS Advisory Opinion, paras. 187, 380). Accordingly, State sovereignty over natural resources is not absolute. Where the exploitation of natural resources poses a risk of significant environmental harm, States are required to regulate such activities consistently with their environmental obligations and, where necessary, cooperate with other States to prevent or minimise that harm.

Accordingly, Bangladesh's obligations under UNCLOS and other environmental treaties cannot be interpreted in isolation from the customary duty to cooperate. Even where Bangladesh exercises its sovereign rights over natural resources, it remains under a concurrent obligation to cooperate with other States in preventing transboundary and global environmental harm, as reaffirmed by the ICJ Advisory Opinion.

Primarily, in *Nicaragua v. United States of America,* the court gave a direction on how this duty becomes customary, stating that 'adoption of this text by states gives them the recognition as *opinio juris'*[9]. As the same obligation appeared in many binding and non-binding instruments relating to the environment, the court declared that these customary rules apply to the environment (Para 140). Specifically, by mentioning Principle 24 of the Stockholm Declaration and Principle 7 of the Rio Declaration, where both recognise co-operation as an essential element in the protection of the environment, the court recognised the customary character of the duty to cooperate (para 140).

This duty, too, is not taken in isolation. In the ICJ Opinion, the court held that the duty to cooperate in customary international law includes the duty to prevent significant harm to the environment (para 141). From *the pulp mills* case, we have observed how uncoordinated individual efforts by States may lead to a destructive result.

Bangladesh's own experience with its two border-sharing states illustrates precisely this risk. With India, the record is one of persistent, unresolved unilateral action despite decades of nominal cooperation. India's construction of the Farakka Barrage in 1975, built to keep the Kolkata port silt-free, reduced dry-season flow of the Ganges into Bangladesh, prompting Bangladesh to raise the issue before the UN General Assembly as early as 1977 (Insights IAS, 2021) Although the 1996

Ganges Water Treaty eventually resolved the dispute on paper, that treaty expires in 2026, and its renewal remains politically uncertain (Observer Research Foundation, 2024). The Teesta River dispute remains entirely unresolved to this day: a draft treaty allocating 37.5% of dry-season flow to Bangladesh was finalised in 2011 but has been blocked for over a decade by opposition from the Chief Minister of West Bengal, causing Bangladesh to lose an estimated 1.5 million tons of rice annually to water shortage[10]. Upstream dam releases by India have also been directly linked to recent flooding in Bangladesh, including the severe floods of August 2024[11].

With Myanmar, cooperation is comparatively minimal. The Joint Rivers Commission's mandate nominally extends to three border rivers shared with Myanmar, but no equivalent water-sharing treaty, joint adaptation mechanism, or environmental cooperation instrument exists between the two states. Bilateral relations remain dominated by the Rohingya displacement crisis rather than environmental or maritime cooperation, leaving the customary duty to cooperate largely unfulfilled on Bangladesh's eastern frontier.

These patterns reflect exactly the danger the Court identified: individually rational upstream or unilateral conduct. A barrage built for a domestic port, a dam operated for domestic hydropower produces, in aggregate, significant and recurring transboundary harm to Bangladesh, precisely because a binding, functioning cooperative mechanism has never been fully realized with either neighbour.

The Court linked the duty to cooperate further with the principle that "the conservation and management of shared resources and the environment are based on shared interests and governed by the principle of good faith", which was derived from the *Legality of the Threat or Use of Nuclear Weapons* case (para 141). Bangladesh shares 38% of the Sundarbans with India (IUCN, 2025, p. 2). Yet, lack of a joint administrative management, limited institutional coordination between the two countries (Islam & Islam, 2026) both the states fails to ensure ecological integrity and climate resilience (Amin & Salekin, 2026). This fragmented coordination can lead to failure of duty to prevent and duty to cooperate under the customary obligation of each state. Again, the Bay of Bengal is shared with both Myanmar and India, Bangladesh is bound to establish shared policy frameworks and remove the ambiguities arising from those shared borders.

3. SCOPE OF THE OBLIGATIONS

The Court opined that the scope of these explicitly mentioned obligations actually arises out of those basic environmental law principles reflecting the nature of environmental protection itself, as the Court states-

"By identifying the core treaties and customary international environmental law, the Court specified the exact obligations concretely, but it did not evade the basic foundational principles of environmental protection and reaffirmed that these are part and parcel to achieve those customary and treaty obligations" (para 146).

One of the core principles is *sustainable development*. Bangladesh is bound by the principle of Sustainable development as identified in the *Gabčíkovo-Nagymaros Project* Case as 'need to reconcile economic development with protection of the environment'[12] If development projects are not aligned with the sustainable development goals allowed by the laws and policy framework, this principle stands violated, and the Rampal Power Station illustrates precisely this failure. The 1,320-megawatt coal-fired plant was sited roughly 14 kilometres from the Sundarbans, the world's largest mangrove forest and a UNESCO World Heritage and Ramsar site, in direct breach of the Department of Environment's own precondition that such projects remain outside a 25-kilometre radius of an ecologically sensitive area. Land was acquired and site clearance granted before an Environmental Impact Assessment had been completed, and the project proceeded despite objections from UNESCO's World Heritage Committee and the IUCN (Human Rights Council 2018). This sequencing, economic commitment preceding, rather than following, environmental assessment, is the inverse of what *Gabčíkovo-Nagymaros* requires: reconciliation demands that environmental protection be built into the development decision, not retrofitted after the decision is effectively irreversible. Rampal therefore stands as a concrete instance where Bangladesh's domestic regulatory practice departed from the reconciliation standard it is otherwise bound to uphold.

Another important principle includes *Intergenerational equity;* a fundamental promise laid in the preamble of the Paris Agreement by stating-

"Climate change is a common concern of humankind, Parties should, when taking action to address climate change, respect, promote and consider their respective obligation on . . . intergenerational equity".

The court vested the responsibility on the present generation to conserve the climate as trustees of future generations, mentioning 'humanity tasked with preserving dignified living conditions and transmitting them to future generations.' (para 156) Therefore, the project plans permission granted to present-generation stakeholders need to be assessed and analysed based on the possibilities and causation of future fundamental harms on soil, air and water that may affect the next generation, who have done nothing wrong, yet may suffer the consequence of the harms caused by the present generations.

Bangladesh's Department of Environment (DoE) needs to move from a reactive to a proactive posture in regulating the projects it approves. Under the updated by the Environment Conservation Rules 2023, industrial projects are classified into Green, Yellow, Orange, and Red categories based on their pollution potential. This classification system, however, has been criticised as fundamentally reactive rather than preventive: scholars note that the Act's own preamble language "improvement of environmental standards" and "mitigation of environmental pollution" signals that it is "cure-oriented" and "deals with the post-harm situations" rather than anticipating future risk. (Naher 2025). The categorization criteria themselves are based on present pollution impact, not on the probability or magnitude of future fundamental harm to soil, air, and water contemplated by the Court's intergenerational trust reasoning.

The Court rules that "Environmental impact assessment is itself closely tied to the principle of due diligence"(para 295). Crucially, the Court held that the principle underlying this obligation is not confined to industrial activity, but applies "generally to proposed activities which may have a significant adverse impact in a transboundary context" (para 296), thereby invoking both substantive and procedural dimensions of the duty to prevent harm (para 295).

This is not a merely theoretical standard: everyday, cumulative conduct, such as the uncontrolled dumping and piling of household and industrial waste along riverbanks, or the direct discharge of untreated waste into rivers, falls squarely within activities capable of generating "significant adverse impact," particularly where those rivers are transboundary in nature, as many of Bangladesh's major rivers are shared with India and Myanmar. These gaps need to be addressed both substantially and procedurally.

Lastly, The Court held that certain obligations under the climate change regime qualify as obligations *erga omnes*, describing climate change as "a common concern of humankind" (para 440). In particular, the Court identified the customary international law- duty to prevent significant transboundary harm as one such obligation *erga omnes* (para 440). The consequence is significant: because these obligations are owed to the international community as a whole, every State party has a legal interest in the protection of the core mitigation obligations enshrined in the climate change treaties, and may invoke the responsibility to another State for its failure to fulfil them (para 441) that is too without distinction according to *The Gambia v Myanmar* case. Therefore, for all the gaps resulting to breach of obligations, any party can invoke responsibility if they cause significant transboundary harm to every alleged violator however the nature and extent of the violation is.

Taken together, the aforementioned obligations and their related principles show that compliance cannot be assessed in isolation. Each obligation must be read within the scope of the principle it is tied to; a state that formally fulfils an obligation while disregarding the principle underlying it has not truly complied; it has only satisfied the letter, not the substance, of that obligation. Such compliance-in-form, therefore, still amounts to a violation, one capable of giving rise to serious consequences. The next chapter deals with the consequences that may arise out of these obligations and their related principles.

4. CONSEQUENCE OF BREACH OF OBLIGATIONS

Issue b) of the Opinion deals with the consequences that arises out of breach of the obligations. As mentioned earlier, the obligations mentioned before are are legally binding; therefore, any 'act or omission' in upholding those obligations shall be a breach of obligations producing certain consequences. The court held that consequences shall "arise upon three stakeholders-

1. States, who are climate vulnerable,
2. states which are specially affected,
3. People and individuals of both present and past generations" (para 107)

The consequences give rise to some further responsibilities. Before stretching this section, a distinction between obligation and responsibility needs to be drawn first.

4.1 Obligation vs Responsibility

The distinction becomes clear by applying the golden rule of reading a judgment established by Lord Halsbury in the Quinn V Leathem case-

" ... every judgment must be read as applicable to the particular facts proved, or assumed to be proved, since the generality of the expressions which may be found there are not intended to be expositions of the whole law, but are governed and qualified by the particular facts of the case in which such expressions are to be found." [Para 28][13].

Although the wording 'obligation' and 'responsibility' both indicate some duties, these duties are different in their nature. One arises out of law; the other is invoked by breach of the former ones. The court in the Opinion stated that "Legal Consequences is to be understood as referring to the legal consequences arising from internationally wrongful acts of States, which are to be ascertained on the basis of the primary rules and the customary rules on State responsibility." (405) and "The rules on State responsibility under customary international law are also applicable to the determination of legal consequences for States that, by their actions or omissions, have breached those obligations."(para 407). These two paragraphs make clear that, in the Court's usage, "responsibility" does not denote a primary duty in itself, but the secondary legal consequence that follows from the breach of an obligation, that is, responsibility presupposes obligation, rather than standing alongside it.

These obligations and responsibilities differ in terms of their binding nature. In the first portion of the chapter, the obligations we have discussed, under treaty law, may become binding to all the parties and may not to some parties that give rise to "common but differentiated responsibilities". For example, Article 4 paragraph

2 of UNFCCC confers legally binding responsibility to developed countries and countries included in Annexure 01 only. However, responsibilities arising out of consequence are always binding upon the parties that contribute to the wrongful Act by any act or omission. The word responsibility thereby reflects and takes its origin from the ILC Report. 2001 on Articles of State Responsibility (ASR), where the responsibility is invoked in the same meaning under Article 02 of ASR.

4.2 Scope of These Consequences:

The court in this case reaffirmed that the application of the rules on State responsibility under customary international law does not differ depending on the category or status of an injured State. (para 109). Therefore, although Bangladesh enjoys some common but differentiated responsibilities, the responsibilities that arise from the consequences of breach of the said obligations are not differentiated and attributed to all the parties alike. Therefore, Bangladesh's status as a vulnerable country does not exempt it from state responsibility.

It is important to note that the Court in *Gabčíkovo-Nagymaros Project* (Hungary vs Slovakia) identified that, "parties to a treaty may accept "obligations of conduct, and obligations of result" under the same instrument (Judgment, I.C.J. Reports 1997, p. 77, para. 135). In the Pulp Mills case, the court defined obligation of conduct as "an obligation to act with due diligence, requiring States parties 'to use all the means at [their] disposal' with a view to fulfilling their international obligations[14]. For example, the obligation to pursue domestic mitigation measures is an obligation of conduct and not an obligation of result (see Responsibilities and Obligations of States with respect to activities in the Area, Advisory Opinion, 1 February 2011, ITLOS Reports 2011, p. 41, para. 110).

On the other hand, Obligation of result requires "a State to guarantee the achievement of a specific, defined outcome, rather than merely to exercise due diligence in pursuing it." By defining the nature of these obligations, the Court conclude that the breach of each obligation depends on the context of the case and their nature. Therefore, the court exemplified that when it's a matter of obligation to conduct, 'taking all measures' is the main attribution to invoke responsibility out of violation. Despite taking all necessary measures, if the objective is not fulfilled, then there shall be no breach of obligation (para 208) as well; not taking measures shall be considered a serious violation. Similarly, in case of an obligation to result, failing "to bring about the result required under the obligation" (para 208) creates responsibilities under this breach despite all measures has been taken. One example of such as an obligation is obligation to "adopt national policies and take corresponding measures on the mitigation of climate change". Therefore, if policies are adopted but not implemented, it doesn't produce the result, making a state climate violator despite taking the measures on paper.

The court also notes that these obligations may co-exist in a single legislation (para 208). In the Court's view, the legal obligations relating to mitigation under Article 4, paragraph 1 that requires parties "to develop, periodically update, publish and make available to COP the national inventories of anthropogenic emissions by sources and removals by sinks of all greenhouse gases" are interconnected obligations of conduct and result.

So, after the Climate Opinion, the threshold for compliance became considerably higher for Bangladesh. As discussed previously, Bangladesh has, in several instances, relied on its status as a climate-vulnerable state to justify the absence of concrete action, treating vulnerability as a shield rather than as a basis for heightened diligence. This is particularly significant in light of the obligation-of-result standard outlined above: where an obligation requires a defined outcome, such as the adoption of policies coupled with corresponding measures, vulnerability cannot excuse a state from failing to produce that outcome. Following the Opinion, Bangladesh can no longer treat its vulnerability as a justification for inaction; instead, it must be read as an aggravating factor that raises, rather than lowers, the standard of diligence expected of it.

The court also shed light on the discretion in the determination of NDC by each state party. The court analysed that this discretionary power is limited and backed by the obligation of due diligence (para 245). Also, the court highlighted the "Paris Rulebook decision" taken on 15 December 2018, decision 4/CMA.1, where it was confirmed that the discretionary power need to be applied in alignment with "[f] airness considerations"and "equity"[15]. Therefore, if the ICJ Opinion is read together with these explanations, it can be said that State cannot use the discretionary power arbitrarily, it can be held liable by invoking consequential responsibilities if NDC's does not align with the global goal set by Paris Agreement. Bangladesh will not be any different.

4.3 The Responsibilities as the Consequences of Breach

The responsibilities arising out of these consequences include, firstly, the duty of performance. The Court stated that "the breach by a State of its international obligations does not extinguish its underlying duty of performance. States have a continuing duty to perform their obligations despite their breaches thereof." The state party is therefore required to continue upholding the very obligation it failed to perform.

However, where the wrongful act arises out of a specific act by the State, the second duty is the duty of cessation and guarantees of non-repetition. The Court opined that "the obligation to put an end to the wrongful act may require a State to revoke all administrative, legislative and other measures that constitute an internationally wrongful act of that State" (para 447).

The most notable finding of the Court was that "the duty of cessation may also require States to employ all means at their disposal to reduce their GHG emissions and take other measures in a manner, and to the extent, that ensures compliance with their obligations."

The third duty is the duty of reparation. According to the *Factory at Chorzów* case, "reparation must, as far as possible, wipe out all the consequences of the illegal act and reestablish the situation which would, in all probability, have existed if that act had not been committed." It is further required that "full reparation" be made, which can be achieved through restitution, compensation, satisfaction, or a combination thereof.

The Advisory Opinion's tripartite framework of responsibility, the duty of performance, the duty of cessation and non-repetition, and the duty of reparation, is not confined to the states most commonly associated with historical emissions. Because the Court characterised the underlying obligations to prevent significant harm and to reduce GHG emissions as erga omnes (para 440), these consequences attach to Bangladesh as much as to any other State party, irrespective of its comparatively low share of historical global emissions. This marks a decisive shift for Bangladesh's legal position: a state that has, until now, engaged with the climate regime principally as a claimant invoking its vulnerability, is now also bound as a respondent-in-waiting under the same responsibility architecture it might otherwise invoke against others. The next section demonstrate that dual legal identity of Bangladesh after the post ICJ Opinion 2025.

5. BANGLADESH'S DUAL LEGAL IDENTITY

The word 'dual legal identity' in this chapter includes two folded legal capacities of Bangladesh-

1. Exercising legal right as a climate-vulnerable country
2. Invoking legal responsibilities as a climate polluter

5.1 Rights of Bangladesh as a Climate Victim

The idea of a "victim" arises from the notion of a party who suffers the wrongdoing of another without having contributed to the wrong itself. Bangladesh and other coastal countries fall squarely within this notion: they bear the consequences of climate change, an existential crisis threatening land, livelihood, and habitability, despite having contributed negligibly to its causes. While all nations experience the adverse effects of rising global temperatures, vulnerability is not evenly distributed; low-lying, geographically exposed states like Bangladesh face disproportionate harm relative to their share of responsibility for causing it.

Adger (2006, p. 268) suggests that "vulnerability is the state of susceptibility to harm from exposure to stresses associated with environmental and social change and from the absence of capacity to adapt."[16] The idea of a having special rights by climate-vulnerable country legally arose from two different articles of the UNFCCC. Article 3, paragraph 1, of the UNFCCC states that "developed country Parties should take the lead in combating climate change and the adverse effects thereof". At the same time, Article 3, paragraph 2, of the UNFCCC states that full consideration should be given to "[t]he specific needs and special circumstances of developing country Parties". Bangladesh, enjoys the rights being a 'least developed country' and 'facing adverse effect' as a climate vulnerable country, making it a climate 'victim'.

In fact, Bangladesh exemplifies this vulnerability in concrete terms. It ranks among the countries most affected by climate change globally, with the Global Climate Risk Index consistently placing it in the top ten most climate-vulnerable nations due to its exposure to cyclones, floods, and sea-level rise (Ecstein, 2021). Roughly a third of its population lives in low-lying coastal zones directly threatened by sea-level rise, and rising salinity is already degrading agricultural land across the southwest.[17] This combination of high physical exposure and comparatively limited adaptive capacity is precisely what Adger's definition of vulnerability captures, and is why Bangladesh is routinely invoked as the paradigmatic case of a UNFCCC Article 3(2) "specific needs and special circumstances" developing country.

Thus, the country is entitled not only to financial aid but also to technological aid. Article 4(9) of the UNFCCC provides that "the Parties shall take full account of the specific needs and special situations of the least developed countries in their actions concerning funding and transfer of technology." The Court held that although the phrases "give full consideration" and "take full account" afford parties some discretion in implementation, they nonetheless impose legally binding obligations, with the COP and its subsidiary bodies tasked with determining the precise scope of that obligation. This discretion in implementation, therefore, does not weaken the obligation itself; it merely locates the mechanism for its elaboration elsewhere. Consequently, Bangladesh's entitlement to funding and technology transfer as an LDC is not aspirational but legally enforceable, giving Bangladesh a legal upper hand in addressing the climate change problem.

This entitlement extends further under the UNFCCC's broader funding architecture. Article 4(4) states that Annex II parties "shall" assist developing country parties that are particularly vulnerable to the adverse effects of climate change in meeting the costs of adaptation to those effects. Read together with Article 4(9), these provisions establish a legally binding obligation on all Annex II parties, and correspondingly, a legally enforceable right for Bangladesh to reparation and climate funding for both mitigation and adaptation.

This claim is further strengthened by the Paris Agreement. Article 9 of the Paris Agreement provides that "[d]eveloped country Parties shall provide financial resources to assist developing country Parties with respect to both mitigation and adaptation". The use of the term "shall" to introduce the obligation indicates the legally binding character of that provision (para 264). As well, Article 11 of the Paris Agreement governs capacity-building activities and encourages co-operation amongst parties to build the capacity of developing countries to implement the Agreement, in particular that of least developed countries and small island developing States. Bangladesh is exercising these rights from the law few decades.

Bangladesh as a climate-vulnerable and least developing country received Multilateral climate finance. Green Climate Fund (GCF) has approved USD 903 million across ten projects in Bangladesh, combining mitigation and adaptation activities which includes climate projects with contributions of $441.2 million from the GCF, $135.5 million from other development partners, national implementing entities, and private banks, and $33 million from the Government of Bangladesh (GCF, 2026). Alongside, Bangladesh has multiple active Global Environment Facility (GEF), GEF-6 and GEF-7 projects, including the flagship "Building Climate Resilient Livelihoods in Vulnerable Landscapes in Bangladesh (BCRL)" project. Besides, the LDCF has funded seven projects in Bangladesh so far. et the irony is stark: despite this considerable financial inflow, Bangladesh has struggled to translate available funds into meaningful implementation of its mitigation and adaptation policies. The bottleneck, in other words, lies not in the availability of climate finance, but in the capacity and governance mechanisms required to convert that finance into concrete outcomes on the ground.

Despite this gap between funding availability and implementation, Bangladesh's underlying vulnerability as a least developed and climate-exposed state remains the basis on which its claims to differentiated treatment rest. This vulnerability, in other words, does not disappear because of implementation shortfalls; it continues to justify the differentiated obligations owed to Bangladesh under international climate law, chief among them the principle of common but differentiated responsibilities.

a) Common but Differentiated Responsibilities

Article 4, paragraph 3, of the Paris Agreement provides that each party's NDCs must reflect its differing national circumstances and capabilities. The Kyoto Protocol's approach required quantified emission limitation and reduction commitments from developed country parties and other parties undergoing the process of transition to a market economy, but not from developing countries. (para 179).

While the UNFCCC and the Paris Agreement establish obligations on all parties, they recognise that such obligations may differ depending on parties' economic situations, their historic contribution to anthropogenic GHG emissions and their capabilities to adapt to and mitigate the adverse impacts of climate change (para 179).

Countries like Bangladesh, are listed under non-Annex I of UNFCCC. Therefore, the obligations of USA as highly emitting countries like the USA are not the same as those of Bangladesh making these obligations country-specific and highly subjective.

The common but differentiated responsibility is further confirmed by the Paris Agreement. Although the common goal of the Paris Agreement is to keep the temperature within the 1.5°C threshold, Article 2 states, "[t]his Agreement will be implemented to reflect equity and the principle of common but differentiated responsibilities and respective capabilities, in the light of different national circumstances". Similarly, Article 4, paragraph 4, provides that developed country parties "should continue taking the lead by undertaking economy-wide absolute emission reduction targets". The word 'absolute' here possess utmost emphasise on uphold the responsibility with positive obligation where no derogation is permitted but only to the developing countries exempting countries like Bangladesh. Developing country parties are expected to "continue enhancing their mitigation efforts, and are encouraged to move over time towards economy-wide emission reduction or limitation targets in the light of different national circumstances".[18] Under this wording of ' continue enhancing' the flexibility and subtle form of negligence towards climate action arises on behalf of countries like Bangladesh.

The Court in the Opinion emphasised, again and again, that implementation of these obligations remains subject to due diligence, requiring even the latter States to take all means at their disposal to protect the climate system in accordance with their capabilities and available resources (para 291). Thus, the slow-paced behaviour of climate-vulnerable countries now warrants serious concern from the world community, making them accountable not to the absolute standard imposed on developed states, but to a due diligence standard that nonetheless demands genuine effort commensurate with whatever capacity and resources they do possess. Add this closing line after the paragraph:

It is precisely this due diligence standard, one that permits flexibility but not inaction, that Bangladesh's own conduct must now be tested against, revealing the extent to which it, too, functions as a climate violator which is discussed in the next section.

5.2 Bangladesh as a Climate Violator

Having examined the international legal obligations Bangladesh has assumed as a party to the UNFCCC and Paris Agreement, and the extent to which it has complied with them, this section turns to a dimension of Bangladesh's climate identity that receives comparatively little attention in the existing literature: Bangladesh is not only a climate-vulnerable victim-state, but also, in absolute and sectoral terms, a meaningful contributor to the very harms it seeks protection from, making it a "climate violator." This duality is central to the "dual legal identity" framework advanced in this chapter.

The preceding sections have already demonstrated this violator identity through the lens of Bangladesh's obligations in the post-ICJ Advisory Opinion era, tracing specific instances of non-compliance, from the absence of enforceable sink-protection duties to the unfunctional MRV platform, and situating them against the heightened due diligence standard the Opinion imposes. What follows extends this finding beyond the obligation-compliance framework, examining Bangladesh's contribution to climate harm through direct empirical indicators, namely its aggregate and sectoral GHG emissions, and practices such as brick kiln operations, that further substantiate its status as a violator independent of any single treaty provision.

Bangladesh's total GHG emissions have risen consistently rather than plateaued. Between 2013 and 2019, emissions including LULUCF rose by 22%, reaching 222 MtCO2e (Enerdata, 2025). NDC 3.0's own business-as-usual projections anticipate continued growth, from 252 MtCO2eq in 2022 to 418.40 MtCO2eq by 2035. This trajectory places Bangladesh's own domestic conduct in tension with the "progression" and "highest possible ambition" standards discussed earlier in this chapter: a state cannot simultaneously invoke its status as an emissions victim while presiding over a rising domestic emissions trajectory of its own making.

This tension becomes still more concrete at the sectoral level. Bangladesh's brick manufacturing industry is the clearest, most quantifiable example of Bangladesh functioning as a climate polluter in its own right, rather than merely a passive recipient of externally caused harm. With over 7,000 active kilns producing approximately 27 billion bricks annually, the sector alone generates 17% of the country's total annual CO2 emissions, 22% of national black carbon emissions, and 11% of national particulate matter emissions (The Business Standard, 2022). In absolute terms, this translates to an estimated 1.8 million tonnes of CO2, 302,200 tonnes of carbon monoxide, 23,300 tonnes of particulate matter, 15,500 tonnes of sulphur dioxide, and 6,000 tonnes of black carbon annually.

Taken together, this evidence, a rising domestic emissions trajectory and a brick kiln sector responsible for nearly a fifth of national CO2 output and over a fifth of national black carbon output, is not merely descriptive. Applying the Advisory

Opinion's own attribution and causation framework, this evidence places Bangladesh squarely within the class of states the Court's reasoning is designed to reach: not only as a claimant seeking protection, but as a contributor whose own conduct falls within the scope of that same framework.

That attribution framework is what allows this evidence to move from mere description to a legal finding of violation. The Court held that "the conduct of any organ of a State must be regarded as an act of that State," extending expressly to conduct such as fossil fuel production, consumption, licensing, and subsidy (para 427). This framework operates on two distinct levels, and Bangladesh's brick kiln sector engages both, establishing Bangladesh's conduct as attributable state action rather than an incidental domestic industry problem.

The first level is direct: state conduct itself. For decades, the Government of Bangladesh has responded to the kiln sector's pollution profile not by mandating emissions standards or phasing out traditional kilns, but by promoting voluntary technology upgrades, such as the hybrid-Hoffman kiln and the Zigzag 2.0 retrofit. This regulatory choice, to rely on incentive rather than mandate, is itself state conduct, and falls squarely within the category of licensing and regulatory action the Court's attribution principal addresses.

The second level does not require the same direct link. Here, the relevant question is not what the state itself did, but what it failed to prevent private actors from doing. The Court addressed precisely this scenario in relation to private conduct: states bear an obligation to regulate such conduct as a matter of due diligence (para 428). Under this second level, attribution does not depend on proving that individual kilns are organs of the state; it depends only on showing that Bangladesh's regulatory apparatus failed to exercise due diligence over an industry whose harms were foreseeable and known. That standard is met here. Traditional, highly polluting kiln designs remain the dominant technology in use nationally despite decades of promotional intervention aimed at replacing them, and the sector's health toll, an estimated 6,000 premature deaths annually (Eil et al. 2020), has been documented and known to government and researchers for well over a decade. The persistence of this technology, in the face of known harm and repeated but non-binding intervention, is what converts Bangladesh's conduct from mere policy shortfall into a due diligence failure attributable to the state.

However, in determining a state's contribution to climate pollution, certain challenges inevitably arise, and the Court addressed these directly. One such challenge concerns quantification: if breach can be established based on a state's total contribution to global GHG emissions, the natural objection is that no reliable method exists to measure each state's share of that total. On this point, the Court's finding that "it is scientifically possible to determine each State's total contribution to global emissions, taking into account both historical and current emissions" (para 429) directly forecloses this objection, and is not a hypothetical standard for Bangladesh; it is a standard Bangladesh's own reporting has already met. The precise sectoral figures cited above (17% of annual CO2, 22% of black carbon, 11% of particulate matter, all attributable to a single named industry) demonstrate that Bangladesh's own domestic data infrastructure, the same BTR1 inventory and MRV platform discussed in section 4, already produces exactly the kind of state-and-sector-specific emissions accounting the Court's evidentiary threshold requires. This cuts against any argument Bangladesh might otherwise raise that its own contribution is too diffuse or complex to quantify.

Causation follows the same pattern. The Court's two-step test asks, first, whether a climatic trend is attributable to anthropogenic climate change generally, and second, to what extent the resulting damage is attributable to a particular State (para 437), applying the Costa Rica v Nicaragua standard of a “sufficiently direct and certain causal nexus.” This standard is directly satisfiable on Bangladesh's brick kiln evidence. The link between brick kiln combustion and specific climatic and health outcomes is not a matter of scientific dispute: peer-reviewed dispersion modelling has already traced Dhaka's regional black carbon concentrations to named kiln clusters in Narayanganj, Gazipur, and Savar, and a randomized controlled trial has directly measured the CO2 and PM2.5 reductions achievable per kiln through operational change. Where the Court's causation standard requires “a sufficiently direct and certain causal nexus” rather than absolute scientific certainty, Bangladesh's own brick sector research already exceeds that threshold.

None of this forecloses the genuine uncertainty the Court acknowledged, that damage may arise from concurrent causes, or that causal science may remain contested in a given case (para 437), and the Court was clear that resolving such uncertainty falls to the adjudicating tribunal on the facts before it, not to the claimant in advance. But this residual uncertainty applies with far greater force to diffuse, multi-state atmospheric mixing than to a case such as Bangladesh's brick kilns, where the polluting source, its emissions volume, its geographic concentration, and its health and climate consequences are already independently documented in the scientific literature. The brick kiln sector therefore stands as a clear illustration that Bangladesh's dual identity is not merely rhetorical: under the Advisory Opinion's own attribution and causation architecture, Bangladesh possesses, indeed has already generated, the evidentiary record capable of grounding a state responsibility claim against itself, not only against others.

Taken together, the obligation-compliance analysis and the empirical evidence assembled in this section converge on the same conclusion. A state does not need to be a major historical emitter to fall within the legal identity of ‘violator’; it needs only to possess the kind of documented, quantifiable, and causally direct contribution that that Bangladesh itself has already produced, whether through unmet obligations or through sectors like brick manufacturing. a state can be simultaneously and legitimately a victim entitled to protection under international climate law, and a violator whose own conduct falls within the scope of the same legal obligations it invokes against others. Having established both dimensions of this dual identity, the chapter now closes by drawing together their combined implications for Bangladesh's position in international climate law.

CONCLUSION

By trying to understand the ICJ Climate Opinion to identify the dual legal identity of Bangladesh, this chapter has produced two meaningful contributions. One, to represent the ICJ Opinion in a well-sorted, understandable manner; secondly, to find out what subtle ways remain to make any climate-vulnerable country a climate violator, as the climate harm is done not only by big industries, but also from household emissions.

In conclusion, this chapter looks to rebut some controversies relating to the Opinion as it has already generated some contrasting scholarly views. While, Some scholars reagrd it as the foundation of a new climate legal order. Others remain unconvinced. Judge Yusuf himself, in his separate opinion, criticised the Opinion for not being sufficiently ambitious. He suggests that it was delivered on "extreme formalism." In the same tone, Fisher argues that the Court's statement that climate change "may impair the effective enjoyment of human rights" (para. 386) is weakened by the Court's own choice of word selection-*may*. He believes it leaves uncertainty as to how transformative the Opinion truly is (Fisher & Gordhan, 2026).

The author would like to have a different view. Reading this Chapter, the author believes, will convince anyone how systematic the Opinion was to deliver the Opinion. From historical analysis to climate litigation backed by scientific claims, the Opinion tried to accumulate and validate all the relevant climate laws so that no one feels ambiguity in understanding which laws are rigorous climate obligations that produce binding accountability. While construing for Bangladesh's dual identity, the paper highlights several perspectives of the court that were unique. Specially, its connection of principles to the core obligations, and answering different questions that created doubt so far, like *lex specelis* and *lex priori*. In doing so, the Court answered many of the doctrinal uncertainties that had previously divided climate stakeholders. It has removed many ambiguities for domestic, regional, and international courts that will produce a meaningful, less time-consuming, clearer path for upcoming climate litigations. The opinion thus created an authoritative framework from which future climate jurisprudence can develop. In that sense, the Opinion functions less as a source of new law than as a constitutional map of existing international climate law, a successful Grund norm to have a basic understanding of source of international environmental law.

Furthermore, the success of the Opinion lies in how it connected the Articles of State Responsibility with a climate perspective and construed the 'act and omission' in a new way. This gave a clearer perspective to measure how the breaches may arise in the case of climate change. The Opinion thus gave rise to the legal lens through which obligations became certain to invoke certain consequences upon States. This chapter, the author shall claim, could only exist because the Opinion, for the first time, gave the author the footing to impose responsibility on a climate-vulnerable state like Bangladesh. The Opinion exposed Bangladesh to legal scrutiny for its own conduct.

It would therefore be inaccurate to suggest that the Advisory Opinion is merely an endlessly and fatally deferred promise of a remedy. The Opinion never purported to solve the practical difficulties of implementation, nor could it. Courts only interpret law, but do not replace political institutions or enforce global environmental governance. The responsibility for implementation remains within the States themselves. The substantial part is done by the Opinion, and it is in the hands of the international Community how they shall invoke these responsibilities whenever a breach occurs.

The author would like to conclude the chapter with a notion that, although Bangladesh has been the main focus of this chapter, Bangladesh's dual legal identity is not unique. Many climate-vulnerable developing countries like Bangladesh, Pakistan, and others simultaneously possess enforceable rights to international assistance while bearing corresponding obligations to prevent environmental harm within their own territories (Haider & Ahmad, 2024). The Opinion therefore presents a universal framework applicable to all States regardless of their level of development. A framework where every State possesses both rights and responsibilities and the balance between the two depends not on political narratives but on context and legal compliance.

Therefore, the chapter vests the accountability measures on the whole world community of the world. Following the status of *erga omnes* as established by the Opinion, the accountability measures can now be adopted by any climate-concerned countries. Binding a state to its required obligations no longer needs the status of 'treaty party' to invoke consequences if the obligation is customary in nature. As presented by Haider et al. (2024), 'the essence of erga omnes obligations is in its power to ensure the inviolability of basic principle norms of the international legal system and to serve the common interests of the international community'. Therefore, this can give rise to a new legal institution to punish environmental crime. Most importantly, as Nody and Amin (2024) argue, safeguarding international environmental law requires not merely new legal principles but a renewed commitment to the environmental rule of law through the reassessment and restructuring of existing legal frameworks. It can be 'establishing technology-enhanced environmental governance systems to improve the monitoring and enforcement of governance processes', leaving no room for violation (Ahmad et al., 2025)

Ultimately, the ICJ Climate Change Advisory Opinion should not be understood as either a promise or a disappointment. As Ahmad et al. (2023) note, "the development of environmental law is a story set against changing societal values and changing conceptions of the relationship between humanity and nature." Therefore, It is a legal standard. It recognises Bangladesh's entitlement to protection while simultaneously holding Bangladesh accountable for its own environmental conduct. The strength of that standard is considerable. Whether it produces meaningful environmental justice, however, will depend less upon the Court itself than upon the willingness of States, and particularly Bangladesh, to transform clarified legal obligations into genuine institutional practice. This chapter thereby concludes with the notion that success in environmental governance ultimately depends upon communities of practice willing to "run against our strongest mental habits" (Douglas, 2013), embracing a culture in which compliance with international environmental law becomes not merely a legal requirement but an institutional norm.

REFERENCES

Adger, W. N. (2006). Vulnerability. *Global Environmental Change, 16*(3), 268–281. Retrieved July 2026, from. DOI: 10.1016/j.gloenvcha.2006.02.006

Ahmad, I., Haider, A., & Ahmad, K. (2025). The impact of judicial activism on environmental protection and justice in Pakistan: A statistical analysis. *Journal of Asian Development Studies, 14*(2), 1–10. Retrieved July 2026, from. DOI: 10.62345/jads.2025.14.2.1

Ahmad, I., Haider, A., & Khan, B. Z. (2023). In the name of nature: The legal frontiers of environmental preservation. *Journal of Asian Development Studies, 12*(4), 401–411. Retrieved July 2026, from. DOI: 10.62345/jads.2023.12.4.30

Amin, M., & Salekin, M. (2026). Sundarbans at risk: A legal and policy analysis of the transboundary management and sustainability challenges. In Dr. Firdous Ahmad Malik & Dr. Moon Moon Hussain, *Human algorithm of inclusive Earthscapes: Weaving education, AI, climate justice and transformative governance into a compassionate future* (pp. 29–49). Alborer (OPC) Pvt. Ltd.

Amin, M. B. (2022). *Emergence of customary international environmental law: A very need to protect the Earth. AMUCM Law Society Review, 1.* Available online at https://www.researchgate.net/publication/362222497_Emergence_of_Customary_International_Environmental_Law_A_Very_Need_to_Protect_the_Earth (accessed on July, 2026)

Application of the Convention on the Prevention and Punishment of the Crime of Genocide (The Gambia v. Myanmar), Preliminary Objections, Judgment, I.C.J. Reports 2022, 477 (International Court of Justice, July 22, 2022)

Belal, A. R., Cooper, S. M., & Roberts, R. W. (2015). Corporate environmental responsibility and accountability: What chance in vulnerable Bangladesh? *Critical Perspectives on Accounting, 33*, 44–58. Retrieved July 2026, from. DOI: 10.1016/j.cpa.2015.01.005

Belal, A. R., & Roberts, R. W. (2010). Stakeholders' perceptions of corporate social reporting in Bangladesh. *Journal of Business Ethics, 97*(2), 311–324. Retrieved July 2026, from. DOI: 10.1007/s10551-010-0511-4

Certain Activities Carried Out by Nicaragua in the Border Area (Costa Rica v. Nicaragua), Compensation Owed by the Republic of Nicaragua to the Republic of Costa Rica, Judgment, I.C.J. Reports 2018, p. 15

Conference of the Parties to the United Nations Framework Convention on Climate Change. (2012). *Decision 1/CP.17: Establishment of an Ad Hoc Working Group on the Durban Platform for Enhanced Action.* In *Report of the Conference of the Parties on its seventeenth session, held in Durban from 28 November to 11 December 2011.*

DNV. (2025). *Understanding and managing the regulatory landscape of ship recycling*. https://www.dnv.com/expert-story/maritime-impact/understanding-and-managing-the-regulatory-landscape-of-ship-recyling/ (accessed on June, 2026)

Douglas, M. (2013). *A very personal method: Anthropological writings drawn from life*. SAGE Publications. https://www.researchgate.net/publication/306032193_A_very_personal_anthropology_of_Mary_Douglas (accessed on June, 2026)

Eil, A., Li, J., Baral, P., & Saikawa, E. (2020). *[Title of dataset]*. Harvard Dataverse. https://doi.org/ (accessed on June, 2026)DOI: 10.7910/DVN/MV9FEH

Fisher, L., & Gordhan, S. (2026). Hope with teeth: Five lessons from the ICJ's climate change advisory opinion. *Journal of Environmental Law*, eqag011. Advance online publication. DOI: 10.1093/jel/eqag011

Haider, A., Ahmad, I., & Yaseen, M. (2024). Jus cogens and the right to self-determination: A study of its peremptory status and erga omnes effects. Pakistan Journal of Law. *Analysis and Wisdom*, *3*(6), 59–72.

Haider, A., Mathlouthi, N., & Ahmad, I. (2024). Beyond the books: Real world challenges in implementing environmental laws in Pakistan. *Remittances Review*, *9*(1), 1383–1405. DOI: 10.33282/rr.vx9il.93

Hassan, M. (2023, July 25). *Environmental impact assessment in Bangladesh: A critical review*. ResearchGate. https://www.researchgate.net/ (accessed on June, 2026)

Hossain, M. S. (2018). *Integrated coastal zone management (ICZM) and marine spatial planning (MSP)* [Conference presentation]. ResearchGate. https://www.researchgate.net/ (accessed on June, 2026)

Hui, Z., Haider, A., & Khan, A. (2025). International trade and plastic waste in oceans: Legal and policy challenges. *Frontiers in Marine Science*, *12*, 1627829. Advance online publication. Retrieved June 2026, from. DOI: 10.3389/fmars.2025.1627829

International Court of Justice. (2025, July 23). *Obligations of States in respect of climate change (Request for Advisory Opinion), Advisory Opinion.* Available online at https://www.icj-cij.org/case/187/advisory-opinions (accessed on July, 2026)

Islam, K. S., Xue, X.-Z., & Rahman, M. M. (2009). Successful integrated coastal zone management (ICZM) program model of a developing country (Xiamen, China)–Implementation in Bangladesh perspective. *Journal of Wetlands Ecology, 2*(1), 35–41. Retrieved July 2026, from. DOI: 10.3126/jowe.v2i1.1854

Islam, M. M., & Islam, A. R. M. T. (2026). Strengthening climate resilience in coastal Bangladesh: Analyzing the role of NGOs in adaptation governance from an intersectionality lens. *Progress in Disaster Science, 29*, 100498. Retrieved July 2026, from. DOI: 10.1016/j.pdisas.2025.100498

Islam, M. R., & Chowdhury, M. (2022). Effectiveness of the coastal and marine conservation initiatives in Bangladesh: Analyzing the drawbacks of the legal, policy, and institutional framework. *Journal of the Indian Ocean Region, 18*(2), 149–172. Retrieved July 2026, from. DOI: 10.1080/19480881.2022.2111050

Khan, M. A. R. (2020). Sustainable coastal zone management: Need for a holistic approach for Bangladesh. *Journal of Management and Sustainability, 10*(2), 112–120. Retrieved July 2026, from. DOI: 10.5539/jms.v10n2p112

Khan, N. A., & Belal, A. R. (1999). The politics of the Bangladesh environmental protection act. *Environmental Politics, 8*(1), 311–317. DOI: 10.1080/09644019908414449

Mahmud, I. (2025). *A human rights critique of Bangladesh's Climate Change Strategy and Action Plan (BCCSAP) & National Adaptation Plan (NAP) (2023–2050): Assessing participation, equity, accountability, and justice in adaptation policy* (Master of Human Rights practicum, University of Manitoba). MSpace Institutional Repository. https://mspace.lib.umanitoba.ca/handle/1993/39126 (accessed on June, 2026)

Maritime Magazine. (2021, December 2). *The Basel Convention's ban on the export of toxic ships to developing countries reestablished.* https://maritimemag.com/en/the-basel-conventions-ban-on-the-export-of-toxic-ships-to-developing-countries-reestablished/ (accessed on June, 2026)

Military and Paramilitary Activities in and against Nicaragua (Nicaragua v. United States of America), Merits, Judgment, I.C.J. Reports 1986, 14 (International Court of Justice, June 27, 1986).

Ministry of Environment. Forest and Climate Change. (2025, September). *Bangladesh's third nationally determined contribution (NDC 3.0).* Government of the People's Republic of Bangladesh. https://unfccc.int/sites/default/files/2025-09/Bangladesh%20Third%20Nationally%20Determined%20Contribution%20%28NDC%203.0%29.pdf (accessed on June, 2026)

Naher, N. (2025, August 15). *Rethinking environmental impact assessment in Bangladesh: Challenges, gaps, and pathways forward.* SSRN. https://doi.org/ (accessed on June, 2026)DOI: 10.2139/ssrn.5393399

Nody, N. I., & Amin, M. B. (2024). *The Post-COVID Environmental Ramifications and a Quest for an Eco-Friendly Legal Hegemony.* BAIUST Academic Journal., Retrieved June 2026, from, DOI: 10.63307/BAJ.5.1.011

Observer Research Foundation. (2024, July 10). *The Teesta water dispute: Geopolitics, myth and economics.* https://www.orfonline.org/research/the-teesta-water-dispute-geopolitics-myth-and-economics (accessed on June, 2026)

Pulp Mills on the River Uruguay (Argentina v. Uruguay), Judgment, I.C.J. Reports 2010, p. 14. https://www.icj-cij.org/case/135 (accessed on June, 2026)

Saleh, M. F. (2015, September 3). *A critical appraisal of Bangladesh Environment Conservation Act, 1995 and Rules, 1997.* Bangladesh Law Digest. https://bdlawdigest.org/bangladesh-environment-conservation-act-1995.html (accessed on June, 2026)

Simpson, G. (2021). The sentimental life of international law: Literature, language, and longing in world politics. Oxford University Press., Retrieved June 24, 2026, from, DOI: 10.1093/oso/9780192849793.001.0001

United Nations. (1992). *United Nations Framework Convention on Climate Change.* https://unfccc.int/resource/docs/convkp/conveng.pdf (accessed on June, 2026)

United Nations. (1998). *Kyoto Protocol to the United Nations Framework Convention on Climate Change.* https://unfccc.int/resource/docs/convkp/kpeng.pdf (accessed on June, 2026)

United Nations. (2015). *Paris Agreement.* https://unfccc.int/sites/default/files/english_paris_agreement.pdf (accessed on June, 2026)

Wewerinke-Singh, M., & Viñuales, J. E. (2026). Bracing for impact: The ICJ Advisory Opinion on climate change and the activation of legal tipping points. *Journal of Environmental Law,* eqag014. Advance online publication. Retrieved June 2026, from. DOI: 10.1093/jel/eqag014

ENDNOTES

1 Article 55, paragraph 4 of the commentary, ILC. Articles on State Responsibility, p. 140

2 Para 135

3 Pulp Mills on the River Uruguay (Argentina v. Uruguay), Judgment, I.C.J. Reports 2010 (I), pp. 55-56, para. 101, and p. 79, para. 197

4 Application of the Convention on the Prevention and Punishment of the Crime of Genocide (Bosnia and Herzegovina v. Serbia and Montenegro), Judgment, I.C.J. Reports 2007 (I), p. 221, para. 430

5 Pulp Mills on the River Uruguay (Argentina v. Uruguay), Judgment, I.C.J. Reports 2010 (I), p. 79, para. 197

6 Ibid para. 101;

7 *Climate Change*, Advisory Opinion, ITLOS Reports 2024, p. 91, para. 239, and p. 137, para. 397.

8 Article 01 UN Charter.

9 (Military and Paramilitary Activities in and against Nicaragua (Nicaragua v. United States of America), Merits, Judgment, I.C.J. Reports 1986, p. 101, para. 191).

10 'Will We Ever Get the Teesta Back?' *The Daily Star* (Dhaka) <https://www.thedailystar.net/opinion/views/news/will-we-ever-get-the-teesta-back-3571031> accessed 3 July 2026, citing International Food Policy Research Institute study data

11 'Unprecedented Floods Fuel Bangladesh-India Tension' *The Diplomat* (24 August 2024)

12 (Gabčíkovo-Nagymaros Project (Hungary/Slovakia), Judgment, I.C.J. Reports 1997, p. 78, para. 140

13 *Quinn v Leathem* [1901] AC 495 (HL), at p. 506

14 Pulp Mills on the River Uruguay (Argentina v. Uruguay), Judgment, I.C.J. Reports 2010 (I), p. 56, para. 101; see paragraphs 132-139).

15 decision 4/CMA.1, Further guidance in relation to the mitigation section of decision 1/CP.21, 15 December 2018, UN doc. FCCC/PA/CMA/2018/3/Add.1, p. 11, Annex I, paragraph 6 (a) and (b)), and how each party has addressed progression, highest possible ambition, common but differentiated responsibilities and respective capabilities in light of different national circumstances (ibid., Annex I, paragraph 6 (c) and (d))

16 W.N. Adger Vulnerability Global Environmental Change, 16 (2006), pp. 268-281

17 World Bank, *Bangladesh: Building Resilience to Climate Change* (2022).

18 Article 4 para 4, Paris Agreement.

Chapter 11

Islamic Environmental Ethics and Climate Justice in Maldives:

Integrating Faith, Law, and Community for Sustainable Climate Resilience

Mohammad Shekaib Alam
http://orcid.org/0009-0005-7991-8184
Villa College, Male', Maldives

Amish Abdullah
http://orcid.org/0009-0001-4791-776X
Villa College, Male', Maldives

Mohd Arsh Shery
http://orcid.org/0009-0008-6277-4040
Villa College, Male', Maldives

ABSTRACT

The Republic of Maldives is facing devastating existential threats from rising sea levels, coastal erosion, and ecosystem degradation. Current technocratic frameworks, which elevate science above religious doctrine and morality, have failed to incorporate Shariah's morally imperative elements, creating a critical governance gap. Islamic environmental ethics (khalifa, mizan, and amanah) therefore provide viable pathways for uniting faith, law, and community action for climate justice. Through qualitative analysis of jurisprudence and policy documents, this study identifies how imam sermons, mosque education, and ethical environmental assess-

DOI: 10.4018/979-8-2600-3678-5.ch011

ments can serve as faith-based drivers for grassroots fulfilment of God's will among vulnerable atoll communities. Recommendations: 1) partnering religious leaders and regulators through collaborative training; 2) implementing tawbah-aligned restorative penalties; and 3) embedding Maqasid al-Shariah curricula within the Maldivian education system thereby establishing a replicable model for climate justice within other Small Island Developing States.

INTRODUCTION

The ecological balance of the planet has been severely disrupted by human activity during the last century. Humans have negatively impacted many of the atmospheric conditions that support life on Earth. Scientific and political solutions alone do not adequately address the multiple moral, cultural, and institutional dimensions of the environmental crisis (Grim & Tucker, 2014). Religious institutions have the large institutional authority and social infrastructure necessary to mobilise large numbers of people toward environmental stewardship, as evidenced by the work of Hulme (2016), Jenkins et al. (2018), and Koehrsen (2021).

The Maldives represent an especially acute test case for this assertion. As a Muslim-majority country, the Maldives consists of a collection of over 1,000 low-lying islands, all averaging 1.5 metres above sea level. The Maldives is now facing imminent threats (from sea levels rising, coastal erosion, ocean warming, and salt-water intrusion) (Klein et al., 2014). Current adaptation strategies by the government (NAPA, MEEW, 2007; Law No. 4/93 (as amended)) lack sufficient support in practice. Consistent evidence provided by enforcement data and independent monitoring indicates a substantial discrepancy between statutory obligation and implementation (Human Rights Watch, 2023, 2025). This discrepancy is not solely rooted in the issue of resource constraints; it is deeper and reflects an absence of normative legitimacy regarding environmental obligations; they are experienced as being externally imposed but not as having any moral foundations (Saniotis, 2012).

In this chapter, the author attempts to fill in the gap through a different approach than that typically taken in the literature today, as the literature is typically very idealistic in nature. This chapter states that Islamic ethical frameworks do not equate to the automatic compliance of individuals, nor do mosques represent inherently trustworthy vehicles for enforcing compliance. This chapter posits that under specific structural conditions of constitutional alignment, institutional accountability, and transparent governance, the integration of khalifa, mizan, and amana into Maldivian environmental law can be both legally justified and empirically grounded.

The author's analysis is conducted through the lens of a scholar who is locally based in the Maldives with an academic interest in proving that faith-based governance reform is possible. Being locally based gives the author an advantage over someone from outside the nation, as the author has a better understanding of the political economy of Maldivian environmental governance, has access to local sources, and has a better sensitivity to community dynamics than would someone from outside of the nation. However, local scholars also have the potential to develop normative biases in how they conduct their research and may skew their research toward being optimistic about the potential for reforms. This chapter attempts to reduce that risk through systematically comparing cases of both success and failure, and highlighting the documented capture dynamics identified by Human Rights Watch (2023, 2025) to be structural barriers rather than peripheral caveats.

The argument will unfold in four parts: (1) a review of comparative evidence of when faith-based environmental governance has been successful and unsuccessful; (2) establishing the constitutional, theoretical, and methodological foundation for the reform argument; (3) applying Islamic ethical concepts to the Maldivian constitutional and statutory context; and (4) proposing a sequenced, conditionally feasible reform architecture.

LITERATURE REVIEW: RELIGION, ECOLOGY, AND ENVIRONMENTAL GOVERNANCE

The scholarly intersection of religion and environmentalism has matured considerably since White's (1967) influential critique of the Judeo-Christian tradition's ecological legacy. Subsequent decades have produced substantial evidence that religious traditions possess both normative resources and institutional capacities relevant to environmental governance, resources that, under the right conditions, translate into measurable behavioural and policy change (Jenkins, 2009; Tucker & Grim, 2017). This literature review restricts its scope to the questions that are most directly relevant to the Maldivian reform agenda: when does faith-based environmental governance succeed, when does it fail, and why?

WHEN FAITH-BASED GOVERNANCE WORKS

The effectiveness of faith-based environmental actions have been documented in multiple settings by multiple empirical social science researchers and the underlying conditions necessary for achieving success have been identified. Veldman et al.'s (2014) global review of religious environmentalism demonstrated that faith communities need an organised structure, not merely a theological motivation, for success. Furthermore, governance structures need to be formally embedded rather than merely articulated. Ramlan's (2019) comparative legal analysis of Singapore, Malaysia, and Indonesia has shown that attempts to operationalise Islamic environmental principles through *fatwas* and mosque sermons in Muslim-majority countries are systematically constrained due to religion-state complexity, and that motivation alone cannot sustain long-term pro-environmental behaviour.

An example of this is the 2011 Majelis Ulama Indonesia (MUI) binding fatwa that declared mining operations that destroy ecosystems as *haram* (Harvard Divinity School, 2024). While the declaration generated immense public discussion regarding the religious legitimacy of such mining operations, compliance with the fatwa was shown to be mainly driven by accompanying statutory enforcement mechanisms rather than the fatwa itself (Gade, 2019; Ramlan, 2019). The presence of religious legitimacy is necessary but not sufficient to achieve compliance; it amplifies statutory compliance rather than replacing it. In Morocco's green mosque programme, compliance with formal government regulations was demonstrated through 40% reductions in energy and water use across participating mosques because the mosque-based implementation was subject to external audit and reporting requirements (Ramlan, 2019). Similarly, Bangladesh's mosque-based cyclone early warning systems have achieved measurable reductions in mortality precisely because imams function as the last-mile component of a coordinated national emergency management system that includes clear protocols and accountability structures (Rahman & Ahmad, 2019).

WHEN FAITH-BASED GOVERNANCE FAILS

According to Koehrsen and Burchardt (2023), there are multiple instances where religiously framed environmental legislation has allowed state actors to shift responsibility to religious institutions within Muslim-majority contexts. This is referred to as "moral outsourcing." This dynamic is especially acute in regions where authority is difficult to separate or differentiate between religious and governmental authorities. The second area of failure is mosque-based programming that has recreated the existing hierarchies of class and gender, thus limiting the ability of women to be heard regarding their environmental knowledge and the adaptive processes they use to increase their climate resilience (Gade, 2019; Koehrsen & Burchardt, 2023). The third area of failure concerns centre-periphery governance: national programmes for faith-based environmental activities in Indonesia and Pakistan often fail to extend to outer islands or remote areas due to the differences between local religious authority structures and national government hierarchies (Ramlan, 2019; Javadi et al., 2020).

In the Maldives, this issue is further complicated due to documented struggles between the central government ministries and local island councils (MEEW, 2007; Human Rights Watch, 2023). Critically, in the Maldives, the moral outsourcing risk is not merely a theoretical risk. Human Rights Watch (2023, 2025) has documented numerous examples of land reclamation and coastal construction interests systematically avoiding environmental enforcement, noting that the ERA lacks the necessary independent institutional capacity to resist political pressure. This chapter's answer to that challenge is structural: ERA capacity reform and anti-capture safeguards must be established as preconditions to activating mosque-based functions, rather than serving as mere aspirational accompaniments to immediate statutory integration.

ISLAMIC ENVIRONMENTAL ETHICS: SCHOLARLY CONTEXT

Scholars interested in environmental ethics from an Islamic perspective have closely scrutinized the concept of stewardship's normative basis. For example, the 2015 Islamic Declaration on Global Climate Change was signed by over 60 nations' scholars and institutions and states that climate action is a religious duty (IFEES, 2019). Nasr (1996) connects the principles of *khalifa* (stewardship), *mizan* (balance), and *amana* (trust) to the metaphysical idea of *tawhid* (unity of creation), which gives stewardship its basis in the universe's order rather than just law; he contends that Islamic stewardship is "an extension of the cosmic order of nature and not just a set of legal regulations governing humans' interaction with other elements of nature" (Nasr, 1996, p. 373). Furthermore, Kamali (2008) and Auda (2008) have situated environmental protection within the *Maqasid al-Shariah* framework, arguing that the preservation of the environment is essential for preserving human life and dignity.

Recent scholarship has paid greater attention to the separation between normative sophistication and practical application. For example, Ramlan (2019) indicates that most Muslim-majority jurisdictions have sought to translate their Islamic environmental ethics into governance through *fatwa* (Islamic legal opinions) and *khutbah* (sermons) rather than formal statutory reform, producing limited durable change. The exceptions, Iran's Article 50 of the 1979 Constitution, Pakistan's Article 227, and Morocco's Green Mosque Regulations, are notable precisely because they moved beyond rhetorical alignment to structural integration.

Shari'ah-compliant green finance and its relevance to the Maldivian reform agenda have been examined in depth through research studies from scholars like Moghul and Safar-Aly (2015). Through their work, they demonstrate how *green sukuk* (Islamic bonds structured to finance environmentally sustainable projects) embed the ethical principles of *khalifa* and *mizan* within formal financial instruments, providing a vehicle to mobilize faith-aligned capital towards climate change adaptation on a large scale. For a Small Island Developing State (SIDS) with severely constrained fiscal space like the Maldives, unlocking these Islamic green finance instruments represents a practically significant and constitutionally resonant dimension of the reform agenda that extends well beyond symbolic rhetoric.

Within the context of faith-based governance, the work of critical Islamic feminist scholars is indispensable to addressing gender exclusion. Wadud's (2006) research demonstrates from within Islamic jurisprudence that the practice of excluding women from mosque governance structures reflects historically contingent patriarchal interpretations rather than Qur'anic injunction, thus creating a distinctly Islamic, not merely secular, basis for the full participation of women in mosque-based environmental bodies. Similarly, Mir-Hosseini (2015) identifies the specific jurisprudential mechanisms through which gender hierarchies are reproduced in Islamic institutional practice, thereby equipping advocates of reform with the language derived from the internal law of Islam needed to challenge resistance within the Maldivian context.

Finally, the scholarship of Barnett and Campbell (2010) provides essential empirical grounding for understanding the particular governance vulnerabilities and power asymmetries that define climate policy in Small Island Developing States. This analysis provides insights that are directly transferable to the Maldivian archipelagic context.

METHODOLOGY AND THEORETICAL FRAMEWORK

The methodology applied in this research is qualitative in nature, being based on critical-doctrinal principles; three different analytical modes are integrated: (1) doctrinal legal analysis of Maldivian constitutional and statutory frameworks; (2) comparative constitutional analysis across Muslim-majority jurisdictions; and (3) a critical review of empirical social science literature on faith-based environmental governance outcomes. Each mode is applied in a defined sequence: doctrinal legal analysis is used to determine *lex lata* (what the law currently requires); then comparative analysis identifies what has worked institutionally elsewhere; and finally, the empirical literature review provides the evidence base for conditional recommendations (*lex ferenda*).

The comparative cases examined include Morocco, Indonesia, Bangladesh, and Pakistan, chosen on the basis of three criteria: (1) their Muslim-majority constitutional status with explicit religion-state provisions; (2) documented attempts to integrate Islamic ethical principles into formal environmental governance; and (3) the availability of independent empirical assessments of outcomes. Malaysia, Jordan, the UAE, and Tunisia were also considered as comparatives, but not selected as primary cases: Malaysia's religion-state constitutional arrangement is significantly more complex and contested (Ramlan, 2019); Jordan and the UAE lack comparative assessments of faith-based *environmental* governance specifically; and Tunisia's constitutional transition after 2011 creates contextual instability that complicates comparison. However, these potential alternatives are not without merit and will be noted in cases where comparative insight can be drawn, but they do not serve as primary analytical anchors.

The chapter instead does not generate primary data, so there are no interviews, community surveys, E.R.A. enforcement records, or evaluations of mosque programming included within it, which is appropriate for a theoretical-doctrinal discussion. However, the analytical conclusions are bounded accordingly. Claims made about the feasibility of proposed reforms are conditional rather than empirically verified. Accordingly, what this chapter provides is theoretical architecture and comparative grounding with respect to an empirical research agenda that urgently needs to be conducted in the context of the Maldives.

The researcher's proximity to the Maldivian context (as noted in the Introduction) shapes both the analytical frame and the potential for normative investment in the reform argument. This chapter will therefore systematically engage with failure cases and documented capture dynamics to manage that risk.

THEORETICAL FRAMEWORK

In selecting the theoretical frameworks for this chapter, two frameworks are integrated: historical institutionalism and climate justice theory. The primary analytical lens used throughout this chapter is historical institutionalism (Hall & Taylor, 1996), which helps to understand why governance design matters more than theological motivation alone. In essence, historical institutionalism argues that institutions (the formal and informal rules that govern how people interact with one another in both a political and social context) shape actors' interests, capacities, and behaviour in ways that are both path-dependent and resistant to change by normative argument alone. When applied to religious governance, this framework explains why *fatwas* without accompanying institutional enforcement mechanisms consistently fail to produce durable compliance (Gade, 2019; Ramlan, 2019): actors embedded in institutional structures respond to incentives and accountability mechanisms, not only to normative persuasion. Gill (2008) applies institutional analysis to religious organisations more specifically and concludes that faith-based institutions do alter their behaviour depending on the political and regulatory environments they operate within; as such, they do not operate purely out of theological motivation. Trejo (2009) similarly demonstrates that the political and regulatory environment in which religious institutions operate is the primary determinant of whether they will function as agents of accountability or as instruments of power consolidation in Latin America. The reforms outlined below with respect to the Maldivian mosque's capacity to function as an effective intermediary for environmental governance depend entirely on this institutional environment, specifically the accountability structures, oversight mechanisms, and incentive systems within which it operates.

The normative framework of *climate justice theory* developed by Schlosberg and Collins (2014) allows one to evaluate both pre-existing arrangements and proposed reforms through the lens of equity. This normative framework consists of three distinct dimensions, which will be referred to throughout this chapter as:

- *Distributive justice*: Who bears the costs of environmental degradation? The communities on outer islands contribute negligibly to global greenhouse gas emissions; however, they bear the most immediate and severe costs of coastal erosion, saltwater intrusion, and ecosystem degradation (Barnett & Campbell, 2010; Klein et al., 2014). A constitutionally grounded reform that increases compliance with environmental law directly addresses this aspect of climate justice.
- *Recognitional justice*: Whose knowledge and voice counts in governance? Currently, the technocratic framework of Law No. 4/93 recognizes scientific and bureaucratic expertise; however, it systematically marginalizes the place-based environmental knowledge of island communities, women, and traditional fishing communities (Okereke, 2010). The proposal for a community monitoring framework below embeds community and women's voices into governance as a means of addressing the recognition component of climate justice.
- *Procedural justice*: Who participates in environmental decision-making? Current practice under Section 5 of Law No. 4/93 does not provide for meaningful community participation in EIA processes. Reforming the legislation to require Islamic ethical assessment criteria will create a procedural basis for communities to use the normative register of *khalifa* and *mizan* to challenge environmentally harmful projects, expanding participatory access for communities that currently have limited formal standing (Schlosberg & Collins, 2014).

ISLAMIC ENVIRONMENTAL ETHICS: FOUNDATIONAL CONCEPTS AND THEIR LEGAL OPERATIONALISATION

Khalifa (Stewardship)

The Islamic sacred texts (the Qur'an) present the human being as *khalifa*, designating humans as God's vicegerents entrusted with the Earth's care (Qur'an 2:30).

Hadith literature frames stewardship as accountability rather than dominion (Khalid, 2002; Nasr, 1996). Contemporary scholars like Fazlun Khalid (2002) align *khalifa* explicitly with climate justice, framing environmental degradation as a form of moral betrayal with both theological and political dimensions.

Of crucial importance to the reform agenda, *khalifa* has pre-modern legal expression in two established doctrines of Islamic jurisprudence. The *hima*, a system of legally protected areas prohibiting resource extraction within designated zones, and the *harim*, inviolable buffer zones surrounding water sources established to prevent contamination and overexploitation, constitute institutions formalised and enforced under classical Islamic law (Ammar, 2001; Izzi Dien, 2000). These doctrines represent a legal precedent within the Islamic tradition itself for statutory environmental protection. Their existence rebuts the objection that integrating Islamic ethics into formal environmental law is a departure from Islamic jurisprudential tradition; it is, rather, a recovery of it.

Mizan (Balance)

Mizan means the cosmic balance that came from the Almighty Allah as shown in the Qur'an (55:7-9) and requires humans to act within ecological and social limits (Nasr 1996; Sarvestani & Shahvali 2008). In terms of climate justice, *mizan* maps directly onto the concept of planetary boundaries and reinforces the normative case for equitable burden-sharing in climate governance (Özdemir 2018). For example, the Maldives contributes less than 0.01% of global greenhouse gas emissions yet faces existential climate risk. As such, *mizan* provides the theological grounding for the distributional justice claim that the costs of climate action must be equitably shared across the global community.

Amana (Trust)

Amana describes the sacred trust bestowed upon humanity to safeguard the Earth (Qur'an 33:72), entailing moral duties to prevent harm, promote ecological care, and uphold intergenerational and social justice (Nasr 1996; Saeed 2008). In the context of Maldivian governance, *amana* has direct implications for accountability: if the State holds the environment in trust for present and future generations (a construction consistent with Article 22 of the Constitution), then failure to enforce environmental law constitutes a breach of constitutional-theological obligation. This framing gives communities legitimate standing, grounded in Islamic jurisprudence, to demand that the State fulfill its constitutional duty.

Maqasid al-Shariah

The *Maqasid al-Shariah* integrates these principles, which are the higher objectives of Islamic law, aimed at preserving religion, life, intellect, lineage, and property (Kamali, 2008). Ibn Ashur's foundational jurisprudential treatise establishes that the preservation of the natural order constitutes a collective public interest (*maslaha amma*) that Islamic law is structurally obligated to protect (Ibn Ashur, 2006). Contemporary scholars, including Auda (2008), have explicitly incorporated environmental protection into this framework, recognizing that ecological preservation is essential to the preservation of human life and dignity. The *Maqasid* framework is significant because it situates environmental protection not as a peripheral concern of Islamic law but as a constitutive element of its core objectives, strengthening the constitutional argument for mandatory integration in Maldivian environmental legislation.

CLIMATE VULNERABILITY AND THE CONSTITUTIONAL FRAMEWORK

Ecological and Socioeconomic Context

The Maldives' vulnerability to climate change is structurally extreme. The Maldives consists of roughly 1,190 coral islands. These islands have an average elevation of just 1.5 metres above sea level. According to projections, an estimated 80% of the land area in the Maldives could be threatened by sea-level rise by 2100 (MEEW, 2007; World Bank, 2024). Coral reef systems provide the ecological foundations for both coastal protection and the tourism economy, which accounts for approximately half of the GDP. However, these systems suffered severe bleaching events in 2016, resulting in lasting damage to their buffering function. Furthermore, when saltwater intrudes into freshwater aquifers, it disproportionately affects women, children, and outer island communities lacking desalination infrastructure (WHO, 2021). This represents a distributional justice concern that sits at the heart of the climate justice framework applied throughout this chapter.

Legal and Constitutional Framework

Constitutional law in the Maldives creates the most fundamental environmental obligation. To begin with, Article 22 of the Constitution of the Republic of the Maldives (2008) imposes a non-derogable obligation on the State to protect and preserve the natural environment, biodiversity, and natural resources for the benefit of present and future generations. This provision is part of a larger constitutional system that establishes an expressly Islamic constitutional order. Articles 9(d) and 10(a) of the same Constitution require every Maldivian citizen to be Muslim and designate Islam as the official religion of the State, establishing the framework within which the Article 22 environmental duty operates.

Through such provisions, the Constitution provides a clear mandate for integrating Islamic ethics into environmental governance. At present, however, the practical integration of this mandate has not been achieved by Law No. 4/93, as amended. Law No. 4/93 is the principal statutory mechanism through which the State operationalises the constitutional obligation established by Article 22. It requires Environmental Impact Assessments (EIAs) for all projects likely to cause environmental harm (Section 5), prohibits illegal waste disposal (Sections 7a-7b), and prescribes penalties for violations (Section 9). The law is administered by the Environmental Regulatory Authority (ERA), which was created by Presidential Directive No. 14/2025 and operates within the purview of the Ministry of Climate Change, Environment and Energy.

Even though the framework appears to be comprehensive, enforcement data reveal systematic failure. As outlined by Human Rights Watch (2023, 2025), many provisions in legislation aimed at protecting the environment are routinely bypassed. For example, land reclamation and coastal construction are two areas where environmental protection provisions are frequently ignored. These failures disproportionately affect outer island communities and other vulnerable populations. This cannot simply be viewed as a problem of resources; it is also a matter of legitimacy. Without community-level normative buy-in, enforcement requires a coercive intensity which is neither feasible nor desirable in the Maldivian context (Saniotis, 2012; Ramlan, 2019).

The Maldives' obligations at the domestic level have been strengthened by the ITLOS Advisory Opinion of 21 May 2024 (Case No. 31). The Opinion found unanimously that anthropogenic greenhouse gas emissions constitute "pollution of the marine environment" as defined under UNCLOS. Furthermore, it established that State parties carry stringent due diligence obligations to prevent, reduce, and control marine environmental harm under Articles 192, 194, 207, 212, 213, and 222 (ITLOS, 2024). As scholarly commentary indicates, the findings of the Opinion went beyond what many anticipated, establishing UNCLOS as an independent legal basis for States' climate obligations and delineating that the due diligence standard is "stringent," proportionate to the "high risk of serious and potentially irreversible harm" associated with climate change and ocean acidification (BIICL, 2024). For the Maldives, a State whose entire territorial existence depends wholly upon the marine environment, the Opinion reinforces the Article 22 constitutional duty and raises the international legal stakes for persistent enforcement failure under Law No. 4/93.

The Governance Gap

The ongoing failure to enforce existing laws relating to the environment in the Maldives demonstrates a structural governance gap that results from a convergence of three separate but interrelated dynamics. The first dynamic is a *constitutional gap*: the Constitution of the Maldives gives the State a mandate to uphold an Islamic constitutional order; however, this mandate has not been operationalised into the substantive provisions of Law No. 4/93. Consequently, there is a disconnect between the constitutional values the State must uphold and those that environmental regulation actually deploys. Secondly, there exists an *institutional gap*: the ERA does not have the distributed presence and community trust necessary to effectively enforce environmental regulation across a dispersed archipelago (Human Rights Watch, 2023). Lastly, there exists a *normative gap*: without cultural and moral resonance, environmental obligations will be treated as administrative or bureaucratic impositions rather than as shared commitments to protect the environment (Saniotis, 2012).

This diagnosis sheds light on what Islamic ethical integration can and cannot achieve. It can address the normative gap by providing a moral framework for environmental obligations that commands genuine local authority. It can partially address the institutional gap by leveraging the distributed presence of mosques across the islands. Islamic ethics cannot, however, substitute for structural reforms in ERA capacity and judicial oversight. A reform agenda treating Islamic ethics as a *replacement* for strengthened formal enforcement would be an analytical error and practically counterproductive.

CRITICAL ANALYSIS: RESOLVING TENSIONS AND CONFRONTING OBSTACLES

Resolving the Constitutional Tension

A significant tension exists within the chapter's core argument: Islamic ethics integration is framed simultaneously as a constitutional mandate (non-negotiable) and as contingent on specific institutional conditions. This tension requires explicit resolution. The constitutional obligation under Articles 9(d), 10(a), and 22 is non-derogable, it does not disappear because conditions are unfavourable. What varies is not the obligation, but the *form of its implementation*. When institutional conditions are absent, specifically, when ERA capacity is insufficient and accountability structures do not exist, the State remains in ongoing constitutional breach. The appropriate remedy, however, is *sequential*: the State must first create the institutional preconditions before activating mosque-based functions within the statutory framework.

The sequencing argument is therefore not a retreat from constitutional obligation, but a specification of it. The constitutional mandate is absolute; the mechanism for discharging it is conditional. The failure to establish institutional preconditions places the State in breach of its constitutional obligation through institutional design failure, a justiciable deficiency that civil society organisations and island communities have standing, under Article 22 and the *amana* principle, to challenge. This resolution transforms conditionality from a limitation of the reform proposal into its strongest structural feature: it holds the State accountable not merely for ultimate outcomes, but for the institutional design choices that determine whether those outcomes are achievable.

The Risk of Political Capture

The moral outsourcing risk is not merely hypothetical within the context of the Maldives. Human Rights Watch (2023, 2025) documents systematic patterns in which development interests have captured both environmental regulatory processes and, to a significant extent, the deployment of religious authority for governance purposes. Any reform proposal that does not address this documented capture dynamic is analytically incomplete.

The proposed safeguards in the reform architectural framework below, namely, independent ERA oversight of mosque-based functions, mandatory civil society representation on review bodies, transparent reporting of enforcement actions, and the constitutional justiciability of Article 22, have been specifically designed to address this dynamic. However, the chapter must be candid: if the ERA itself is subject to political capture, then oversight of mosque functions by the ERA alone will not prevent moral outsourcing. Therefore, the multi-stakeholder Policy Review Committee proposed below will consist of civil society organisations, representatives of the outer island communities, and independent legal practitioners, actors whose structural interests are in opposition to those of the development interests that have historically captured regulatory institutions. Membership must be appointed through open, advertised selection processes subject to Majlis (parliamentary) confirmation, rather than by executive appointment, to prevent the appointment process itself from being captured.

Centre-Periphery Tensions

In the Maldives, ongoing tensions exist between central government authority and island-level local councils. These tensions carry direct implications for a reform model premised on distributed faith-based implementation. Competing jurisdictional claims between the Ministry of Climate Change, Environment and Energy, island councils, and local elites over environmental regulation, land use, and enforcement authority mean that mosque-based monitoring and reporting mechanisms could, in the wrong institutional context, become instruments of local power consolidation rather than accountability (MEEW, 2007; Human Rights Watch, 2023). Therefore, the reform design proposed below attempts to address this by embedding mosque-based functions within a clearly defined national governance architecture, with ERA oversight and transparent reporting requirements applying uniformly across all islands.

Gender and Recognitional Justice

Mosque-based environmental programmes have so far focused on male leadership within the religious community, with very few formal roles for women. This creates a failure both of recognitional justice and of practical governance capacity, as women and outer island communities typically possess the most immediate and locally specific environmental knowledge (Gade, 2019; Koehrsen & Burchardt, 2023). Furthermore, the resistance to women's participation in mosque governance structures is not theologically mandated. According to Wadud (2006), the exclusion of women from governance structures reflects historically contingent patriarchal interpretations rather than Qur'anic injunction. Mir-Hosseini (2015) identifies the specific jurisprudential mechanisms through which gender hierarchies are reproduced, thus equipping reformers with the internal Islamic legal language, not merely secular equality arguments, needed to challenge resistance in the Maldivian context. As a result, the reform proposal outlined below grounds its gender participation requirements in Islamic jurisprudence as well as in the empirical evidence that inclusive governance produces better environmental outcomes than hierarchically restricted approaches (Gade, 2019).

POLICY RECOMMENDATIONS: A SEQUENCED REFORM ARCHITECTURE

The following recommendations are grounded in the analysis above, consistently distinguish lex lata from lex ferenda, and are sequenced to reflect the constitutional tension resolution: accountability infrastructure first, statutory integration second.

Phase 1 (Immediate): Establishing Institutional Preconditions

1. ERA Capacity Reform and Anti-Capture Safeguards

Prior to the enactment of mosque-based functions through legislation, the ERA must undergo structural reform to achieve institutional independence from political capture. The following changes are needed: a) providing statutory security of tenure for ERA leadership, removing them from direct executive dismissal; b) establishing independent funding mechanisms not subject to annual executive discretion; c) requiring mandatory public reporting of all enforcement decisions and all cases where enforcement was not pursued; and d) creating a civil society advisory panel with formal standing to challenge decisions made by the ERA. If these changes do not take place, additional legislative integration of mosque-based functions will reproduce the capture dynamics documented by Human Rights Watch (2023, 2025).

2. Multi-Stakeholder Policy Review Committee

Establish a permanent multi-sectoral Islamic Environmental Ethics and Law Review Committee. Membership will include individuals from several sectors: Islamic scholarship, environmental law practitioners, ERA officials, outer island community representatives, civil society organisations, gender advocacy groups, and international comparative law experts. All members must be selected through an open, advertised selection process subject to Majlis confirmation (rather than executive appointment), to structurally prevent appointment capture by the interests that have historically bypassed environmental law. The Committee's annual reports should be tabled in the Majlis as a matter of constitutional accountability under Article 22.

Phase 2 (Following Institutional Reform): Statutory Integration

3. Statutory Incorporation of Islamic Ethical Principles Into Law No. 4/93

Lex ferenda: Amend Law No. 4/93 to incorporate explicit references to *khalifa*, *mizan*, and *amana* in the Act's operative provisions, not only in its preamble:

1. Section 5 (EIA requirements) should be amended to require that EIA reports address the project's consistency with *khalifa* and *mizan* as constitutional values mandated by Articles 9(d), 10(a), and 22. This creates procedural justice by giving communities formal standing to challenge projects through an Islamic ethical lens (Schlosberg & Collins, 2014).

2. Add a definitional section to provide statutory definitions of *khalifa*, *mizan*, and *amana* drawing on the *Maqasid al-Shariah* framework and the jurisprudential literature of Kamali (2008) and Ibn Ashur (2006). This will provide for legal certainty and prevent the arbitrary application of the law.
3. The preamble of Law No. 4/93 already states that the Maldivian environment should be conserved as a heritage for future generations, making it consistent with the inter-generational stewardship principle of *amana*. Amending the operative provisions to reflect this founding intent represents a coherent statutory development, not a radical departure.

4. Graduated Enforcement With Mosque-Administered Restorative Sanctions, Subject to Independent Oversight

Lex ferenda: Amend Section 9 of Law No. 4/93 to establish a graduated penalty framework consisting of the following:

- **Monetary penalties**: The statutory range of penalties will remain unchanged (Rf 5 to Rf 100,000,000) but must be calibrated to violation severity and persistence.
- **Custodial sanctions**: For egregious or repeat offenders, custodial sanctions must be created, requiring express new legislative authority.
- **Restorative sanctions**: Offenders will be required to fund ERA-supervised rehabilitation programmes, with mosque institutions serving as community implementation partners for environmental education, ecosystem restoration, and community engagement framed as *tawbah* (repentance).

A phased pilot across 10-15 islands, with independent evaluation against defined outcome metrics, is recommended before full national implementation. Mosque institutions participating in restorative sanction administration will require investment in trained personnel and programme coordination capacity.

The following safeguards are non-negotiable and must accompany statutory enactment:

- All mosque-based programmes must be subject to independent ERA audit with publicly reported performance metrics.
- Religious institutions participating in sanction administration must have no financial interest in the penalties imposed.
- Clear grievance and appeal mechanisms must be available to all affected individuals and communities.

- The multi-stakeholder oversight committee must have formal authority to suspend mosque-based functions if accountability requirements are not met.

5. Participatory Environmental Monitoring With Religious Institutions as Intermediaries

Establish a structured community-based environmental monitoring system where mosques act as trusted intermediaries, not enforcement bodies, for reporting environmental violations:

- A standardised mobile-enabled reporting platform for community members to log violations with GPS coordinates and photographic evidence.
- Imam-facilitated community environmental committees with legally mandated gender-equitable composition in each island. The Maldivian context presents documented barriers to women's participation in mosque governance structures, including the absence of women from *khutbah* delivery and mosque administrative roles, that reflect patriarchal jurisprudential interpretations rather than Qur'anic requirements (Wadud, 2006). Because the proposed community environmental committees are constituted under the statutory framework rather than as mosque governance bodies, their gender composition requirements are enforceable as a matter of law. Any threshold adopted should be informed by comparative legislative consultation and be sufficient to prevent tokenism (Mir-Hosseini, 2015).
- A clear protocol for transmitting community monitoring data to the ERA, with defined response timelines and public outcome reporting.

Phase 3 (Sustained): Education and Finance Integration

6. Mandatory Islamic Environmental Ethics Curriculum

All schools, *madrasas*, and mosque-based education programmes will be required to have compulsory modules on Islamic environmental ethics. The curricula should address: the *Maqasid al-Shariah* framework; citizens' rights under Law No. 4/93 and Article 22; climate science and its relationship to Islamic cosmological principles; and practical prescriptions for individual and community environmental responsibility (Kamali, 2008; Auda, 2008). An important consideration when developing the curriculum is that it must involve Islamic scholars, environmental law practitioners, climate scientists, and critically, youth and women's community representatives from outer islands, whose localised environmental knowledge should be incorporated as pedagogically legitimate.

7. Islamic Green Finance Integration

The reform architecture is not sustainable without addressing its funding dimension. Green *sukuk*, Shariah-compliant financial instruments structuring returns from environmentally sustainable assets, represent a constitutionally resonant mechanism for mobilising climate adaptation finance in a Muslim-majority SIDS (Moghul & Safar-Aly, 2015). The Islamic identity of the Maldives' Constitution establishes a framework to draw faith-aligned capital from Gulf Cooperation Council and Southeast Asian Islamic finance markets to fund climate adaptation infrastructure. The statutory incorporation of *khalifa*, *mizan*, and *amana* into Law No. 4/93 would create the legal framework within which green *sukuk* issuance could be constitutionally grounded, linking the statutory reform agenda to Islamic finance instruments capable of addressing the SIDS climate finance gap at scale.

CONCLUSION

By examining the statutory integration of Islamic environmental ethics into Maldivian environmental law through the lens of constitutional obligation, comparative evidence, historical institutionalist theory, and a critical assessment of the conditions under which faith-based governance succeeds or fails, this chapter has articulated a central analytical contribution to resolving the tension between constitutional mandate and conditional feasibility. The constitutional obligation is absolute, but the mechanism for discharging it is conditional. Furthermore, the failure to establish these institutional preconditions is itself a justiciable constitutional deficiency, placing the State in breach not only for enforcement failure but for institutional design failure.

Applied through the three dimensions of climate justice, the proposed reform framework creates a pathway to address distributive justice by reducing locally imposed environmental harm on the communities least responsible for climate change; recognitional justice by embedding women's and island communities' voices in governance structures; and procedural justice by establishing statutory standing for communities to utilize Islamic ethical criteria to challenge potentially environmentally harmful projects (Schlosberg & Collins, 2014; Barnett & Campbell, 2010).

What the comparative evidence firmly supports is this: where Islamic ethical integration is constitutionally grounded, structurally embedded into formal governance architecture, and supported by robust accountability mechanisms, it can generate normative conditions for environmental compliance that technocratic regulation alone cannot achieve (Ramlan, 2019; Gade, 2019). The constitutional framework of the Maldives, Articles 9(d), 10(a), and 22 operating in conjunction, provides an exceptionally strong foundation for exactly this kind of integration. The ITLOS

Advisory Opinion of May 2024 reinforces the urgency of reform and establishes a stringent due diligence standard that existing Maldivian enforcement practice demonstrably fails to meet (ITLOS, 2024; BIICL, 2024).

That said, the proposed reform is conditional, not simple; the ability of Islamic ethical integration to strengthen Maldivian environmental governance depends on whether it is preceded by ERA independence reform, accompanied by anti-capture safeguards, and evaluated against measurable outcomes. If these conditions are not satisfied, the reform risks producing exactly the outcome it is designed to prevent: a governance system in which moral language substitutes for, rather than demands, institutional performance.

Future research should prioritise empirical community-level studies of Islamic environmental ethics operationalisation in the Maldives, independent evaluation of existing faith-based climate programmes, and comparative analysis of graduated enforcement frameworks in Muslim-majority legal systems.

REFERENCES

Abedi-Sarvestani, A., & Shahvali, M. (2008). Environmental ethics: Toward an Islamic perspective. *American-Eurasian Journal of Agricultural & Environmental Sciences, 3*(4), 609–617. https://www.idosi.org/aejaes/jaes3(4)/15.pdf

Auda, J. (2008). *Maqasid al-Shariah as philosophy of Islamic law: A systems approach.* International Institute of Islamic Thought. DOI: 10.2307/j.ctvkc67tg

Barnett, J., & Campbell, J. (2010). *Climate change and small island states: Power, knowledge and the South Pacific.* Routledge.

British Institute of International and Comparative Law (BIICL). (2024). *Reflections on the ITLOS Advisory Opinion.* https://www.biicl.org/documents/184_reflections_on_the_itlos_advisory_opinion_final.pdf

Constitution of the Islamic Republic of Pakistan. (1973). Art. 227.

Constitution of the Republic of Maldives. (2008). Arts. 9(d), 10(a), 22.

Environmental Protection and Preservation Act of Maldives (Act No. 4/1993). (1993). Republic of Maldives.

Gade, A. M. (2019). *Muslim environmentalisms: Religious and social foundations.* Columbia University Press. DOI: 10.7312/gade19104

Gill, A. (2008). *The political origins of religious liberty.* Cambridge University Press.

Green Climate Fund. (2021). *Building Climate Resilient Safer Islands (BCRSI) project: Maldives.* https://www.greenclimate.fund/project/fp151

Grim, J., & Tucker, M. E. (2014). *Ecology and religion.* Island Press. DOI: 10.2307/jj.41003348

Hall, P. A., & Taylor, R. C. R. (1996). Political science and the three new institutionalisms. *Political Studies, 44*(5), 936–957. DOI: 10.1111/j.1467-9248.1996.tb00343.x

Harvard Divinity School Religion and Public Life Project. (2024). *Environmental fatwas of Indonesia.* https://rpl.hds.harvard.edu/religion-context/case-studies/climate-change/environmental-fatwas-indonesia

Hulme, M. (2016). *Weathered: Cultures of climate.* SAGE Publications.

Human Rights Watch. (2023). *Maldives: Environmental protection laws bypassed.* https://www.hrw.org/news/2023/10/17/maldives-environmental-protection-laws-bypassed

Human Rights Watch. (2025). *World report 2025: Maldives*. https://www.hrw.org/world-report/2025/country-chapters/maldives

Ibn Ashur, M. T. (2006). *Treatise on Maqasid al-Shariah* (El-Mesawi, M. E., Trans.). International Institute of Islamic Thought. DOI: 10.2307/j.ctvkc673b

International Tribunal for the Law of the Sea (ITLOS). (2024). *Request for an Advisory Opinion submitted by the Commission of Small Island States on Climate Change and International Law* (Case No. 31). Advisory Opinion of 21 May 2024. https://www.itlos.org/fileadmin/itlos/documents/cases/31/Advisory_Opinion/C31_Adv_Op_21.05.2024_orig.pdf

Islamic Foundation for Ecology and Environmental Sciences (IFEES). (2019). *Applying Islamic environmental ethics*. https://ifees.org.uk/wp-content/uploads/2019/12/applying_islamic_environmental_ethics.pdf

Izzi Dien, M. Y. (2000). *The environmental dimensions of Islam.*

Javadi, H., Mirjalili, S., & Hajizadeh, A. (2020). Faith-based environmental organisations and water governance in Pakistan. *Water Alternatives*, *13*(2), 401–421. https://www.water-alternatives.org/index.php/alldoc/articles/vol13/v13issue2

Jenkins, W., Berry, T., & Kreider, L. B. (2018). Religion and climate change. *Annual Review of Environment and Resources*, *43*(1), 85–108. DOI: 10.1146/annurev-environ-102017-025855

Jenkins, W. J. (2009). Religious responses to environmental challenges. In Gottlieb, R. (Ed.), *The Oxford handbook of religion and ecology* (pp. 45–64). Oxford University Press.

Kamali, M. H. (2008). *Shari'ah law: An introduction*. Oneworld Publications.

Khalid, F. (2002). Islam and the environment. In Munn, T. (Ed.), *Encyclopaedia of global environmental change* (Vol. 5, pp. 332–339). Wiley.

Klein, R. J. T., Midgley, G. F., Preston, B. L., Alam, M., Berkhout, F. G. H., Dow, K., & Shaw, M. R. (2014). Adaptation opportunities, constraints, and limits. In *Climate change 2014: Impacts, adaptation, and vulnerability* (Part A, pp. 899–943). Cambridge University Press. https://doi.org/DOI: 10.1017/CBO9781107415379.021

Koehrsen, J. (2021). Does religion promote environmental sustainability? Exploring the role of religion in local energy transitions. *Social Compass*, *68*(1), 3–19. DOI: 10.1177/0037768620982244

Koehrsen, J., & Burchardt, M. (2023). Faith-based organisations and climate justice: Between moral outsourcing and institutional transformation. *Global Environmental Politics*, *23*(1), 75–95. DOI: 10.1162/glep_a_00643

Ministry of Environment. Energy and Water (MEEW). (2007). *National Adaptation Programme of Action: The Republic of Maldives*. UNFCCC. https://unfccc.int/resource/docs/napa/mdv01.pdf

Ministry of Environment and Energy. (2015). *Maldives Climate Change Policy Framework 2015–2025*. Republic of Maldives.

Mir-Hosseini, Z. (2015). Muslim legal tradition and the challenge of gender equality. In Mir-Hosseini, Z., Al-Sharmani, M., & Rumminger, J. (Eds.), *Men in charge? Rethinking authority in Muslim legal tradition* (pp. 1–24). Oneworld Publications.

Moghul, U. F., & Safar-Aly, S. H. K. (2015). Green sukuk: The introduction of Islam's environmental ethics to contemporary Islamic finance. *Georgetown International Environmental Law Review*, *27*(1), 1–60.

Nasr, S. H. (1996). *Religion and the order of nature*. Oxford University Press. DOI: 10.1093/acprof:oso/9780195108231.001.0001

Okereke, C. (2010). Climate justice and the international regime. *Wiley Interdisciplinary Reviews: Climate Change*, *1*(3), 462–474. DOI: 10.1002/wcc.52

Özdemir, İ. (2018). The concept of mizan and ecological ethics in Islam. *International Journal of Islamic Thought*, *13*(1), 58–73. DOI: 10.24035/ijit.13.2018.006

Presidency of the Republic of Maldives. (2019). *Climate Smart Resilient Islands initiative*. https://presidency.gov.mv/Press/Article/22213

Presidency of the Republic of Maldives. (2025). *Presidential Directive No. 14/2025: Renaming of the Environmental Protection Agency to the Environmental Regulatory Authority*. https://presidency.gov.mv/Press/Article/34548

Rahman, A., & Ahmad, N. (2019). Mosque-based disaster preparedness in Bangladesh: Institutional integration and community outcomes. *International Journal of Disaster Risk Reduction*, *35*, 101066. DOI: 10.1016/j.ijdrr.2019.01.021

Ramlan, S. (2019). *Religious Law for the Environment: Comparative Islamic Environmental Law in Singapore, Malaysia, and Indonesia* (NUS Centre for Asian Legal Studies Working Paper 19/03). National University of Singapore Faculty of Law. https://law.nus.edu.sg/wp-content/uploads/2020/04/CALS-WPS-1903.pdf

Regulation on the Preparation of Environmental Impact Assessment Report (Regulation No: 2012/R-27). (2012). Ministry of Housing and Environment, Republic of Maldives.

Saeed, A. (2008). Environmental ethics and Islam. *Islamic Studies*, *47*(3), 313–340.

Saniotis, A. (2012). Muslims and ecology: Fostering Islamic environmental ethics. *Contemporary Islam*, *6*(2), 155–171. DOI: 10.1007/s11562-011-0173-8

Schlosberg, D., & Collins, L. B. (2014). From environmental to climate justice: Climate change and the discourse of environmental justice. *Wiley Interdisciplinary Reviews: Climate Change*, *5*(3), 359–374. DOI: 10.1002/wcc.275

Trejo, G. (2009). Religious competition and ethnic mobilization in Latin America: Why the Catholic Church promotes indigenous rights in Mexico. *The American Political Science Review*, *103*(3), 323–342. DOI: 10.1017/S0003055409990025

Tucker, M. E., & Grim, J. (2017). *Religion and ecology: Developing a planetary ethic*. Columbia University Press.

United Nations Development Programme (UNDP). (2022). *Maldives country programme document 2022–2026*. https://www.undp.org/maldives/publications/country-programme-document-maldives-2022-2026

Veldman, R. G., Jenkins, W. J., & Chapple, C. K. (2014). *How the world's religions are responding to climate change: Social scientific investigations*. Harvard University Press.

Wadud, A. (2006). *Inside the gender jihad: Women's reform in Islam*. Oneworld Publications.

White, L.Jr. (1967). The historical roots of our ecological crisis. *Science*, *155*(3767), 1203–1207. DOI: 10.1126/science.155.3767.1203

World Bank. (2024). *Key Highlights: Maldives Country Climate and Development Report*. https://www.worldbank.org/en/country/maldives/publication/key-highlights-maldives-country-climate-and-development-report

World Health Organization (WHO). (2021). *Climate change and health in small island developing states*. https://lancetcountdown.org/wp-content/uploads/2024/03/Climate-Change-and-Health-in-Small-Island-Developing-States_February-2022.pdf

Compilation of References

Aayog, N. I. T. I. (2021). *SDG India Index and Dashboard 2020–21: Partnerships in the decade of action*. Government of India. https://www.niti.gov.in/reports-sdg

Abedi-Sarvestani, A., & Shahvali, M. (2008). Environmental ethics: Toward an Islamic perspective. *American-Eurasian Journal of Agricultural & Environmental Sciences*, *3*(4), 609–617. https://www.idosi.org/aejaes/jaes3(4)/15.pdf

Achterfeldt, S., & Traverso, M. (2026). Classification of the EU taxonomy for sustainable activities among other sustainable finance taxonomies worldwide. *Clean Technologies and Environmental Policy*, *28*(4), 88. Advance online publication. DOI: 10.1007/s10098-025-03384-6

Adger, W. N. (2006). Vulnerability. *Global Environmental Change*, *16*(3), 268–281. Retrieved July 2026, from. DOI: 10.1016/j.gloenvcha.2006.02.006

Adil., ... (2025, February 12). *Climate Risk Index 2025 | Germanwatch e.V.* https://www.germanwatch.org/en/93013

Afzal, J., & Haider, A. (2025). Environmental and Resource Protection Law by Ke Zhou, Luozhi Yi, Xinjian Su and Youhai Sun. *International Journal of Law and Legal Advancement*, *1*(2). Advance online publication. DOI: 10.64060/IJLLA.v1i2.3

Ahmad, I., Haider, A., & Ahmad, K. (2025). The impact of judicial activism on environmental protection and justice in Pakistan: A statistical analysis. *Journal of Asian Development Studies*. https://www.researchgate.net/publication/391771576

Ahmad, I., Haider, A., & Ahmad, K. (2025). The impact of judicial activism on environmental protection and justice in Pakistan: A statistical analysis. *Journal of Asian Development Studies*, *14*(2), 1–10. Retrieved July 2026, from. DOI: 10.62345/jads.2025.14.2.1

Ahmad, I., Haider, A., & Zeb, B. (2023). In the name of nature: The legal frontiers of environmental preservation. *Journal of Asian Development Studies, 12*(4), 401–411. DOI: 10.62345/jads.2023.12.4.30

Alexander, A., & Orlando, E. (2025). The Triple Alignment Model for Effective Sustainable Supply Chain Regulations. *Academy of Management Proceedings, 2025*(1). DOI: 10.5465/AMPROC.2025.19814abstract

Ali, K., & Tintawi, G. (2026). Governing environmental decisions with artificial intelligence: Algorithmic sustainability for accountability and lifecycle impact, a policy review. *Preprints*. DOI: 10.20944/preprints202602.1746.v3

Ali, K., Tintawi, G., Bassma, M. K., & Haider, A. (2026). Algorithmic sustainability: Governance conditions for AI-driven environmental decision-making. *Preprints.org*. https://doi.org/DOI: 10.20944/preprints202602.0733.v2

Almaqtari, F. A., Elsheikh, T., Abdelkhair, F., & Mazrou, Y. S. A. (2023). The impact of corporate environmental disclosure practices and board attributes on sustainability: Empirical evidence from Asia and Europe. *Heliyon, 9*(8), e18453. DOI: 10.1016/j.heliyon.2023.e18453 PMID: 37560679

Alsaleh, M., & Wang, X. (2025). The integration of ecosystem-based management in EU fisheries governance: Challenges and opportunities. *Economics of Governance, 26*(4), 527–558. DOI: 10.1007/s10101-025-00338-y

Amin, M. B. (2022). *Emergence of customary international environmental law: A very need to protect the Earth. AMUCM Law Society Review, 1*. Available online at https://www.researchgate.net/publication/362222497_Emergence_of_Customary_International_Environmental_Law_A_Very_Need_to_Protect_the_Earth (accessed on July, 2026)

Amin, M., & Salekin, M. (2026). Sundarbans at risk: A legal and policy analysis of the transboundary management and sustainability challenges. In Dr. Firdous Ahmad Malik & Dr. Moon Moon Hussain, *Human algorithm of inclusive Earthscapes: Weaving education, AI, climate justice and transformative governance into a compassionate future* (pp. 29–49). Alborer (OPC) Pvt. Ltd.

Animal Defenders International v. United Kingdom (2013). European Court of Human Rights.

Application of the Convention on the Prevention and Punishment of the Crime of Genocide (The Gambia v. Myanmar), Preliminary Objections, Judgment, I.C.J. Reports 2022, 477 (International Court of Justice, July 22, 2022)

Ascui, F., & Lovell, H. (2011). As frames collide: Making sense of carbon accounting. *Accounting, Auditing & Accountability Journal, 24*(8), 978–999. DOI: 10.1108/09513571111184724

Asghar Leghari vs. Federation of Pakistan. (2025). *Asghar Leghari vs. Federation of Pakistan | UNEP Law and Environment Assistance Platform*. https://leap.unep.org/en/countries/pk/national-case-law/asghar-leghari-vs-federation-pakistan

Asio, M. V. S., Cvitanovic, C., Lau, J., & Diedrich, A. (2026). Collaborative governance in marine and coastal management: A systematic review. *Ocean and Coastal Management, 272*, 108020. DOI: 10.1016/j.ocecoaman.2025.108020

Assche, A. V., & Narula, R. (2022). Internalization strikes back? Global value chains and the rising costs of effective cascading compliance. *Economia e Politica Industriale, 50*(1), 161–173. DOI: 10.1007/s40812-022-00237-x

Auda, J. (2008). *Maqasid al-Shariah as philosophy of Islamic law: A systems approach*. International Institute of Islamic Thought. DOI: 10.2307/j.ctvkc67tg

Backes, J. G., & Traverso, M. (2021). Application of life cycle sustainability assessment in the construction sector: A systematic literature review. *Processes (Basel, Switzerland), 9*(7), 1248. DOI: 10.3390/pr9071248

Bais, B., Nassimbeni, G., & Orzes, G. (2024). Global Reporting Initiative: Literature review and research directions. *Journal of Cleaner Production, 471*, 143428. DOI: 10.1016/j.jclepro.2024.143428

Bangladesh Environmental Lawyers Association (BELA) v. Bangladesh, Writ Petition No. 7260 of 2008 (High Court Division of the Supreme Court of Bangladesh 2009).

Barbieri, N., Consoli, D., Napolitano, L., Perruchas, F., Pugliese, E., & Sbardella, A. (2022). Regional technological capabilities and green opportunities in Europe. *The Journal of Technology Transfer, 48*(2), 749–778. DOI: 10.1007/s10961-022-09952-y

Barnett, J., & Campbell, J. (2010). *Climate change and small island states: Power, knowledge and the South Pacific*. Routledge.

Bartl, M. (2022). Towards the imaginary of collective prosperity in the European Union (EU): Reorienting the corporation. *European Law Open, 1*(4), 957–986. DOI: 10.1017/elo.2023.5

Basurto, X., Virdin, J., Smith, H., & Juskus, R. (2017). *Strengthening governance of small-scale fisheries: An initial assessment of theory and practice*. Nicholas Institute for Environmental Policy Solutions.

Belal, A. R., Cooper, S. M., & Khan, N. A. (2015). Corporate environmental responsibility and accountability: What chance in vulnerable Bangladesh? *Critical Perspectives on Accounting*, *33*, 44–58. DOI: 10.1016/j.cpa.2015.01.005

Belal, A. R., & Roberts, R. W. (2010). Stakeholders' perceptions of corporate social reporting in Bangladesh. *Journal of Business Ethics*, *97*(2), 311–324. Retrieved July 2026, from. DOI: 10.1007/s10551-010-0511-4

Berkes, F. (2018). Evolution of co-management: Role of knowledge generation, bridging organizations and social learning. *Journal of Environmental Management*, *90*(5), 1692–1702. DOI: 10.1016/j.jenvman.2008.12.001 PMID: 19110363

Billedeau, D. B., & Wilson, J. (2025). Misalignment in the automotive supply Chain: Sustainability commitments of automotive manufacturing firms and their suppliers. *Journal of Cleaner Production*, *520*, 146122. DOI: 10.1016/j.jclepro.2025.146122

Birnie, P., Boyle, A., & Redgwell, C. (2009). *International law and the environment* (3rd ed.). Oxford University Press.

Black, J. (2001). Decentring regulation: Understanding the role of regulation and self-regulation in a 'post-regulatory' world. *Current Legal Problems*, *54*(1), 103–146. DOI: 10.1093/clp/54.1.103

Black, J., & Baldwin, R. (2012). When risk-based regulation aims low: Approaches and challenges. *Regulation & Governance*, *6*(1), 2–22. DOI: 10.1111/j.1748-5991.2011.01124.x

Blažo, O. (2025). The Mirage of Territoriality: EU Law's Extraterritorial Reach Between Protectionism and Altruism. *Comenius Bratislava Legal Forum, 1*, 545. DOI: 10.46282/bpf.2025.33

Bodansky, D. (2010). *The art and craft of international environmental law*. Harvard University Press.

Bodansky, D. (2016). The Paris Climate Change Agreement: A new hope? *The American Journal of International Law*, *110*(2), 288–319. DOI: 10.5305/amerjintelaw.110.2.0288

Boffo, R., & Patalano, R. (2020). *ESG investing: Practices, progress and challenges*. OECD Publishing. https://www.oecd.org/content/dam/oecd/en/publications/reports/2020/09/esg-investing-practices-progress-and-challenges_8a73d683/b4f71091-en.pdf

Boiral, O. (2013). Sustainability reports as simulacra? A counter-account of A and A+ GRI reports. *Accounting, Auditing & Accountability Journal, 26*(7), 1036–1071. DOI: 10.1108/AAAJ-04-2012-00998

Bookman, S. (2024). Demystifying Environmental Constitutionalism. *Environmental Law (Northwestern School of Law), 54*(1), 1–77. https://www.jstor.org/stable/48778824

Boström, M., Jönsson, A. M., Lockie, S., Mol, A. P. J., & Oosterveer, P. (2014). Sustainable and responsible supply chain governance: Challenges and opportunities. *Journal of Cleaner Production, 107*, 1–7. DOI: 10.1016/j.jclepro.2014.11.050

Both, C. M., & Wilhelm, M. (2025). Supply chain sustainability risk management in the era of mandatory due diligence: A literature review. *Journal of Purchasing and Supply Management, 32*(2), 101083. DOI: 10.1016/j.pursup.2025.101083

Bounfour, T., & Chatelain, L. (2025). A (So Far Unlikely) Day in Court: An Overview of the First Judicial Decisions under the French Duty of Vigilance Law. *Business and Human Rights Journal, 10*(2-3), 434–439. DOI: 10.1017/bhj.2025.22

Boyle, A. (2012). Human rights and the environment: Where next? *European Journal of International Law, 23*(3), 613–642. DOI: 10.1093/ejil/chs054

Bragança, L., Mateus, R., & Koukkari, H. (2010). Building sustainability assessment. *Sustainability (Basel), 2*(7), 2010–2023. DOI: 10.3390/su2072010

Brander, M., Gillenwater, M., & Ascui, F. (2018). Creative accounting: A critical perspective on the market-based method for reporting purchased electricity (scope 2) emissions. *Energy Policy, 112*, 29–33. DOI: 10.1016/j.enpol.2017.09.051

British Institute of International and Comparative Law (BIICL). (2024). *Reflections on the ITLOS Advisory Opinion*. https://www.biicl.org/documents/184_reflections_on_the_itlos_advisory_opinion_final.pdf

Broadstock, D. C., Chan, K., Cheng, L. T., & Wang, X. (2020). The role of ESG performance during times of financial crisis: Evidence from COVID-19 in China. *Finance Research Letters, 38*, 101716. Advance online publication. DOI: 10.1016/j.frl.2020.101716

Bueno, N., Bernaz, N., Holly, G., & Martín-Ortega, O. (2024). The EU Directive on Corporate Sustainability Due Diligence (CSDDD): The Final Political Compromise. *Business and Human Rights Journal, 9*(2), 294–300. DOI: 10.1017/bhj.2024.10

Buhmann, K., & Feld, L. (2024). Shifting boundaries: A transnational legal perspective on the EU Corporate Sustainability Due Diligence Directive. *Transnational Legal Theory, 15*(3), 370–396. DOI: 10.1080/20414005.2024.2426961

Casado Coca v. Spain (1994). European Court of Human Rights.

Castrillón-Mendoza, R., Rey-Hernández, J. M., Castrillón-Mendoza, L., & Rey-Martínez, F. J. (2024). Sustainable building tool by energy baseline: Case study. *Applied Sciences (Basel, Switzerland), 14*(20), 9403. DOI: 10.3390/app14209403

Certain Activities Carried Out by Nicaragua in the Border Area (Costa Rica v. Nicaragua), Compensation Owed by the Republic of Nicaragua to the Republic of Costa Rica, Judgment, I.C.J. Reports 2018, p. 15

Cherednychenko, O. O. (2023). Private enforcement of EU law. In *Edward Elgar Publishing eBooks* (p. 19). Edward Elgar Publishing., DOI: 10.4337/9781802208030.00010

Chu, H., Liu, N., Tsang, A., & Zeng, X. (2025). Environmental Risk, Environmental Action, and Firm Value: Evidence From Climate-Linked Contracts. *Australian Accounting Review, 35*(3), 215–240. Advance online publication. DOI: 10.1111/auar.70010

Ciacchi, S., Faure, M. G., & Oded, S. (2026). Navigating the EU's Corporate Sustainability Due Diligence Directive: Strategic Implications for US Companies. *Michigan Business & Entrepreneurial Law Review, 34*, 34. Advance online publication. DOI: 10.36639/mbelr.15.1.navigating

Code de la consommation. (2016). *Article L121-1*. https://www.legifrance.gouv.fr/codes/article_lc/LEGIARTI000032227301/2016-07-01

Cohn, M. (2001). Fuzzy Legality in Regulation: The Legislative Mandate Revisited. *Law & Policy, 23*(4), 469–497. DOI: 10.1111/1467-9930.00121

Conference of the Parties to the United Nations Framework Convention on Climate Change. (2012). *Decision 1/CP.17: Establishment of an Ad Hoc Working Group on the Durban Platform for Enhanced Action*. In *Report of the Conference of the Parties on its seventeenth session, held in Durban from 28 November to 11 December 2011.*

Constitution of Kenya. (2010). https://new.kenyalaw.org/akn/ke/act/2010/constitution/eng@2010-09-03

Constitution of the Islamic Republic of Pakistan. (1973). Art. 227.

Constitution of the Republic of Maldives. (2008). Arts. 9(d), 10(a), 22.

Convention on access to information, public participation in decision-making and access to justice in environmental matters (Aarhus Convention). (1998). United Nations Economic Commission for Europe

Corfu Channel Case (U.K. v. Alb.), Judgment, 1949 I.C.J. 4 (Apr. 9).

Corino, C. (2000). Environmental due diligence. *European Environmental Law Review, 9*(4), 120–124. DOI: 10.54648/268317

Council Directive (EU) 2022/2464 of 14 December 2022 as regards corporate sustainability reporting (CSRD), 2022 O.J. (L 322).

Council Directive (EU) 2024/1760 of 24 May 2024 on corporate sustainability due diligence (CS3D), 2024 O.J. (L 1760).

Council Directive (EU) 2026/470 amending Directives 2006/43/EC, 2013/34/EU, (EU) 2022/2464 and (EU) 2024/1760 ("Omnibus I"), 2026 O.J.

Council of Europe. (2017). *Protecting the right to freedom of expression under the European Convention on Human Rights: A handbook for legal practitioners.* Council of Europe.

Court of Justice of the European Union. (2000). Germany v European Parliament and Council (Tobacco Advertising I) (Case C-376/98), ECLI:EU:C:2000:544. https://eur-lex.europa.eu/legal-content/EN/TXT/?uri=CELEX:61998CJ0376.

Court of Justice of the European Union. (2004). Omega Spielhallen- und Automatenaufstellungs-GmbH v Oberbürgermeisterin der Bundesstadt Bonn (Case C-36/02), ECLI:EU:C:2004:614. https://eur-lex.europa.eu/legal-content/EN/TXT/?uri=CELEX:62002CJ0036

Court of Justice of the European Union. (2006). Germany v European Parliament and Council (Tobacco Advertising II) (Case C-380/03), ECLI:EU:C:2006:772. https://eur-lex.europa.eu/legal-content/EN/TXT/?uri=CELEX:62003CJ0380.

Court of Justice of the European Union. (2010). Vodafone Ltd and Others v Secretary of State for Business, Enterprise and Regulatory Reform (Case C-58/08), ECLI:EU:C:2010:321. https://eur-lex.europa.eu/legal-content/EN/TXT/?uri=CELEX:62008CJ0058.

Crawford, J. (2019). *Brownlie's principles of public international law* (9th ed.). Oxford University Press. DOI: 10.1093/he/9780198737445.001.0001

Cremasco, C., & Boni, L. (2024). Is the European Union (EU) sustainable finance disclosure regulation (SFDR) effective in shaping sustainability objectives? An analysis of investment funds' behaviour. *Journal of Sustainable Finance & Investment, 14*(4), 1018–1019. DOI: 10.1080/20430795.2022.2124838

Crovini, C., Højrup, E. M., Jensen, M., Sørensen, K., Jensen, S., Castellano, N. G., D'Onza, G., Melis, A., Luca, F. D., & Maffei, M. (2025). *The Integration of Sustainability into the Risk Management Process and the Impact of the Energy Crisis and Green Transition: Insights from a Danish Energy Trading Company. In the SIDREA series in accounting and business administration.* Springer International Publishing., DOI: 10.1007/978-3-031-76618-3_16

Cuomo, F., Mallin, C., & Zattoni, A. (2015). Corporate Governance Codes: A Review and Research Agenda [Review of *Corporate Governance Codes: A Review and Research Agenda*]. *Corporate Governance An International Review, 24*(3), 222. Wiley. DOI: 10.1111/corg.12148

Das, J. (2023). Blue economy, blue growth, social equity and small-scale fisheries: A global and national review. *Studies in Social Science Research, 4*(1), 38–58. DOI: 10.22158/sssr.v4n1p38

De Cristofaro, T., & Gulluscio, C. (2023). In search of double materiality in non-financial reports: First empirical evidence. *Sustainability (Basel), 15*(2), 924. DOI: 10.3390/su15020924

de Freitas, J. T. (2024). Prioritization under Value Chain Due Diligence. *European Journal of Risk Regulation, 1*(4), 985–998. Advance online publication. DOI: 10.1017/err.2024.19

Debnath, R., Ebanks, D., Mohaddes, K., Roulet, T., & Alvarez, R. M. (2023). Do fossil fuel firms reframe online climate and sustainability communication? A data-driven analysis. *npj. Climate Action, 2*(1), 47. DOI: 10.1038/s44168-023-00086-x PMID: 38694952

Dehalwar, K., & Sharma, S. N. (Eds.). (2026). *Deltas resilience: Nature-based solutions for sustainable development in India*. Springer Nature. DOI: 10.1007/978-3-032-07240-5

Demirci, A., Şimşek, E., & Demirci, S. (2026). Weak representation, fragmented governance: Evidence from four fisheries conflicts in Türkiye. *Marine Policy, 192*, 107176. DOI: 10.1016/j.marpol.2026.107176

Demirgüç-Kunt, A., Klapper, L., Singer, D., & Ansar, S. (2022). *The Global Findex Database 2021: Financial inclusion, digital payments, and resilience in the age of COVID-19*. World Bank. https://doi.org/DOI: 10.1596/978-1-4648-1897-4

Dietz, T., Ostrom, E., & Stern, P. C. (2003). The struggle to govern the commons. *Science, 302*(5652), 1907–1912. DOI: 10.1126/science.1091015 PMID: 14671286

Dill, A. (2025). The role of corporate governance in macro-prudential regulation of systemic risk. SSRN *Electronic Journal*. DOI: 10.2139/ssrn.5072748

Dispute Concerning Delimitation of the Maritime Boundary Between Bangladesh and Myanmar in the Bay of Bengal (Bangl. v. Myan.), Judgment, 2012 ITLOS Rep. 4 (Mar. 14).

Dixon, M., Grilli, G., Stewart, B. D., Bark, R. H., & Ferrini, S. (2024). The importance of rebuilding trust in fisheries governance in post-Brexit England. *Marine Policy, 161*, 106034. DOI: 10.1016/j.marpol.2024.106034

DNV. (2025). *Understanding and managing the regulatory landscape of ship recycling*. https://www.dnv.com/expert-story/maritime-impact/understanding-and-managing-the-regulatory-landscape-of-ship-recyling/ (accessed on June, 2026)

Douglas, M. (2013). *A very personal method: Anthropological writings drawn from life*. SAGE Publications. https://www.researchgate.net/publication/306032193_A_very_personal_anthropology_of_Mary_Douglas (accessed on June, 2026)

Downie, J., & Stubbs, W. (2013). Evaluation of Australian companies' scope 3 greenhouse gas emissions assessments. *Journal of Cleaner Production, 56*, 156–163. DOI: 10.1016/j.jclepro.2011.09.010

Eakpisankit, A., & Pirzada, K. (2025). Comparative environmental disclosure practices: Evidence from the UK and the USA. *Cogent Business & Management, 12*(1), 2534093. Advance online publication. DOI: 10.1080/23311975.2025.2534093

Eil, A., Li, J., Baral, P., & Saikawa, E. (2020). *[Title of dataset]*. Harvard Dataverse. https://doi.org/ (accessed on June, 2026)DOI: 10.7910/DVN/MV9FEH

Elbel, J., Böse-O'Reilly, S., & Hrzic, R. (2023). A European Union corporate due diligence act for whom? Considerations about the impact of a European Union due diligence act on artisanal and small-scale cobalt miners in the Democratic Republic of Congo. *Resources Policy, 81*, 103241. DOI: 10.1016/j.resourpol.2022.103241

Environmental Protection and Preservation Act of Maldives (Act No. 4/1993). (1993). Republic of Maldives.

European Commission. (2020). *Fisheries innovations under the European Maritime and Fisheries Fund 2014–2020*. https://oceans-and-fisheries.ec.europa.eu/document/download/0eb81df7-8d30-4eb0-8e00-2fb187f29859_en

European Commission. (2022). *Proposal for a directive on corporate sustainability due diligence* (COM/2022/71 final). Publications Office of the European Union.

European Commission. (2023). *The common fisheries policy*. https://wodnesprawy.pl/en/common-fisheries-policy-evaluation-consultations/

European Commission. (2023, March 22). *Proposal for a directive on substantiation and communication of explicit environmental claims (Green Claims Directive), COM(2023) 166 final.*

European Commission. (n.d.). *Green claims*. https://environment.ec.europa.eu/topics/circular-economy-topics/green-claims_en

European Financial Reporting Advisory Group. (2024). *European sustainability reporting standards (ESRS) 2024*. https://www.efrag.org

European Parliament and Council of the European Union. (2022). Directive (EU) 2022/2464 of the European Parliament and of the Council of 14 December 2022, as regards corporate sustainability reporting. *Official Journal of the European Union, L, 322*, 15–80. https://eur-lex.europa.eu/eli/dir/2022/2464/oj

European Parliament and Council. (2005). Directive 2005/29/EC concerning unfair business-to-consumer commercial practices. *Official Journal of the European Union, L, 149*, 22–39.

European Parliament and of the Council Directive (EU) 2022/2464 of 14 December 2022. OJ L 322, 16.12.2022, p. 15–80. https://eur-lex.europa.eu/eli/dir/2022/2464/oj

European Parliament. (2024). *EU oceans and fisheries policy: Latest developments and future challenges*. https://oceans-and-fisheries.ec.europa.eu/policy_en

Evans, J., & Silbey, S. S. (2021). Co-Opting Regulation: Professional Control Through Discretionary Mobilization of Legal Prescriptions and Expert Knowledge. *Organization Science, 33*(5), 2041–2064. DOI: 10.1287/orsc.2021.1525

Federal Trade Commission. (2012). *Guides for the use of environmental marketing claims* (16 C.F.R. Part 260). https://www.ftc.gov/sites/default/files/attachments/press-releases/ftc-issues-revised-green-guides/greenguides.pdf

Felbermayr, G., Friesenbichler, K. S., Gerschberger, M., Klimek, P., & Meyer, B. (2024). Designing EU Supply Chain Regulation. *Inter Economics, 59*(1), 28–34. DOI: 10.2478/ie-2024-0007

Feld, L. (2026). Leading the Way or Crossing the Line? The Extraterritorial Dimension of the EU Directive on Corporate Sustainability Due Diligence. *Business and Human Rights Journal*, *1*(1), 23–45. Advance online publication. DOI: 10.1017/bhj.2025.10038

Fisher, L., & Gordhan, S. (2026). Hope with teeth: Five lessons from the ICJ's climate change advisory opinion. *Journal of Environmental Law*, eqag011. Advance online publication. DOI: 10.1093/jel/eqag011

Foerster, A., Laundau, I., & House, M. (2025). Sustainable Finance and Corporate Sustainability Due Diligence: Interactions, Complementarities and Accountability Logics. *The University of New South Wales Law Journal*, *48*(3). Advance online publication. DOI: 10.53637/GUXD7806

Food and Agriculture Organization. (2022). *Guidebook for evaluating fisheries co-management effectiveness*. DOI: 10.4060/cc2228en

Food and Agriculture Organization. (2024). *The state of world fisheries and aquaculture 2024*. https://openknowledge.fao.org/items/ceda292f-ac05-4254-b8b5-695156007fab

Ford, J., & Nolan, J. (2020). Regulating transparency on human rights and modern slavery in corporate supply chains: The discrepancy between human rights due diligence and the social audit. *Australian Journal of Human Rights*, *26*(1), 27–45. DOI: 10.1080/1323238X.2020.1761633

Friede, G., Busch, T., & Bassen, A. (2015). ESG and financial performance: Aggregated evidence from more than 2,000 empirical studies. *Journal of Sustainable Finance & Investment*, *5*(4), 210–233. DOI: 10.1080/20430795.2015.1118917

Friesenbichler, K. S., Felbermayr, G., Gerschberger, M., Klimek, P., & Meyer, B. (2024). Designing EU Supply Chain Regulation. SSRN *Electronic Journal*. DOI: 10.2139/ssrn.4917535

Frost, J., Gambacorta, L., Huang, Y., Shin, H. S., & Zbinden, P. (2019). BigTech and the changing structure of financial intermediation. *Economic Policy*, *34*(100), 761–799. DOI: 10.1093/epolic/eiaa003

Gade, A. M. (2019). *Muslim environmentalisms: Religious and social foundations*. Columbia University Press. DOI: 10.7312/gade19104

Gelcich, S., Martínez-Harms, J., Tapia-Lewin, S., Vásquez-Lavín, F., & Ruano-Chamorro, C. (2019). Co-management of small-scale fisheries and ecosystem services. *Conservation Letters*, *12*(4), e12637. DOI: 10.1111/conl.12637

Gill, A. (2008). *The political origins of religious liberty*. Cambridge University Press.

Gill, G. N. (2016). Environmental justice in India: The National Green Tribunal and expert members. *Transnational Environmental Law*, *5*(1), 175–205. DOI: 10.1017/S2047102515000278

Global Alliance for Buildings and Construction and United Nations Environment Programme. (2024). *Global Status Report for Buildings and Construction: Beyond foundations—Mainstreaming sustainable solutions to cut emissions from the buildings sector* (I. Hamilton, H. Kennard, O. Rapf, J. Amorocho, S. Steuwer, J. Kockat, & Z. Toth, Trans.; Nos. 978-92-807-4131-5). DOI: 10.59117/20.500.11822/45095

Global Reporting Initiative. (2023). *GRI 3: Material topics 2021*. https://www.globalreporting.org/standards/

Global Sustainable Investment Alliance. (2023). *Global sustainable investment review 2022*. https://www.gsi-alliance.org/wp-content/uploads/2023/12/GSIA-Report-2022.pdf

Gomber, P., Koch, J., & Siering, M. (2017). Digital finance and FinTech: Current research and future research directions. *Journal of Business Economics*, *87*(5), 537–580. DOI: 10.1007/s11573-017-0852-x

Gomez, M. L., & Çalı, B. (2025). *FROM LITIGATION TO IMPLEMENTATION: FRAMING SMART REMEDIES IN RIGHTS-BASED CLIMATE LITIGATION*.

Government of Balochistan & United Nations Development Program. (2024). *Balochistan Climate Change Policy*. UNDP. https://www.undp.org/pakistan/publications/balochistan-climate-change-policy

Green Climate Fund. (2021). *Building Climate Resilient Safer Islands (BCRSI) project: Maldives*. https://www.greenclimate.fund/project/fp151

Green, S., & Temper, L. (2026). The health case for a fossil fuel advertising ban. *Canadian Medical Association Journal*, *198*(5), E176–E180. DOI: 10.1503/cmaj.250022 PMID: 41663133

Grim, J., & Tucker, M. E. (2014). *Ecology and religion*. Island Press. DOI: 10.2307/jj.41003348

Gugkang, A. S., Tanakinjal, G. H., & Hamid, M. R. A. (2025). The Challenges of Implementing Socially Sustainable Supply Chain Management to Enhance Fair Labor Practices for Small-Scale Farmers in Malaysia. *Global Conference on Business and Social Sciences Proceedings*, *17*(1), 56. DOI: 10.35609/gcbssproceeding.2025.1(56)

Gutterman, A. (2023). Certifications and Rating Systems for Assessing Social and Environmental Impact. *ResearchGate*. https://www.researchgate.net/publication/370631183_Certifications_and_Rating_Systems_for_Assessing_Social_and_Environmental_Impact

Guye, V., Meyfroidt, P., & zu Ermgassen, E. (2025). The need for improved public transparency in the era of due diligence regulations. *SPIRE - Sciences Po Institutional Repository*. https://hal.science/hal-05420553

Habermas, J. (1996). *Between facts and norms*. Polity Press. DOI: 10.7551/mitpress/1564.001.0001

Haider, A. (2025). Review of the 26th Amendment to the Constitution of Pakistan and the fundamental right of a clean and healthy environment: A case study of new Article 9A. https://www.researchgate.net/publication/392602711

Haider, A., Ahmad, I., & Yaseen, M. (2024). Jus cogens and the right to self-determination: A study of its peremptory status and erga omnes effects. Pakistan Journal of Law. *Analysis and Wisdom*, *3*(6), 59–72.

Haider, A., Mathlouthi, N., & Ahmad, I. (2024). Beyond the books: Real world challenges in implementing environmental laws in Pakistan. *Remittances Review*, *9*(1), 1383–1405. DOI: 10.33282/rr.vx9il.93

Hall, P. A., & Taylor, R. C. R. (1996). Political science and the three new institutionalisms. *Political Studies*, *44*(5), 936–957. DOI: 10.1111/j.1467-9248.1996.tb00343.x

Handyside v. United Kingdom (1976). European Court of Human Rights.

Harrison, J. (2017). *Saving the oceans through law*. Oxford University Press. DOI: 10.1093/law/9780198707325.001.0001

Harvard Divinity School Religion and Public Life Project. (2024). *Environmental fatwas of Indonesia*. https://rpl.hds.harvard.edu/religion-context/case-studies/climate-change/environmental-fatwas-indonesia

Hassan, M. (2023, July 25). *Environmental impact assessment in Bangladesh: A critical review*. ResearchGate. https://www.researchgate.net/ (accessed on June, 2026)

Healy, P. M., & Palepu, K. G. (2001). Information asymmetry, corporate disclosure, and the capital markets: A review of the empirical disclosure literature. *Journal of Accounting and Economics*, *31*(1–3), 405–440. DOI: 10.1016/S0165-4101(01)00018-0

Higgins, P., Short, D., & South, N. (2013). Protecting the planet: A proposal for a law of ecocide. *Crime, Law, and Social Change*, *59*(3), 251–266. DOI: 10.1007/s10611-013-9413-6

Hill, S., Nemes, N., Montgomery, A., Scanlan, S., McNally, B., Tubiello, F., Aronczyk, M., Wood, T., Smith, T., & Kaupa, C. (2025). Testing the greenwashing assessment framework. *Ecology and Society*, *30*(2), 31. DOI: 10.5751/ES-16106-300231

Hossain, M. S. (2018). *Integrated coastal zone management (ICZM) and marine spatial planning (MSP)* [Conference presentation]. ResearchGate. https://www.researchgate.net/ (accessed on June, 2026)

Hu, B., & Xu, Q. (2025). Environmental regulation penalties and corporate environmental information disclosure. *International Review of Economics & Finance*, *102*, 104344. DOI: 10.1016/j.iref.2025.104344

Hui, Z., Haider, A., & Khan, A. (2025). International trade and plastic waste in oceans: Legal and policy challenges. *Frontiers in Marine Science*, *12*, 1627829. Advance online publication. DOI: 10.3389/fmars.2025.1627829

Hulme, M. (2016). *Weathered: Cultures of climate*. SAGE Publications.

Human Rights Committee. (2011). *General comment No. 34: Article 19: Freedoms of opinion and expression* (CCPR/C/GC/34). United Nations.

Human Rights Committee. (2018). *General comment No. 36: Article 6 (Right to life)*. United Nations.

Human Rights Committee. (2019). *General comment No. 36 on Article 6 of the International Covenant on Civil and Political Rights, on the right to life* (CCPR/C/GC/36). United Nations.

Human Rights Watch. (2023). *Maldives: Environmental protection laws bypassed*. https://www.hrw.org/news/2023/10/17/maldives-environmental-protection-laws-bypassed

Human Rights Watch. (2025). *World report 2025: Maldives*. https://www.hrw.org/world-report/2025/country-chapters/maldives

Ibn Ashur, M. T. (2006). *Treatise on Maqasid al-Shariah* (El-Mesawi, M. E., Trans.). International Institute of Islamic Thought. DOI: 10.2307/j.ctvkc673b

Indian Ocean Memorandum of Understanding on Port State Control. (2025). *Annual report 2024*. Indian Ocean MoU Secretariat.

Intergovernmental Panel on Climate Change. (2022). *Climate change 2022: Impacts, adaptation and vulnerability. Contribution of Working Group II to the Sixth Assessment Report*. Cambridge University Press.

Intergovernmental Panel on Climate Change. (2023). *Synthesis report of the IPCC sixth assessment report (AR6).* IPCC. https://www.ipcc.ch/report/ar6/syr/

International Court of Justice. (1997). *Case concerning the Gabcikovo-Nagymaros Project (Hungary v. Slovakia), judgment of 25 September 1997. ICJ Reports, 1997*, 7.

International Court of Justice. (2025, July 23). *Obligations of States in respect of climate change (Request for Advisory Opinion), Advisory Opinion.* Available online at https://www.icj-cij.org/case/187/advisory-opinions (accessed on July, 2026)

International Covenant on Civil and Political Rights. (1966). United Nations.

International Finance Corporation. (2021). *Sustainable bonds: Bridging the gap for gender equality.* World Bank Group. https://www.ifc.org/en/insights-reports/2021/sustainable-bonds-to-bridge-the-gender-gap

International Maritime Organization. (2020). *Fourth IMO GHG study 2020.*

International Maritime Organization. (2023). *2023 IMO strategy on reduction of GHG emissions from ships* (MEPC.377(80)).

International Maritime Organization. (2025). *Entry into force of the Hong Kong International Convention for the safe and environmentally sound recycling of ships.*

International Tribunal for the Law of the Sea (ITLOS). (2024). *Request for an Advisory Opinion submitted by the Commission of Small Island States on Climate Change and International Law* (Case No. 31). Advisory Opinion of 21 May 2024. https://www.itlos.org/fileadmin/itlos/documents/cases/31/Advisory_Opinion/C31_Adv_Op_21.05.2024_orig.pdf

International Tribunal for the Law of the Sea. (2024). *Advisory opinion on climate change and international law obligations under the United Nations Convention on the Law of the Sea* (Case No. 31).

Iqbal, N. (2022). *Supreme Court orders avoidance of term 'competent authority'.* Dawn. https://www.dawn.com/news/1671469

Islamic Foundation for Ecology and Environmental Sciences (IFEES). (2019). *Applying Islamic environmental ethics.* https://ifees.org.uk/wp-content/uploads/2019/12/applying_islamic_environmental_ethics.pdf

Islam, K. S., Xue, X.-Z., & Rahman, M. M. (2009). Successful integrated coastal zone management (ICZM) program model of a developing country (Xiamen, China)–Implementation in Bangladesh perspective. *Journal of Wetlands Ecology*, *2*(1), 35–41. Retrieved July 2026, from. DOI: 10.3126/jowe.v2i1.1854

Islam, M. M., & Islam, A. R. M. T. (2026). Strengthening climate resilience in coastal Bangladesh: Analyzing the role of NGOs in adaptation governance from an intersectionality lens. *Progress in Disaster Science*, *29*, 100498. Retrieved July 2026, from. DOI: 10.1016/j.pdisas.2025.100498

Islam, M. R., & Chowdhury, M. (2022). Effectiveness of the coastal and marine conservation initiatives in Bangladesh: Analyzing the drawbacks of the legal, policy, and institutional framework. *Journal of the Indian Ocean Region*, *18*(2), 149–172. Retrieved July 2026, from. DOI: 10.1080/19480881.2022.2111050

Izzi Dien, M. Y. (2000). *The environmental dimensions of Islam.*

Janik, A., & Ryszko, A. (2025). Greenwashing in sustainability reporting: A systematic literature review of strategic typologies and content-analysis-based measurement approaches. *Sustainability (Basel)*, *18*(1), 17. DOI: 10.3390/su18010017

Járvás, T., Illés, B., & Tóth, Á. B. (2025). Uncertainty factors in supply chains. *Advanced Logistic Systems -. Theory into Practice*, *19*(1), 53–60. DOI: 10.32971/als.2025.005

Javadi, H., Mirjalili, S., & Hajizadeh, A. (2020). Faith-based environmental organisations and water governance in Pakistan. *Water Alternatives*, *13*(2), 401–421. https://www.water-alternatives.org/index.php/alldoc/articles/vol13/v13issue2

Jenart, C. (2024). The Belgian Climate Case and rising (sea) levels: Human rights, separation of powers and multi-level constitutionalism. *Maastricht Journal of European and Comparative Law*, *31*(3), 445–461. DOI: 10.1177/1023263X241287408

Jenkins, W. J. (2009). Religious responses to environmental challenges. In Gottlieb, R. (Ed.), *The Oxford handbook of religion and ecology* (pp. 45–64). Oxford University Press.

Jenkins, W., Berry, T., & Kreider, L. B. (2018). Religion and climate change. *Annual Review of Environment and Resources*, *43*(1), 85–108. DOI: 10.1146/annurev-environ-102017-025855

Jentoft, S. (2007). Limits of governability: Institutional implications for fisheries and coastal governance. *Marine Policy*, *31*(4), 360–370. DOI: 10.1016/j.marpol.2006.11.003

Jentoft, S., Chuenpagdee, R., Bugeja, A., & Isaacs, M. (2022). *Blue justice: Small-scale fisheries in a sustainable ocean economy.* Springer., DOI: 10.1007/978-3-030-89624-9

Jesenko, K. (2018). Freedom of expression in environmental cases before the European Court of Human Rights. *ERA Forum*, 19, 295–301.

Jian, Z., Xu, G., & Li, X. (2025). Intersection of ESG performance and corporate innovation: A perspective of vertical linkages within the industrial chain. *Frontiers in Sustainability*, *6*, 1642317. Advance online publication. DOI: 10.3389/frsus.2025.1642317

Jo-Ellen Parry. (2016). *Review of Current and Planned Adaptation Action in Pakistan | International Institute for Sustainable Development*. https://www.iisd.org/publications/report/review-current-and-planned-adaptation-action-pakistan

Johannsen, K. (2025). An empty victory? Shell v. Milieudefensie et al. 2024, the legal obligations of carbon majors, and the prospects for future climate litigation action. *Review of European, Comparative & International Environmental Law, 34*(1).

Joshi, A. B. (2012). Corporate environmental responsibility: A liability or challenge. SSRN *Electronic Journal*. https://doi.org/DOI: 10.2139/ssrn.2157398

Kalimo, H., Ville, F. D., & Happersberger, S. (2023). The Unilateral Turn in EU Trade Policy? The Origins and Characteristics of the EU's New Trade Instruments. *European Foreign Affairs Review*, *28*(Special Issue), 15–34. DOI: 10.54648/EERR2023012

Kamali, M. H. (2008). *Shari'ah law: An introduction*. Oneworld Publications.

Kashyap, A. (2025). *Sustainable global trade: Balancing economic growth with environmental accountability*. *SSRN*. DOI: 10.2139/ssrn.5550500

Kathmandu School of Law. (2022). *Project Bajho: A Study of the Climate Change Laws and Policies in South Asian Countries*. https://ksl.edu.np/resource/home/resource_detail/50

Kauffman, C. M., & Martin, P. L. (2021). *The rights of nature: A legal revolution that could save the world*. Yale University Press.

Khalid, F. (2002). Islam and the environment. In Munn, T. (Ed.), *Encyclopaedia of global environmental change* (Vol. 5, pp. 332–339). Wiley.

Khan, I. (2021, April 13). *Disinformation and freedom of opinion and expression: Report of the Special Rapporteur on the promotion and protection of the right to freedom of opinion and expression*. United Nations. https://digitallibrary.un.org/record/3925306

Khan, M. A. R. (2020). Sustainable coastal zone management: Need for a holistic approach for Bangladesh. *Journal of Management and Sustainability*, *10*(2), 112–120. Retrieved July 2026, from. DOI: 10.5539/jms.v10n2p112

Khan, M. I. (2025). *Deforestation*. Forest Degradation, and Flood Risk in Pakistan.

Khan, N. A., & Belal, A. R. (1999). The politics of the Bangladesh environmental protection act. *Environmental Politics*, *8*(1), 311–317. DOI: 10.1080/09644019908414449

Khare, P. (2026). Integrating the SDGs Into Corporate Governance: A Legal Imperative for Global Sustainability. *IGI Global Scientific Publishing*, 391-416. https://doi.org/DOI: 10.4018/979-8-3373-4192-7.ch015

Kiewisch, M. (2025). Navigating the EU Corporate Sustainability Due Diligence Directive: How Multinational Enterprises Approach Regulatory Familiarization in the Chocolate Sector. *Regulation & Governance*, rego.70042. Advance online publication. DOI: 10.1111/rego.70042

Kingston, E., & Carnovale, M. (2026). Social Upgrading in Global Supply Chains: Reframing Corporate Responsibilities. *Business Ethics Quarterly*, *1*, 1–28. Advance online publication. DOI: 10.1017/beq.2025.10103

Kitzes, J. (2013). An introduction to environmentally-extended input–output analysis. *Resources*, *2*(4), 489–503. DOI: 10.3390/resources2040489

Klein, R. J. T., Midgley, G. F., Preston, B. L., Alam, M., Berkhout, F. G. H., Dow, K., & Shaw, M. R. (2014). Adaptation opportunities, constraints, and limits. In *Climate change 2014: Impacts, adaptation, and vulnerability* (Part A, pp. 899–943). Cambridge University Press. https://doi.org/DOI: 10.1017/CBO9781107415379.021

Koehrsen, J. (2021). Does religion promote environmental sustainability? Exploring the role of religion in local energy transitions. *Social Compass*, *68*(1), 3–19. DOI: 10.1177/0037768620982244

Koehrsen, J., & Burchardt, M. (2023). Faith-based organisations and climate justice: Between moral outsourcing and institutional transformation. *Global Environmental Politics*, *23*(1), 75–95. DOI: 10.1162/glep_a_00643

Koos, S. (2024). Civil Law, Conflict of Laws, and Extraterritoriality in the European Supply Chain Due Diligence Law. *Hasanuddin Law Review*, *10*(2), 144. DOI: 10.20956/halrev.v10i2.5535

Kotzé, L. J. (2019). *Global environmental constitutionalism in the Anthropocene*. Hart Publishing.

Krebs, D. (2022). Environmental due diligence obligations in home state law with regard to transnational value chains. *Springer International Publishing*, 245-350. https://doi.org/DOI: 10.1007/978-3-031-13264-3_7

Krüger, P. (2015). Corporate goodness and shareholder wealth. *Journal of Financial Economics, 115*(2), 304–329. DOI: 10.1016/j.jfineco.2014.09.008

Kübek, G., & Mancini, I. (2023). EU Trade Policy between Constitutional Openness and Strategic Autonomy. *European Constitutional Law Review, 19*(3), 518–547. DOI: 10.1017/S1574019623000226

Kulovesi, K., Oberthür, S., van Asselt, H., & Savaresi, A. (2024). The european climate law: Strengthening EU procedural climate governance? *Journal of Environmental Law, 36*(1), 23–42. DOI: 10.1093/jel/eqad034

Kyoto Protocol to the United Nations Framework Convention on Climate Change, December 11, 1997, 2303 U.N.T.S. 148.

Lachenmeier, D. W., & Montagnon, C. (2024). Convention on Biological Diversity (CBD) and the Nagoya Protocol: Implications and Compliance Strategies for the Global Coffee Community. *Foods, 13*(2), 254. DOI: 10.3390/foods13020254 PMID: 38254555

Landgericht Köln. (2025, March 21). *84 O 29/24* (ECLI:DE:LGK:2025:0321.84O29.24.00) [Judgment]. Justiz NRW. https://nrwe.justiz.nrw.de/lgs/koeln/lg_koeln/j2025/84_O_29_24_Urteil_20250321.html

Lawrence, J., Rasche, A., & Kenny, K. (2018). Sustainability as opportunity: Unilever's sustainable living plan. In *Managing Sustainable Business* (pp. 435–455). Springer Netherlands., DOI: 10.1007/978-94-024-1144-7_21

Leary, D. (2023). International maritime organization (IMO). *Yearbook of International Environmental Law, 34*(1). Advance online publication. DOI: 10.1093/yiel/yvaf050

Lenzen, M., Moran, D., Kanemoto, K., & Geschke, A. (2013). Building Eora: A global multi-region input–output database at high resolution. *Economic Systems Research, 25*(1), 20–49. DOI: 10.1080/09535314.2013.769938

Lingens v. Austria (1989). European Court of Human Rights.

Lin, W. L.. (2025). The impact of greenwashing: Risks and implications for corporate performance and stakeholder trust. In *Proceedings of the International Conference on Environmental, Social, and Governance (ICESG 2024)* (p. 79). DOI: 10.2991/978-2-38476-358-0_8

Loi n° 2017-399 du 27 mars 2017 relative au devoir de vigilance des sociétés mères et des entreprises donneuses d'ordre, Journal Officiel de la République Française [J.O.], Mar. 28, 2017.

Lotfi, M., & Walker, H. W. (2026). *Beyond paper compliance: why supply chains struggle to confront forced labor. ORCA Online Research @Cardiff.* Cardiff University.

Luciano Lliuya v. RWE AG, Case No. I-5 U 15/17 (Oberlandesgericht Hamm May 28, 2025).

Lyon, T. P., & Montgomery, A. W. (2015). The means and end of greenwash. *Organization & Environment*, *28*(2), 223–249. DOI: 10.1177/1086026615575332

M.C. Mehta v. Union of India, AIR 1987 SC 1086 (India).

Mahaseth, H., & Goyal, S. (2021). Asghar Leghari and Environmental Justice. *Transformative Climate Change Litigation Judgements One Step at a Time.*, *1*, 124–129.

Mahmood Mamdani. (2018, January 5). *Citizen and Subject | Princeton University Press*. https://press.princeton.edu/books/paperback/9780691180427/citizen-and-subject

Mahmud, I. (2025). *A human rights critique of Bangladesh's Climate Change Strategy and Action Plan (BCCSAP) & National Adaptation Plan (NAP) (2023–2050): Assessing participation, equity, accountability, and justice in adaptation policy* (Master of Human Rights practicum, University of Manitoba). MSpace Institutional Repository. https://mspace.lib.umanitoba.ca/handle/1993/39126 (accessed on June, 2026)

Maiju, V. (2025). Drivers and barriers for lower-tier suppliers' sustainability compliance in emerging countries : a narrative review. In *Theseus (Ammattikorkeakoulujen)*. https://www.theseus.fi/handle/10024/906481

Majekodunmi, A. O., Edohen, A., & Conteh, J. Y. (2025). Regulatory Divergence and Security Implementation: Compliance-Driven Security Architecture in Multi-Jurisdictional Financial Organizations. *Research Journal in Civil Industrial and Mechanical Engineering*, *2*(2), 53–71. DOI: 10.61424/rjcime.v2i2.336

Malik. (2025, November 16). The Promise of Article 9A - Bridging constitutional rights and environmental governance. *Human Rights Commission of Pakistan*. https://hrcp.wittywebsolutions.com/the-promise-of-article-9a-bridging-constitutional-rights-and-environmental-governance/

Mangiuc, M.-D., & Kagitci, M. (2025). CSRD as a Paradigm Shift in Sustainability Reporting: From Voluntary Practice to Legal Requirement. *Amfiteatru Economic, 27*(70), 1193. DOI: 10.24818/EA/2025/70/1193

Marco, D. D., & Cagliano, R. (2026). Toward Sustainable Organizations: Routines to Embed Sustainability in Corporate Culture. *Creativity and Innovation Management, 35*(3), 590–611. Advance online publication. DOI: 10.1111/caim.70047

Maritime Magazine. (2021, December 2). *The Basel Convention's ban on the export of toxic ships to developing countries reestablished.* https://maritimemag.com/en/the-basel-conventions-ban-on-the-export-of-toxic-ships-to-developing-countries-reestablished/ (accessed on June, 2026)

Markt Intern Verlag GmbH v. Germany (1989). European Court of Human Rights.

Marušić, B. (2025). The role of Article 114 TFEU in balancing fundamental rights and product safety in the AI Act. *Papers from the International Scientific Conference of the European Studies Department Jean Monnet Center of Excellence, Faculty of Philosophy at Sofia University "St Kliment Ohridski," 12*, 277. DOI: 10.60054/PEU.2025.12.277-289

Matthews, H. S., Hendrickson, C. T., & Weber, C. L. (2008). The importance of carbon footprint estimation boundaries. *Environmental Science & Technology, 42*(16), 5839–5842. DOI: 10.1021/es703112w

Mayer, B. (2018). *The international law on climate change.* Cambridge University Press. DOI: 10.1017/9781108304368

Mercier, V. (2026). The Protection of Human Dignity Through the European Directive on Corporate Sustainability Due Diligence. *Bulletin of the Transilvania University of Braşov Series VII Social Sciences • Law*, 129. DOI: 10.31926/but.ssl.2025.18.67.3.14

Merkl-Davies, D., & Brennan, N. (2007). Discretionary disclosure strategies in corporate narratives: Incremental information or impression management? *Journal of Accounting Literature, 26*, 116–196.

Merriam-Webster. (n.d.). *Intentionality.* In *Merriam-Webster.com dictionary.* https://www.merriam-webster.com/dictionary/intentionality

Michalina, D., Mederly, P., Diefenbacher, H., & Held, B. (2021). Sustainable urban development: A review of urban sustainability indicator frameworks. *Sustainability (Basel), 13*(16), 9348. DOI: 10.3390/su13169348

Milieudefensie et al. v. Royal Dutch Shell plc, Hague District Court, C/09/571932/HA ZA 19-379 (2021).

Milieudefensie et al. v. Royal Dutch Shell plc., (2021). Case No. C/09/571932/HA ZA 19-379 (District Court of The Hague, 2021).

Military and Paramilitary Activities in and against Nicaragua (Nicaragua v. United States of America), Merits, Judgment, I.C.J. Reports 1986, 14 (International Court of Justice, June 27, 1986).

Milnitz, D., & Marchi, J. J. (2025). TEXTILE AND APPAREL SUPPLY CHAIN: STRUCTURE, COMPLEXITY, AND GLOBAL TRANSFORMATION TRENDS. In *Seven Editora eBooks*. DOI: 10.56238/sevened2025.036-005

Ministry of Climate Change. (2025). *Pakistan's Third National Communication on Climate Change (TNC)—Climate Change Laws of the World*. https://climate-laws.org/documents/pakistans-third-national-communication-on-climate-change-tnc_e262

Ministry of Corporate Affairs. (2013). *The Companies Act, 2013* (Act No. 18 of 2013). Government of India.

Ministry of Environment and Energy. (2015). *Maldives Climate Change Policy Framework 2015–2025*. Republic of Maldives.

Ministry of Environment. Energy and Water (MEEW). (2007). *National Adaptation Programme of Action: The Republic of Maldives*. UNFCCC. https://unfccc.int/resource/docs/napa/mdv01.pdf

Ministry of Environment. Forest and Climate Change. (2025, September). *Bangladesh's third nationally determined contribution (NDC 3.0)*. Government of the People's Republic of Bangladesh. https://unfccc.int/sites/default/files/2025-09/Bangladesh%20Third%20Nationally%20Determined%20Contribution%20%28NDC%203.0%29.pdf (accessed on June, 2026)

Miret-Pastor, L., Molina-García, A., García-Aranda, C., & Herrera-Racionero, P. (2022). Social vulnerability and adaptive capacity in small-scale fisheries under climate change. *Aquaculture and Fisheries*, *7*(6), 593–604. DOI: 10.1016/j.aaf.2022.03.004

Mir-Hosseini, Z. (2015). Muslim legal tradition and the challenge of gender equality. In Mir-Hosseini, Z., Al-Sharmani, M., & Rumminger, J. (Eds.), *Men in charge? Rethinking authority in Muslim legal tradition* (pp. 1–24). Oneworld Publications.

Moghul, U. F., & Safar-Aly, S. H. K. (2015). Green sukuk: The introduction of Islam's environmental ethics to contemporary Islamic finance. *Georgetown International Environmental Law Review*, *27*(1), 1–60.

Montgomery, A. W., Lyon, T. P., & Barg, J. (2024). No end in sight? A greenwash review and research agenda. *Organization & Environment*, *37*(2), 221–256. DOI: 10.1177/10860266231168905 PMID: 39091586

Morgera, E. (2020). Introduction - In Corporate Environmental Accountability in International Law. *Oxford University Press.* 1-23. https://doi.org/DOI: 10.1093/oso/9780198738046.003.0001

Müller and Others v. Switzerland (1988). European Court of Human Rights.

Müller Gómez, J. (2024). Federalism, Decentral Governance, and Joint Decision-Making: Bad News for the Implementation of International Environmental Agreements? *Publius*, *54*(2), 407–433. DOI: 10.1093/publius/pjae003

Mu, Y. (2025). Divergence in HRDD Legislation: A Comparative Study of France, Germany, and the United States. *Lecture Notes in Education. Psychology and Public Media*, *107*(1), 27. DOI: 10.54254/2753-7048/2025.LD25578

Naher, N. (2025, August 15). *Rethinking environmental impact assessment in Bangladesh: Challenges, gaps, and pathways forward.* SSRN. https://doi.org/ (accessed on June, 2026)DOI: 10.2139/ssrn.5393399

Nairobi Convention for the Protection, Management and Development of the Marine and Coastal Environment of the Western Indian Ocean, June 21, 1985.

Nasr, S. H. (1996). *Religion and the order of nature*. Oxford University Press. DOI: 10.1093/acprof:oso/9780195108231.001.0001

Neubauer et al. v. Germany, (2021). 1 BvR 2656/18, Federal Constitutional Court of Germany, BVerfGE 157, 30

Ng, A. W., Yorke, S. M., & Nathwani, J. (2022). Enforcing double materiality in global sustainability reporting for developing economies: Reflection on Ghana's oil exploration and mining sectors. *Sustainability (Basel)*, *14*(16), 9988. DOI: 10.3390/su14169988

NGO Shipbreaking Platform. (2026). *Annual list of ships dismantled worldwide, 2025.*

Niehaus, M. (2023). Constitution, Constitutionalisation and Constitutionalism. In Niehaus, M. (Ed.), *Global Climate Constitutionalism "from below": The Role of Climate Change Litigation for International Climate Lawmaking* (pp. 71–104). Springer Fachmedien., DOI: 10.1007/978-3-658-43191-4_3

Nody, N. I., & Amin, M. B. (2024). *The Post-COVID Environmental Ramifications and a Quest for an Eco-Friendly Legal Hegemony*. BAIUST Academic Journal., Retrieved June 2026, from, DOI: 10.63307/BAJ.5.1.011

Obligations of States in Respect of Climate Change, Advisory Opinion, 2025 I.C.J. Rep. 187 (July 23).

Observer Research Foundation. (2024, July 10). *The Teesta water dispute: Geopolitics, myth and economics*. https://www.orfonline.org/research/the-teesta-water-dispute-geopolitics-myth-and-economics (accessed on June, 2026)

Ochoa, A. G., Khare, P., & Setia, S. (2024). Expanding Environmental Education in Honduras: A Holistic Approach for Sustainable Development. *IGI Global Scientific Publishing*, 153-170. https://doi.org/DOI: 10.4018/979-8-3693-2577-3.ch010

OECD. (2021). *Gender and the environment: Building evidence and policies to achieve the SDGs*. OECD Publishing., DOI: 10.1787/3d32ca39-

OECD. (2022). *Policy guidance on market practices to strengthen ESG investing and finance a climate transition* (OECD Business and Finance Policy Papers). OECD Publishing. https://doi.org/DOI: 10.1787/bf84ff64-en

OECD. (2025). Measuring the uptake and impact of due diligence for responsible supply chains. In *OECD Publishing eBooks*. DOI: 10.1787/52ef761f-en

Okereke, C. (2010). Climate justice and the international regime. *Wiley Interdisciplinary Reviews: Climate Change, 1*(3), 462–474. DOI: 10.1002/wcc.52

Özdemir, İ. (2018). The concept of mizan and ecological ethics in Islam. *International Journal of Islamic Thought, 13*(1), 58–73. DOI: 10.24035/ijit.13.2018.006

Paris Agreement to the United Nations Framework Convention on Climate Change, December 12, 2015, T.I.A.S. No. 16-1104.

Paris Judicial Court. (2025, October 23). *Greenpeace France and Others v. TotalEnergies SE and TotalEnergies Électricité et Gaz France* (machine-translated English version). ClientEarth. https://www.clientearth.org/media/1dmfkuvo/gpf-vs-totalenergies_english_machine-translated.pdf

Parroco, V. (2026). The right to an effective remedy and due diligence legislation in the EU. *Archivio Istituzionale Della Ricerca (Alma Mater Studiorum Università Di Bologna)*. https://hdl.handle.net/11585/1038050

Partzsch, L. (2025). Hardening sustainability: Supply chain laws complement cocoa and coffee certifications. *Forest Policy and Economics, 181*, 103661. DOI: 10.1016/j.forpol.2025.103661

Peel, J., & Osofsky, H. M. (2015). *Climate change litigation: Regulatory pathways to cleaner energy*. Cambridge University Press. DOI: 10.1017/CBO9781139565851

Peters, A., Gleß, S., Thomale, C., & Weller, M.-P. (2023). Business and Human Rights: Towards a 'Smart Mix' of Regulation and Enforcement. *Zeitschrift für ausländisches öffentliches Recht Und Völkerrecht / Heidelberg Journal of International Law, 83*(3), 415. DOI: 10.17104/0044-2348-2023-3-415

Petersmann, E.-U. (2024). European Economic and Environmental Constitutionalism as Driver for UN and WTO Sustainable Development Reforms. In G. Z. Capaldo (Ed.), *The Global Community Yearbook of International Law and Jurisprudence 2023: Global Law, Politics, Ethics, Justice* (p. 0). Oxford University Press. DOI: 10.1093/oso/9780197795392.003.0004

Pinsky, M. L., Fenichel, E., Fogarty, M., Levin, S., McCay, B., St. Martin, K., Selden, R. L., & Young, T. (2020). Fish and fisheries in hot water: What is happening and how do we adapt? *Population Ecology, 63*(1), 17–26. DOI: 10.1002/1438-390X.12050

Pítzen, S., Kivimaa, P., Hildén, M., & Mosoni, C. (2025). From informal coordination to formalized policy integration: Options for strengthening climate change adaptation in EU finance and trade policy. *Frontiers in Climate, 7*, 1543732. Advance online publication. DOI: 10.3389/fclim.2025.1543732

Portillo Cáceres v. Paraguay (2019). Human Rights Committee, Communication No. 2751/2016.

Poseidon Principles. (2019). *Poseidon Principles: A global framework for responsible ship finance*.

Post, R. (2012). *Democracy, expertise, and academic freedom*. Yale University Press.

Presidency of the Republic of Maldives. (2019). *Climate Smart Resilient Islands initiative*. https://presidency.gov.mv/Press/Article/22213

Presidency of the Republic of Maldives. (2025). *Presidential Directive No. 14/2025: Renaming of the Environmental Protection Agency to the Environmental Regulatory Authority*. https://presidency.gov.mv/Press/Article/34548

Preston, B. J. (2021). The influence of the Paris Agreement on climate litigation: Causation, corporate governance, and costs. *Journal of Environmental Law, 33*(1), 1–30. DOI: 10.1093/jel/eqaa020

Pulp Mills on the River Uruguay (Arg. v. Uru.), Judgment, 2010 I.C.J. 14 (Apr. 20).

Pulp Mills on the River Uruguay (Argentina v. Uruguay), Judgment, I.C.J. Reports 2010, p. 14. https://www.icj-cij.org/case/135 (accessed on June, 2026)

Quirico, O., & Baber, W. (Eds.). (2024). *Implementing Climate Change Policy: Designing and Deploying Net Zero Carbon Governance*. Cambridge University Press., DOI: 10.1017/9781009341493

Radzi, H. (2023). An overview of environmental, social and governance (ESG) and company performance. In *European Proceedings of Social and Behavioural Sciences* (pp. 1111–1122). DOI: 10.15405/epsbs.2023.11.90

Rahman, A., & Ahmad, N. (2019). Mosque-based disaster preparedness in Bangladesh: Institutional integration and community outcomes. *International Journal of Disaster Risk Reduction, 35*, 101066. DOI: 10.1016/j.ijdrr.2019.01.021

Rajamani, L. (2007). *The Right to Environmental Protection in India: Many a Slip between the Cup and the Lip?* https://www.climate.ox.ac.uk/publication/1041472/dimensions

Rajamani, L., & Ghosh, S. (2020). India. In Rajamani, L., & Peel, J. (Eds.), *The Oxford handbook of international environmental law* (2nd ed., pp. 977–994). Oxford University Press.

Ram Mohan, M. P., Reynaers Kini, E., & Prasad, S. (2024). A worthy roadmap for global climate change litigation. *Georgetown Journal of International Law*, •••, 55.

Ramlan, S. (2019). *Religious Law for the Environment: Comparative Islamic Environmental Law in Singapore, Malaysia, and Indonesia* (NUS Centre for Asian Legal Studies Working Paper 19/03). National University of Singapore Faculty of Law. https://law.nus.edu.sg/wp-content/uploads/2020/04/CALS-WPS-1903.pdf

Ranjan, D., Singh, S., Singh, A., Misra, V., Singh, C., & Kumar, D. (2026). Recent advancements in sustainable aquaculture: Innovative techniques and future prospects. *Frontiers in Aquaculture, 5*, 1770106. DOI: 10.3389/faquc.2026.1770106

Rashid, H. (2025). Evaluating the efficacy of multi-tier supplier audits in enhancing supply chain reliability. *International Journal of Multidisciplinary Research and Growth Evaluation, 6*(1), 1918–1923. DOI: 10.54660/.IJMRGE.2025.6.1-1918-1923

Razzak, M. (2024). The UN guiding principles on business and human rights: A commentary. *Australian Journal of Human Rights, 30*(1), 125–130. DOI: 10.1080/1323238X.2024.2349322

Regulation (EU) 2020/852 of the European Parliament and of the Council of 18 June 2020 on the establishment of a framework to facilitate sustainable investment. (2020). *Official Journal of the European Union, L 198*, 13–43.

Regulation (EU) 2021/1119 of the European Parliament and of the Council of 30 June 2021 establishing the framework for achieving climate neutrality. (2021). *Official Journal of the European Union, L 243*, 1–17.

Regulation on the Preparation of Environmental Impact Assessment Report (Regulation No: 2012/R-27). (2012). Ministry of Housing and Environment, Republic of Maldives.

Repas, L. (2024). Accountability Framework initiative and Canopy: Statement of Alignment. *Canopy*. https://canopyplanet.org/news/accountability-framework-initiative-and-canopy-statement-of-alignment

Request for an Advisory Opinion Submitted by the Commission of Small Island States on Climate Change and International Law, Advisory Opinion, 2024 ITLOS Rep. Case No. 31 (May 21).

Reserve Bank of India. (2025). *Report on trend and progress of banking in India 2024–25*. https://www.rbi.org.in/Scripts/AnnualPublications.aspx?head=Trend%20and%20Progress%20of%20Banking%20in%20India

Responsibilities and Obligations of States Sponsoring Persons and Entities with Respect to Activities in the Area, Advisory Opinion, 2011 ITLOS Rep. 10 (Feb. 1).

Rezarta, T., & Richard, T. (2025). Navigating the Future: Assessing the Impact of Due Diligence Rules in the European Union on Business, Environment, and Workers. *Università Degli Studi Di Bari*. DOI: 10.15162/2612-6583/2390

Rio Declaration on Environment and Development. (1992). United Nations.

Rocha, H. M. B. C. C. L. (2016). *Social work and environment: Ecological sustainability of vulnerable communities (Serviço social e ambiente: A sustentabilidade ecológica das comunidades vulneráveis)* (Doctoral dissertation, ISCTE – Instituto Universitário de Lisboa). https://hdl.handle.net/10071/11964

Rodríguez-Gómez, S., Sánchez, I., Arco-Castro, M. L., & Pérez, M. V. L. (2025). The Effect of Institutional Pressures and Business Incentives on the Level of Corporate Compliance System Implementation. *Corporate Social Responsibility and Environmental Management, 33*(1), 18–38. DOI: 10.1002/csr.70155

Rościszewski-Dodgson, M. J., & Cirella, G. T. (2026). Quota constraints, market pressures, and structural change in Polish Baltic fisheries. *Marine Policy, 189*, 107120. DOI: 10.1016/j.marpol.2026.107120

Rothwell, D. R., & Stephens, T. (2016). *The international law of the sea* (2nd ed.). Hart Publishing.

Ruggie, J. (2011). Guiding principles on business and human rights: Implementing the United Nations "Protect, Respect and Remedy" Framework (Vol. A). UN Human Rights Council., HRC/17/31.

Saeed, A. (2008). Environmental ethics and Islam. *Islamic Studies, 47*(3), 313–340.

Safronova, A. (2025). Experts faced with a drying wetland: History of fisheries in the Volga River delta, 1930–1962. *Water History, 17*(2), 201–228. DOI: 10.1007/s12685-024-00357-2

Saleh, M. F. (2015, September 3). *A critical appraisal of Bangladesh Environment Conservation Act, 1995 and Rules, 1997*. Bangladesh Law Digest. https://bdlawdigest.org/bangladesh-environment-conservation-act-1995.html (accessed on June, 2026)

Salih, R. H., Khinger, I. K., Ubaid, H. R., Aldabagh, D., & Alani, S. S. (2025). Legal Perspectives on Corporate Accountability in Environmental Sustainability. *Environment and Social Psychology, 10*(11). Advance online publication. DOI: 10.59429/esp.v10i11.3992

Salminen, J., Sobel-Read, K., Viljanen, M., & Eller, K. H. (2022). Digital Platforms as Second-Order Lead Firms: Beyond the Industrial/Digital Divide in Regulating Value Chains. *European Review of Private Law/Revue Européenne de Droit Privé/Europäische Zeitschrift für Privatrecht, 30*, 1059. DOI: 10.54648/ERPL2022049

Sands, P., Peel, J., Fabra, A., & MacKenzie, R. (2018). *Principles of international environmental law* (4th ed.). Cambridge University Press. DOI: 10.1017/9781108355728

Saniotis, A. (2012). Muslims and ecology: Fostering Islamic environmental ethics. *Contemporary Islam, 6*(2), 155–171. DOI: 10.1007/s11562-011-0173-8

Scheper, C. (2026). Due Diligence Legislation and the Politics of Implementation: The Case of the German Supply Chain Law. *Regulation & Governance*, rego.70134. Advance online publication. DOI: 10.1111/rego.70134

Schilling-Vacaflor, A., & Lenschow, A. (2021). Hardening foreign corporate accountability through mandatory due diligence in the European Union? New trends and persisting challenges. *Regulation & Governance, 17*(3), 677–693. DOI: 10.1111/rego.12402

Schlosberg, D., & Collins, L. B. (2014). From environmental to climate justice: Climate change and the discourse of environmental justice. *Wiley Interdisciplinary Reviews: Climate Change, 5*(3), 359–374. DOI: 10.1002/wcc.275

Schmidt, S. K. (2018). The European Court of Justice and the Policy Process. In *Oxford University Press eBooks*. Oxford University Press., DOI: 10.1093/oso/9780198717775.001.0001

Science Based Targets Initiative. (2021). *Science-based target setting manual (Version 1.1)*. https://sciencebasedtargets.org/resources

Sea Cargo Charter. (2020). *Sea Cargo Charter: Aligning shipping with climate goals.*

Securities and Exchange Board of India. (2023, July 12). *BRSR Core – Framework for assurance and ESG disclosures for value chain* (Circular No. SEBI/HO/DDHS/DDHS-RACPOD2/P/CIR/2023/120). https://www.sebi.gov.in/legal/circulars/jul-2023/brsr-core-framework-for-assurance-and-esg-disclosures-for-value-chain_73854.html

Sellare, J., Börner, J., Brugger, F., Garrett, R., Günther, I., Meemken, E., Pelli, E. M., Steinhübel, L., & Wuepper, D. (2022). Six research priorities to support corporate due diligence policies. *Nature*, *606*(7916), 861–863. DOI: 10.1038/d41586-022-01718-8 PMID: 35764799

Setzer, J., & Higham, C. (2023). *Global trends in climate change litigation: 2023 snapshot*. Grantham Research Institute on Climate Change and the Environment, London School of Economics and Political Science. https://www.lse.ac.uk/granthaminstitute/wp-content/uploads/2023/06/Global_trends_in_climate_change_litigation_2023_snapshot.pdf

Sharma v. Minister for the Environment, [2021] FCA 560 (Federal Court of Australia).

Sharma, S. N., & Dehalwar, K. (2025). Transit-oriented development and economic development. *Transportation in Developing Economies*, *11*(2), 23. DOI: 10.1007/s40890-025-00245-1

Shekhawat, V. & Khare, P. (2026). Analyzing the Impact of International Law and Sanctions Mechanisms Towards Sustainable Outcomes. *IGI Global Scientific Publishing*, 327-348. https://doi.org/DOI: 10.4018/979-8-3373-1707-6.ch010

Shipbreaking Platform, N. G. O. (2024). *Bangladesh*. https://shipbreakingplatform.org/our-work/the-problem/bangladesh/

Shulman, S. (2024). Corporate Sustainability Due Diligence. *Columbia Journal of Environmental Law*, *49*(2), 479–523. DOI: 10.52214/cjel.v49i2.12633

Simataa, C. B., Persendt, F. C., & Gomez, C. (2025). Illegal, unreported and unregulated (IUU) fishing in Africa: A systematic review. *Aquaculture, Fish and Fisheries*, *5*(4), e70107. DOI: 10.1002/aff2.70107

Simpson, G. (2021). The sentimental life of international law: Literature, language, and longing in world politics. Oxford University Press., Retrieved June 24, 2026, from, DOI: 10.1093/oso/9780192849793.001.0001

Sinnig, J., & Zetzsche, D. A. (2025). The EU's Corporate Sustainability Due Diligence Directive: From Disclosure to Mandatory Prevention of Adverse Sustainability Impacts in Supply Chains. *European Journal of Risk Regulation, 1*(2), 628–652. Advance online publication. DOI: 10.1017/err.2024.100

Sivaramakrishnan. (1999). *Modern Forests | Stanford University Press*. https://www.sup.org/books/asian-studies/modern-forests

Small Industries Development Bank of India. (2024). *Annual report 2023–24*. https://www.sidbi.in/annualreport/AnnualReport202324/index.php

Sneideriene, A., & Legenzova, R. (2025). Greenwashing prevention in environmental, social, and governance (ESG) disclosures: A bibliometric analysis. *Research in International Business and Finance, 74*, 102720. DOI: 10.1016/j.ribaf.2024.102720

Sørensen, K. E. (2025). Applying CSDDD in Corporate Groups. *European Company and Financial Law Review, 22*(2), 129–164. DOI: 10.1515/ecfr-2025-0006

State of the Netherlands v. Urgenda Foundation, Supreme Court of the Netherlands, ECLI:NL:HR:2019:2006 (2019).

Stiglitz, J. E., Sen, A., & Fitoussi, J.-P. (2009). *Report by the Commission on the Measurement of Economic Performance and Social Progress*. European Commission.

Stop Ecocide Foundation. (2021). *Independent expert panel for the legal definition of ecocide: Commentary and core text.*

Stop Ecocide International. (2021). *Independent expert panel for the legal definition of ecocide: Commentary and core text*. Stop Ecocide Foundation. https://www.stopecocide.earth/legal-definition

Subhash Kumar v. State of Bihar, AIR 1991 SC 420 (India).

Suchman, M. C. (1995). Managing legitimacy: Strategic and institutional approaches. *Academy of Management Review, 20*(3), 571. DOI: 10.2307/258788

Sunny, S. A., Cottle, G. W., & Coffman, C. D. (2025). Moving the Needle: Evaluating Firm Responsiveness to the Issue of Global Climate Change. *Academy of Management Proceedings, 2025*(1). DOI: 10.5465/AMPROC.2025.14769abstract

Sunstein, C. R. (1993). *Democracy and the problem of free speech*. Free Press.

Supreme Court of Pakistan. (2025, January 25). PLR 2025 SC 5 = PLD 2025 SC 36 = 2024 SCP 372. *PakistanKanoon.Com*. https://pakistankanoon.com/pakistan-law-reporter/plr-2025-sc-5/

Tanhuanpää, T., Altarriba, E., & Rahiala, S. (2025). Transition towards Fossil-Free energy sources in the shipping sector: A Sustainability Index Analysis of methanol and liquefied Biogas. In *Lecture notes in intelligent transportation and infrastructure* (pp. 571–587). DOI: 10.1007/978-3-031-82818-8_43

Tathavadekar, V. P. (2025). Optimizing Supply Chain Sustainability: Leveraging the Circular Economy for Effective Resource Management and Reduced Environmental Impact. *Advances in Consumer Research. Association for Consumer Research (U. S.)*, *2*(4), 602. DOI: 10.61336/acr/25-04-01

Tinger, M., Walton, S., & O'Kane, P. (2024). The climate impact of misinformation, disinformation, and greenwashing in advertising. In Vandenberg, K. M., & Tinger, M. (Eds.), *Rethinking advertising: Ethics and effectiveness* (pp. 68–69). Palgrave Macmillan.

Trail Smelter Arbitration (U.S. v. Can.), 3 R.I.A.A. 1905 (1941).

Trejo, G. (2009). Religious competition and ethnic mobilization in Latin America: Why the Catholic Church promotes indigenous rights in Mexico. *The American Political Science Review*, *103*(3), 323–342. DOI: 10.1017/S0003055409990025

Tucker, M. E., & Grim, J. (2017). *Religion and ecology: Developing a planetary ethic*. Columbia University Press.

Tysiachniouk, M., & Vorbrugg, A. (2025). The transformation of the Forest Stewardship Council (FSC) Certification Scheme in Russia after the full-scale invasion of Ukraine. *Forest Policy and Economics*, *178*, 103559. DOI: 10.1016/j.forpol.2025.103559

Ullah, Z., Saleem, M. M., & Haider, A. (2026). State's strict liability for managed forest fires under Pakistan's tort jurisprudence. https://www.researchgate.net/publication/400576266

UN Human Rights Office of the High Commissioner. (2024, May 3). *Freedom of expression and media crucial to tackling climate crisis: International freedom of expression rapporteurs*. United Nations. https://www.ohchr.org/en/statements-and-speeches/2024/05/freedom-expression-and-media-crucial-tackling-climate

UN Human Rights Office of the High Commissioner. (2024, May 3). *Joint declaration on the climate crisis and freedom of expression*. United Nations.

United Nations Conference on the Human Environment. (1972). *Declaration of the United Nations Conference on the Human Environment (Stockholm Declaration)* (U.N. Doc. A/CONF.48/14/Rev.1).

United Nations Convention on the Law of the Sea, Dec. 10, 1982, 1833 U.N.T.S. 3.

United Nations Development Programme (UNDP). (2022). *Maldives country programme document 2022–2026*. https://www.undp.org/maldives/publications/country-programme-document-maldives-2022-2026

United Nations Development Programme. (2024). *UNDP annual report 2023*. https://www.undp.org/publications/undp-annual-report-2023

United Nations Economic Commission for Europe. (1998). *Convention on access to information, public participation in decision-making and access to justice in environmental matters*. Aarhus Convention.

United Nations Economic Commission for Europe. (2014). *The Aarhus Convention: An implementation guide* (2nd ed.). United Nations Economic Commission for Europe.

United Nations Environment Programme. (1995). *Convention for the protection of the marine environment and the coastal region of the Mediterranean (consolidated text)*. UNEP Mediterranean Action Plan.

United Nations Environment Programme. (2021). *Making peace with nature: A scientific blueprint to tackle climate, biodiversity and pollution crises*. DOI: 10.18356/9789280738377

United Nations Framework Convention on Climate Change, May 9, 1992, 1771 U.N.T.S. 107.

United Nations General Assembly. (2022). Resolution 76/300: The human right to a clean, healthy and sustainable environment (Vol. A). United Nations., RES/76/300.

United Nations Human Rights Committee. (2022). *Billy et al. v. Australia: Views adopted by the Committee under Article 5(4) of the Optional Protocol* (CCPR/C/135/D/3624/2019).

United Nations Human Rights Council. (2025). *The imperative of defossilizing our economies: Report of the Special Rapporteur on the promotion and protection of human rights in the context of climate change, Elisa Morgera* (A/HRC/59/42). United Nations. https://docs.un.org/en/A/HRC/59/42

United Nations. (1992). *Rio Declaration on Environment and Development* (U.N. Doc. A/CONF.151/26, Vol. I).

United Nations. (1992). *Rio declaration on environment and development.*

United Nations. (1992). *United Nations Framework Convention on Climate Change.* https://unfccc.int/resource/docs/convkp/conveng.pdf (accessed on June, 2026)

United Nations. (1998). *Kyoto Protocol to the United Nations Framework Convention on Climate Change.* https://unfccc.int/resource/docs/convkp/kpeng.pdf (accessed on June, 2026)

United Nations. (2015). *Paris Agreement.* https://unfccc.int/sites/default/files/english_paris_agreement.pdf (accessed on June, 2026)

United Nations. (2015). *Transforming our world: The 2030 agenda for sustainable development.* https://sdgs.un.org/2030agenda

United Nations. (n.d.). *Greenwashing – the deceptive tactics behind environmental claims | United Nations.* https://www.un.org/en/climatechange/science/climate-issues/greenwashing

Urgenda Foundation v. State of the Netherlands, ECLI:NL:HR:2019:2006 (Supreme Court of the Netherlands, 2019).

Uvarova, O., & Bernaz, N. (2025). Heightened Human-Rights Due Diligence as Risk Governance: Food-Retail Supply Chains in Wartime Ukraine. *European Journal of Risk Regulation, 1.* Advance online publication. DOI: 10.1017/err.2025.10049

van Oijen, J. C. F., Wallenburg, I., Bal, R., & Grit, K. (2020). Institutional work to maintain, repair, and improve the regulatory regime: How actors respond to external challenges in the public supervision of ongoing clinical trials in the Netherlands. *PLoS One, 15*(7), e0236545. Advance online publication. DOI: 10.1371/journal.pone.0236545 PMID: 32735568

Van Santen, G. (2024). Fisheries governance: The search for effective management and the illusion of control. *Global Perspectives, 5*(1), 94988. Advance online publication. DOI: 10.1525/gp.2024.94988

Veldman, R. G., Jenkins, W. J., & Chapple, C. K. (2014). *How the world's religions are responding to climate change: Social scientific investigations.* Harvard University Press.

Vellore Citizens Welfare Forum v. Union of India, AIR 1996 SC 2715 (India).

Verein gegen Tierfabriken Schweiz (VgT) v. Switzerland (No. 2) (2009). European Court of Human Rights.

Verein KlimaSeniorinnen Schweiz and Others v. Switzerland [GC], App. No. 53600/20 (ECtHR Apr. 9, 2024).

VgT Verein gegen Tierfabriken v. Switzerland (2001). European Court of Human Rights.

Vidhi Centre for Legal Policy. (2024). *Holding the state accountable.* https://vidhilegalpolicy.in/

Villiers, C. (2022). New Directions in the European Union's Regulatory Framework for Corporate Reporting, Due Diligence and Accountability: The Challenge of Complexity. *European Journal of Risk Regulation*, *13*(4), 548–566. DOI: 10.1017/err.2022.25

Viola, P. (2022). The Ongoing Foundation of the Constitutionalisation of Climate. In Viola, P. (Ed.), *Climate Constitutionalism Momentum: Adaptive Legal Systems* (pp. 41–82). Springer International Publishing., DOI: 10.1007/978-3-030-97336-0_3

Vizzuality. (2026). *Pakistan Deforestation Rates & Statistics | GFW.* https://www.globalforestwatch.org/dashboards/country/PAK?category=undefined

Wadud, A. (2006). *Inside the gender jihad: Women's reform in Islam.* Oneworld Publications.

Wahyuni, W., Syamsidar, S., Indasari, F., & Bualemo, L. E. G. (2024). Corporate environmental accountability: A qualitative study of stakeholder expectations and corporate practices. *Advances: Jurnal Ekonomi Bisnis*, *2*(6), 323–336. DOI: 10.60079/ajeb.v2i6.352

Wang, E. L. (2024). Environmental Constitutionalism in China: A Constitution without Constitutionalism? *Journal of Environmental Law*, *36*(2), 187–202. DOI: 10.1093/jel/eqae012

Wang, Q., & Li, G. (2024). Research on the effect of corporate environmental responsibility on corporate sustainability and the mediator effect of corporate environmental strategy. *SAGE Open*, *14*(3), 21582440241266115. Advance online publication. DOI: 10.1177/21582440241266115

Wang, X. (2023). The 1972 Stockholm Conference and China's diplomatic response. *Cultura e Scuola*, *6*(2), 146–152. DOI: 10.1177/20966083231184691

Watts, R. L. (with Queen's University & Queens's University). (2008). *Comparing federal systems* (Third edition). McGill-Queen's University Press.

Wewerinke-Singh, M., & Viñuales, J. E. (2026). Bracing for impact: The ICJ Advisory Opinion on climate change and the activation of legal tipping points. *Journal of Environmental Law*, eqag014. Advance online publication. Retrieved June 2026, from. DOI: 10.1093/jel/eqag014

White, L.Jr. (1967). The historical roots of our ecological crisis. *Science*, *155*(3767), 1203–1207. DOI: 10.1126/science.155.3767.1203

Wilhelm, M. (2024). Mandatory due diligence legislation: A paradigm shift for the governance of sustainability in global value chains? *Journal of International Business Policy*, *7*(4), 459–465. DOI: 10.1057/s42214-024-00193-4

Woetzel, J., Madgavkar, A., Sneader, K., Tonby, O., Lin, D.-Y., Lydon, J., Sha, S., Krishnan, M., Ellingrud, K., & Gubieski, M. (2018). *The power of parity: Advancing women's equality in Asia Pacific*. McKinsey Global Institute. https://www.mckinsey.com/mgi

Women, U. N. (2022). *Progress on the Sustainable Development Goals: The gender snapshot 2022*. https://www.unwomen.org/en/digital-library/publications/2022/09/progress-on-the-sustainable-development-goals-the-gender-snapshot-2022

Wong, Y. A. (2025). A study on the effectiveness of Patagonia's sustainability-driven brand strategy. *Advances in Economics. Management and Political Sciences*, *180*(1), 302–306. DOI: 10.54254/2754-1169/2025.23726

World Bank. (2022). *Open Knowledge Repository*. https://openknowledge.worldbank.org/entities/publication/614ddc2b-ca31-53c9-b59c-6bf12a56d336

World Bank. (2023). *Women Entrepreneurs Finance Initiative (We-Fi)*. https://www.worldbank.org/en/programs/women-entrepreneurs

World Bank. (2024). *Key Highlights: Maldives Country Climate and Development Report*. https://www.worldbank.org/en/country/maldives/publication/key-highlights-maldives-country-climate-and-development-report

World Bank. (2024). *Women, business and the law 2024*. https://www.worldbank.org/en/publication/women-business-and-the-law

World Commission on Environment and Development. (1987). *Our common future*. Oxford University Press.

World Health Organization (WHO). (2021). *Climate change and health in small island developing states*. https://lancetcountdown.org/wp-content/uploads/2024/03/Climate-Change-and-Health-in-Small-Island-Developing-States_February-2022.pdf

World Resources Institute and World Business Council for Sustainable Development. (2004). *The greenhouse gas protocol: A corporate accounting and reporting standard* (Revised Edition). World Resources Institute.

Yadav, K., Dehalwar, K., & Sharma, S. N. (2025). Factors affecting first and last mile accessibility. *GeoJournal, 90*(6), 298. DOI: 10.1007/s10708-025-11546-8

Yadav, K., Dehalwar, K., & Sharma, S. N. (2026). Exploring the environmental determinants of mode choice in first and last mile connectivity. *Innovative Infrastructure Solutions, 11*(4), 204. DOI: 10.1007/s41062-026-02614-0

Yale Center. (2024). *2024 Environmental Performance Index—Pakistan.* Environmental Performance Index. https://epi.yale.edu/country/2024/PAK

Yanuardi, Y., Vijge, M. J., & Biermann, F. (2022). Social-ecological reflexivity of extractive industry governance? The case of the Extractive Industries Transparency Initiative in Indonesia. *Environmental Policy and Governance, 32*(5), 426–437. DOI: 10.1002/eet.1988

Yuan, L., & Mähönen, J. (2024). Can Integrate a Sustainable Business Model and Global Value Chains Revive the Value Chain's Sustainable Growth? *Circular Economy and Sustainability, 4*(4), 2957–2980. Advance online publication. DOI: 10.1007/s43615-024-00352-y

Zhang, H., & Haider, A. (2025). China's first indoor smoking air pollution case: Remarkable but not justifiable. *Chinese Journal of Environmental Law*. https://www.researchgate.net/publication/397598122

Zhang, H., Haider, A., & Khan, A. (2025a). Biodiversity v. development: The Supreme Court's verdict in M.K. Ranjitsinh v. Union of India. https://www.researchgate.net/publication/395949557

Zhang, H., Haider, A., & Khan, A. (2025b). International trade and plastic waste in oceans: Legal and policy challenges. *Frontiers in Marine Science*. https://www.researchgate.net/publication/394575729

Zhu, L. (2023). Some thoughts on application of the polluter pays principle for controlling marine greenhouse gas emissions. *Marine Policy, 158*, 105877. DOI: 10.1016/j.marpol.2023.105877

Zöllmer, J. (2026). The end of negative integration within the internal market? Reassessing the EU's rigidity in the free movement of goods. *Figshare*. DOI: 10.6084/m9.figshare.31119056

Zumbansen, P. (2025). Devaluing Sustainability: Financialized Disclosure Governance and Transparency in Modern Slavery and Climate Change. *Journal of Law and Political Economy*, *5*(2). Advance online publication. DOI: 10.5070/LP60065156

About the Contributors

Aftab Haider is a dedicated legal scholar and practitioner whose work operates at the critical intersection of environmental law, international law, and human rights. Currently pursuing a PhD in International Law at the Southwest University of Political Science and Law in China, he is exploring how global legal frameworks can address modern challenges like climate change, organized crime, and digital governance. Academic Profile & Research Aftab's academic journey is marked by excellence, having earned both his Bachelor and Master of Laws from Abdul Wali Khan University, Mardan, where he achieved a perfect 4.00 CGPA in both programs and was a Bronze Medalist for his Master's degree. His research is prolific, with numerous publications in high-impact journals covering diverse topics such as: Environmental Justice: Legal challenges of plastic waste in oceans and the implementation of environmental laws in developing nations. Global Governance: The geopolitical impact of BRICS and the legalities of the China-Pakistan Economic Corridor (CPEC). Emerging Technology: The intersection of AI, cybersecurity law, and data sovereignty. He has also expanded his expertise through international certifications from the University of Pennsylvania and LUND University. Professional & Editorial Leadership Beyond academia, Aftab is an active International Arbitrator at the Chongqing Arbitration Commission and has extensive experience as a Lawyer with the KP Bar Council in Pakistan. He is a respected voice in the scholarly community, serving as: Reviewer for top-tier journals like PLOS ONE and Juris. Managing Editor for the Pakistan Journal of Law, Analysis and Wisdom. Editorial Board Member for several international publications across China, Indonesia, and Hong Kong. Achievements & Engagement: Aftab's contributions have been recognized through prestigious awards, including the Chongqing Mayor Scholarship and multiple Fauji Foundation scholarships. He is a frequent speaker at international forums, recently presenting at the Sustainable Justice Law Conference in Saudi Arabia and serving as a plenary speaker in Indonesia. A multilingual professional, he is fluent in English, Urdu, and Pashto, with intermediate proficiency in Chinese.

Whether through his legal advocacy or his role as a Youth Ambassador for Asian Cultural Heritage, Aftab remains committed to using law as a tool for sustainable development and global cooperation.

***

Amish Abdullah is a legal academic, researcher, and practitioner focused on constitutional governance, human rights, and international policy. He serves as an Expert with the World Bank Group and Rapporteur for Oxford International Organizations (OXIO), and is a Senior Lecturer in the UWE Bristol–Villa College partnership in the Maldives. He also works with the Westminster Foundation for Democracy, leads the Maldives chapter of Street Law Inc., and brings prior Supreme Court-level advocacy experience to his current research and pro bono work on comparative constitutional law, administrative justice, and migrant workers' rights.

Mohammad Shekaib Alam is an Assistant Professor in the Faculty of Shariah and Law at Villa College, Maldives. He holds a Ph.D. and a Master's degree in Political Science from the International Islamic University Malaysia (IIUM), along with formal academic training in Islamic Jurisprudence, Islamic Studies, and Arabic Language and Literature from institutions in Malaysia and India. This interdisciplinary background enables him to integrate Islamic legal thought with contemporary political, legal, and governance analysis. Dr. Alam's scholarship focuses on Islamic law and governance, minority rights, constitutionalism, and political theory, with particular attention to South and Southeast Asia, while remaining globally engaged. His research addresses Siyasah Shariah, pluralism, waqf reform, interfaith relations, Islamic finance, and ethical governance, including emerging debates on digital accountability, artificial intelligence, and environmental ethics within Islamic moral frameworks. He has published in international peer-reviewed journals, contributed to edited volumes and conference proceedings, and presented widely at international academic forums. His work emphasizes rigorous interdisciplinary dialogue and the application of normative frameworks to pressing global challenges.

Khuloud Ali is an Assistant Professor of Architecture and Sustainability, Architectural Engineer, and Environmental Energy Designer specializing in AI-driven Frameworks in Architecture and Sustainable Development, Algorithmic Sustainability, Intelligent Building Systems, and Climate-Responsive Design. Her professional affiliations include the United Nations Environment Programme (UNEP) Global Alliance for Buildings and Construction (GlobalABC) Sufficiency Hub, the International Union of Architects (UIA), the Federation of Arab Engineers

(FAE), the Maghreb Urban Hub, the Organization for Women in Science for the Developing World (OWSD), the International Urban Professionals Forum (IUPF), the Saudi Green Building Forum (SGBF), and the Global Ambassadors of Sustainability (GAoS). Dr. Khuloud is a UN COP30 speaker, a guest peer reviewer at Wiley, and a Guest Editor at the World Journal of Women and Sustainable Development (WJWSD), actively contributing to global academic and professional discourse. Dr. Ali is currently a postdoctoral researcher at the Federal University of Rio de Janeiro (UFRJ). She holds a Doctor of Science in Architecture from UFRJ, a Master of Science in Environmental Design and Energy Efficiency in Buildings from Cairo University, and a Bachelor's degree in Architectural Engineering from Damascus University. Her research spans environmental performance, sociocultural sustainability, and adaptive design systems. Her doctoral work examined Palm Jumeirah in Dubai through the lens of the UN 2030 Agenda, proposing strategies to reconcile rapid urbanization with cultural identity and ecological resilience. In 2017, she co-founded IMAGINE Studios (Intelligent Modeling for Architecture & Green Interactive Net-Zero Engineering Lab, Architecture Studio Delivering Applied Research, Design, Drafting, Simulation, and Visualization; ROR: 051dks184) and currently leads its Research and Development Division, advancing AI-driven architecture and sustainable building innovation. Since 2021, Dr. Ali has served as an Assistant Professor at Centro Universitário Gama e Souza (UNIGAMA) in Rio de Janeiro, teaching Environmental Comfort, Sustainable Design, Lighting, Building Systems, and Architectural Design, while supervising final-year research projects. Her teaching approach frames the design studio as an experimental laboratory where sustainability, computational intelligence, and architectural creativity converge.

Mehenaj Binte Amin is a Lecturer in Law at Green University of Bangladesh. She also serves as a lawyer at the District and Sessions Judge Court, Dhaka. As an early-career researcher, she has demonstrated strong research capability, having received the Best Paper Award at the 2025 Virtual Conference organised by B.S. Abdur Rahman Crescent Institute of Science and Technology in collaboration with the Jammu and Kashmir Economic Association, held on 12–13 November 2025. Her research interests lie in environmental justice and international law. She also served as 'Editors' in several peer reviewed journals and as a 'coach'to different moot court competitions, including the 10th round of Prestigious Jessup Moot Court Competition. With her commitment to contributing greater knowledge to the existing body of scholarship on climate justice and international environmental law, she continues to pursue research at the intersection of doctrinal legal analysis and pressing global environmental challenges facing climate-vulnerable states such as Bangladesh.

Kanika Arora is Associate Professor at Lovely Professional University with over 13 years of academic and industry experience. She holds a Ph.D. in Management and works at the intersection of entrepreneurship, sustainability, ESG frameworks, Ethical AI, and digital transformation in emerging economies. Her research includes 10+ peer-reviewed publications, international book chapters with IGI Global and Routledge, and two Indian patents in AI-based systems. A UGC-NET qualifier and Google Certified Data Analyst, she mentors postgraduate scholars with a focus on research ethics, publication quality, sustainable entrepreneurship, and policy-driven innovation.

Mohamad Khaled Bassma earned a Bachelor of Architecture (2007) and a Master of Science in Architecture, specializing in Building Science and Technology. He is an architectural professional with a strong technical orientation toward the performance, feasibility, and economic viability of building systems. He is a member of the International Union of Architects (UIA) and the Federation of Arab Engineers (FAE). His research and professional interests focus on intelligent and responsive façade systems, building economics, and the integration of technological innovation with cost efficiency and constructability in contemporary architecture.

Pranjal Khare is a distinguished corporate lawyer and is currently working as a law faculty at O.P. Jindal Global University, India. He specializes in Property Law, Corporate laws, Intellectual Property Rights, and International law. Mr. Pranjal is a highly regarded scholar and has published on numerous issues related to international affairs, economy, politics, data & privacy, artificial intelligence and IP disputes. He is also a frequent speaker at legal conferences and has served as a consultant to numerous law firms and corporations.

Aswathy Prakash G is an Associate Professor at the Saveetha School of Law (SIMATS), Chennai, India. She teaches a wide array of legal subjects, including Constitutional Law, Cyber Law, the Law of Torts, and Jurisprudence, among others. Dr. Aswathy holds a Ph.D. in Law from Saveetha School of Law, a Master's degree in Corporate Law from Osmania University, Hyderabad, India, and two Post-Graduate Diplomas from NALSAR University, Hyderabad, India. Her research primarily focuses on AI and emerging technologies, Intellectual Property Rights (IPR), Family Law, Consumer Protection, and children's digital rights, with her work extending into various issues within cyberspace. She has published numerous academic papers on diverse legal topics.

Helena Rocha, PhD in Social Work, is an Assistant professor at ISCTE-University Institute of Lisbon in the Department of Political Science and Public Policies and deputy director of the Soft Skills Lab. Integrated researcher at CIES,

Centre for Research and Studies in Sociology, linked to national and international research projects, namely 3 from Marie Curie Actions. She was pionner in Ecology and sustainability in Social Work creating the EcoSocial Model for intervention. Author of papers and communications at national and international congresses, in the areas of social work theory and methodology, environment, sustainability, community intervention, ethics, human rights, social policies and Well-being, education and soft skills. Member of the Editorial Board of national/international journals. Member of Inclusive Policy Lab UNESCO. Award of the Development Networks Award - Project "Promotion of Education for Global Citizenship - UN17" (ISCTE-IUL / COI Foundation).

Abraham S is a Research Scholar in the School of Law with academic and research interests in Maritime Law, International Law, Environmental Law, Sustainable Development, and Blue Economy Governance. He has authored and co-authored several research articles and book chapters published in peer-reviewed and indexed journals and edited volumes. His research focuses on maritime governance, environmental sustainability, women seafarers, and emerging legal challenges in global trade and policy. He has actively participated in and presented papers at various national and international conferences, seminars, and workshops, contributing to interdisciplinary legal scholarship and policy-oriented research

Mohd Arsh Shery is a Lecturer and LL.M. Programme Coordinator at Villa College, Maldives, and the Faculty of Business and Law at UWE Bristol (Partnership Programme). His academic expertise intersects law, technology, environmental law, and Islamic jurisprudence. He holds an LL.M. from the Tata Institute of Social Sciences (TISS) and a B.A.LL.B. (Hons.) from Aligarh Muslim University. As an active researcher and editorial leader, he serves as a Rapporteur for Oxford International Organizations (OXIO), an Expert Contributor for the World Bank's B-READY project, and an Editorial Board Member for the Manchester Journal of Transnational Islamic Law & Practice. He also serves as Lead Editor and peer reviewer for IGI Global Scientific Publishing. A recipient of the Oxford–UNESCO Certificate in AI, Justice, and the Rule of Law, his research is widely published in Scopus-indexed journals and by premier publishers including Springer Nature.

Ghayth Tintawi is an Assistant Lecturer in Architecture and Sustainability, an Architectural Engineer, and an Environmental Energy Designer, as well as a BIM and ICF Specialist. His research and professional work focus on AI-driven Architecture and Sustainable Development, Algorithmic Sustainability, Intelligent Building Systems, Climate-responsive Design, and Carbon Emissions Mitigation. His professional affiliations include the International Union of Architects (UIA), the European Association for Architectural Education (EAAE), the United Nations Global Marketplace (UNGM), the Federation of Arab Engineers (FAE), the Research Data Alliance (RDA), the Sustainability Professionals of Saudi Arabia (SPSA), the Global Ambassadors of Sustainability (GAoS), and the Saudi Green Building Forum (SGBF). Mr. Tintawi holds a Master's in Architecture and Urban Culture (MAUC) from the Universitat Politècnica de Catalunya, a Master of Science (M.Sc.) in Environmental Design, Renewable Energy, and Energy Efficiency in Buildings from Cairo University, a Bachelor of Science in Architectural Engineering, and a MicroMaster Certificate in Project Management from the Rochester Institute of Technology, USA. This multidisciplinary background enables the integration of architectural design, environmental systems, and project management within a unified framework. He is the Founder and Project Manager of IMAGINE Studios (Intelligent Modeling for Architecture & Green Interactive Net-Zero Engineering) and contributes to its Research and Development Division, advancing AI-driven architecture, environmental simulation, and high-performance building design. The studio operates at the intersection of applied research, design, visualization, and digital innovation, delivering architectural and real estate projects across multiple international contexts. His professional experience includes roles in architectural design, project management, and real estate development, including work with Real Estate Mogul Group in New York, where he coordinated and delivered complex architectural projects. In academia, Tintawi has served as an Assistant Lecturer and Academic Coordinator at MSA University in Cairo and Damascus University. He has taught and supervised architectural design studios, sustainable building systems, and urban planning courses, supporting the development of emerging architects and researchers.

C.A.M. Al Wahi is an emerging legal researcher and maritime law scholar from Bangladesh with research interests in international environmental law, maritime governance, climate justice, ocean policy and sustainable development. He is currently pursuing the Master of Laws (LL.M.-Professional) degree at Bangladesh University of Professionals and completed his LL.B. from Bangladesh Maritime University. Alongside his academic pursuits, he serves as an Apprentice Lawyer at the District and Sessions Judge's Court, Dhaka, and works as Grants and Funding Coordinator at the Bangladesh Centre for Ocean Law and Policy. He has also contributed as a reviewer and editor for several academic journals and legal research platforms. His scholarly work primarily focuses on maritime environmental protection, decarbonization of shipping, marine pollution governance, blue economy sustainability, and climate accountability in the Bay of Bengal region. He has authored and co-authored multiple research articles published in national and international journals, including works on maritime greenhouse gas reduction initiatives, shipping decarbonization, marine environmental governance and maritime security law. Al Wahi has presented research papers at national and international conferences on maritime law, cyber security legislation, environmental sustainability and ocean governance. His broader research agenda seeks to strengthen legal and institutional frameworks for sustainable maritime governance and environmental accountability in South Asia.

Serdar Yener currently works as an academic staff in Sinop University; a punctual individual with a proven track record of success and reliability in roles such as academic staff, researcher, project coordinator, and leader. He has participated in different organizations like UN, NATO, EU, academy as public affairs officer, academic staff, coordinator/director. After serving 19 years as army officer in Turkish Army, he has joined Sinop University, Turkey in 2016. He has carried out many projects under UN, EU, University in different areas. He worked as a public affairs officer under civil-military cooperation of NATO in Afghanistan in 2003. He contributed as coordinator of many social projects for the improvement of social life in Afghanistan. He has carried out many social projects in Lebanon under the umbrella of the UN. He got his PhD degree in Business and carried out many researches. He has more than 25 WOS/SCOPUS articles/books. He worked as Project Coordinator in employee issues, environment and marine litter-related projects.

Index

A

B

C

D

E

F

G

H

I

K

L

M

N

O

P

S

U

www.ingramcontent.com/pod-product-compliance
Lightning Source LLC
LaVergne TN
LVHW080906300926
849162LV00003B/578
* 9 7 9 8 2 6 0 0 3 6 7 8 5 *